SOIL PHYSICAL PROPERTIES

Standard Methods of Laboratory and Field Investigations

ABOUT THE AUTHORS

Dr. S K Gupta, Born (1949), is an Agricultural Engineer with specialization in Soil and Water Resources Engineering. He obtained B. Tech (Agril. Engg.) from the Punjab Agricultural University (PAU), Ludhiana, in 1970 and Master of Agricultural Engineering from the same University in 1976. He obtained Ph.D. in Civil Engineering from Jawahar Lal Nehru Technological University, Hyderabad, Telangana, India in 1984. Dr. Gupta is Fellow of the National Academy of Agricultural Sciences, Indian National Academy of Engineering and Indian Society of Agricultural Engineering. Since his joining CSSRI in 1971, besides holding scientific positions, he has been Head, Division of Drainage and Water Management, Head, Indo-Dutch Network Project, Head, Division of Irrigation and Drainage Engineering and Project Coordinator (AICRP on Management of Salt Affected Soils and Use of Saline Water in Agriculture). After superannuation, Dr. Gupta worked as Emeritus Scientist (ICAR) and INAE Distinguished Professor at the same institute. Dr. Gupta has been engaged in conducting researches in surface and subsurface drainage for land reclamation, irrigation water management, hydrology of salt affected soils, leaching and use of saline water in agriculture. He has published more than 150 research papers in high impact factor journals and published 15 books. Currently, he is an INAE visiting professor at ICAR-NDRI, Karnal and an independent Researcher pursuing his own interests in research and writing books. Dr. Gupta has been consultant to WAPCOS-Louis Berger, MoWR, MoRD, UP-Dutch Tube Well Project, ActionAid International, Synergics Hydro India-Oromia Water Works Enterprise, Ethiopia and others. He has been bestowed with Rafi Ahmad Kidwai Award by ICAR. Besides he is recipient of several awards from NAAS, ISAE, IE (India), CBIP and MoWR among other organizations. He is the Chief Editor of Journal of Water Management of the Indian Society of Water Management.

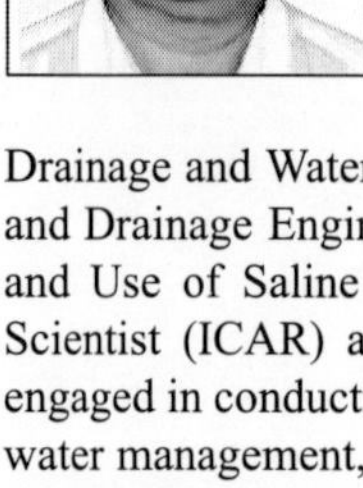

Dr. I.C. Gupta, (b. 1941), was awarded Ph.D. on a subject thesis of environmental soil science by CAZRI/University of Jodhpur in 1968. He was *Officer In-charge, Public Health Engineering Laboratory, Jodhpur before joining* at Central Soil Salinity Research Institute where he worked from 1970 to 1986 (Principal Scientist). Dr. I.C. Gupta served Central Arid Zone Research Institute from 1986 to 2001 where he worked as Head, Regional Research Station, Bikaner and Head, Division of Natural Resources and Environment. A Fellow of Indian Water Works Association, United Writers Association of India & Indian Society of Salinity Research Scientists, Dr. Gupta was elected Vice-Chairman of Arid Zone Research Association of India. He has been working as Secretary of Indian Society of Salinity Research Scientists & Chief Editor Current Agriculture since 1977. Dr. Gupta had been conferred several awards notable being Glaxo Scientific Honours Club, Dr. Rajendra Prasad Puraskar, First Prize, Outstanding Book Award by ISAE, Dr. Gorakh Prasad Puraskar First Prize, S.P. Unvala Memorial Prize by IWWA and Life Achievement Award by IARS. He was nominated as member of the 'Environment Conservation Council of Rajasthan' and of 'Special Committee on Environment Impacts' of INCID, Ministry of Water Resources, Govt. of India. He has about 250 publications including 30 books and two inventions (*Jaltripti* and *Jalshuddhi*). Dr. Gupta was appointed Ph.D. examiner by ten universities. He was also nominated for one of the most INTRIGUING PEOPLE of 2003 by American Biographical Institute, North Carolina, USA, after close scrutiny of biographical archives of over one million individuals.

SOIL PHYSICAL PROPERTIES

Standard Methods of Laboratory and Field Investigations

S.K. Gupta

Ph.D.

Ex- Head, Division of Irrigation and Drainage Engineering
Ex-Project Coordinator, Use of Saline Water in Agriculture
ICAR-Central Soil Salinity Research Institute, Karnal

I.C. Gupta

Ph.D.

Formerly Officer In-charge,
Public Health Engineering Laboratory, Jodhpur
Principal Scientist, CSSRI, Karnal
Head, Regional Research Station, Bikaner
Head, Division of Natural Resources and Environment
ICAR-Central Arid Zone Research Institute, Jodhpur

SCIENTIFIC® PUBLISHERS

Published by
SCIENTIFIC PUBLISHERS (INDIA)
5 A, New Pali Road, P.O. Box 91
Jodhpur 342 001 (INDIA)
E-mail: info@scientificpub.com
Website: http://www.scientificpubonline.com

ISBN: 978-93-90495-84-9 (HB)
eISBN: 978-93-90495-13-9 (E-book)

Printed in India

Preface

Finite global land resource is experiencing tremendous pressure, as the global population nears 10 billion mark. Coupled with rising consumption levels, nations are worried and exploring ways and means to meet the food, energy and water demand from this limited resource. Although only a decade away from the enunciated sustainable development goal of achieving a land degradation neutral world by 2030, land degradation is not showing any signs of slowing. As such, concerted efforts are being made to keep track of the soil health through regular monitoring by national and International organizations. In the Indian context, Indian Council of Agricultural Research and Soil and Land Use Survey of India, under the Ministry of Agriculture and Farmers' Welfare assisted by several central and state government agencies look after this activity. A scheme on 'Soil Health Cards' (SHC) has recently been implemented in mission mode. Millions of SHCs have been prepared and distributed to the farming community. While the effort is laudable, some concerns on improving the contents of these cards have been expressed at various forums. One of key recommendation is to include few soil physical properties that determine the overall soil health/quality. It is emerging that lack of facilities and trained manpower in the existing laboratories is withholding any action on this timely and important recommendation. Nevertheless, to sustain our efforts to assess and maintain the soil health and its quality, these deficiencies in our set-up have to be removed on war footing. Realization and urgency of this need is the prime motivating force for preparing this book in this format.

The current book is divided into 15 chapters, each chapter describing one or a group of inter-related physical properties. The chapter 1 begins with the general sampling procedures to draw disturbed and undisturbed samples, their pre-processing and storage. Next 12 chapters describe physical measurements such as examination of soil profile, soil colour, soil densities and pore space, texture, structure, dispersion ratio, soil moisture constants, water content determinations, infiltration rate, hydraulic conductivity, drainable pore space, soil aeration, engineering properties such as compaction, soil consistency, bearing capacity, angle of repose and relief, modulus of rupture, groundwater levels, and thermal properties including temperature. A chapter on miscellaneous properties includes soil cracks, repellency, surface area, liquid solid contact angle, and coefficient of linear extensibility. Soil carbon, which is the main governing factor of the soil physical state, is included as a separate chapter. It was an uphill task to select and include important determinations in one book, since the numbers of such determinations are huge. Therefore, it was decided to include most determinations needed to monitor the soil health and those in the course curricula of undergraduate and post graduate programs in agricultural universities of India and abroad. As a result, this book has emerged as a comprehensive compendium on the subject. Some of the salient features of this compendium are:

- It is written in an easy to understand language, each measurement procedure being explained in a step-wise manner.
- The procedures are supported by illustrations wherever necessary.
- Solved examples are included at several places to illustrate the calculations for easy understanding.
- In most cases, more than one method of determination is included. It may enable the researchers to choose the most appropriate method of measurement depending upon the required accuracy or instruments available with them.
- Measurements techniques with modern instruments requiring huge investments are not included, as manufacturer's recommendations on their use override the general procedures.
- A glossary of terms related to soil physics is included for ready reference.

Scientific studies in soil physical properties began in Europe in early nineteenth century. While Darcy's law was published in 1856, Wollny considered being the father of soil physics in Europe founded one of the first journal specifically addressed to the soil physics. Since then many researchers across nations have contributed to the development of this branch of soil science. The current book heavily draws from direct and indirect contributions of many national and international organizations and individuals. We scrupulously attempted to include most contributions by referring the work in the references. But, it proved extremely difficult and almost impossible to name each and everyone's contribution in this small brief. We take this opportunity to thankfully acknowledge all sources quoted and unquoted referred during preparation of this publication. We place on record our appreciation for all the contributors who have painstakingly researched and developed the measurement procedures. Finally, we like to thank our respective families and friends for their support.

We believe that this publication is a useful addition to the literature and will be of great benefit to the students, teachers, researchers, field practitioners, and NGOs engaged in agricultural sciences, engineering and in the basic streams of sciences and environment. The publication will be of immense use to technicians engaged in central and state laboratories associated with soil, water and plant analysis work. Since analytical techniques are developing and improving, this manual will be subject to revision at regular intervals. Any helpful suggestions and feedback from users for incorporation in future editions will be highly appreciated.

S.K. Gupta
I.C. Gupta

Contents

Chapter **1**

Soil Sampling Techniques

Soil is an extremely complex substance. Its testing for physical, chemical and biological properties can only unravel this complexity. The whole process of testing comprises the following four distinct phases.

- Sample collection
- Field and laboratory investigations and tests
- Interpretation of results
- Correlations and recommendations

Clearly, correct assessment of the soil's properties requires besides standard measurement techniques, precise sampling methods and strategies to collect representative samples. The first and foremost requirement to develop a sound sampling plan is to clearly identify and state the sampling objectives. A sound sampling and monitoring plan must be capable of meeting the stated objectives of soil testing. This chapter discusses general sampling plans to draw undisturbed or disturbed soil samples for truly characterizing the soil physical properties.

PHYSICAL PROPERTIES

Physical properties of the soil unravel the supporting capability of the soil, movement, retention and availability of water and nutrients to plants, ease of root penetration, and flow of heat and air though the soil. Physical properties also influence many chemical and biological properties of the soil. In fact, physical properties of the soil such as soil texture, type of clay, depth to bedrock, drainage class, aggregate stability, infiltration, organic matter content etc besides a host of chemical and biological properties have a great bearing on the soil quality for its suitability for agricultural, environmental and engineering uses. Management dependent soil physical properties get altered because of human interventions. Such alterations can either improve or degrade the soil quality in relation to its proposed use. According to a recent estimate, around 90 million hectares (M ha) area in India show one or the other kind of physical constraint (Indoria et al., 2017, Table 1.1).

Sampling for Physical Properties Tests

Analysis methods to determine chemical and biological properties have been refined and even automated to minimize human errors. Contrarily, little efforts have been made to achieve the same kind of accuracy in assessing the physical properties of the soils. We still make assumptions such as: soil is homogeneous, soil particles are spherical, and soil pores are composed of a bundle of capillary tubes etc. Moreover, laboratory test results of many physical properties obtained from disturbed and/or undisturbed samples compared to *in-situ* field investigations vary by one or even more than one order of magnitude. Although *in-situ* field assessment provides the most accurate information of the physical state of the soil for devising management practices; but it might entail huge investments and human resource.

The next best option to truly characterize the soil's physical properties is to conduct tests on undisturbed soil samples. Physical tests on disturbed soil samples should be the last option.

Table 1.1. *Problem soil areas affected by various soil physical constraints in India*

Physical constraint	Area affected (M ha)	Affected states
Shallow depth	26.40	Andhra Pradesh, Maharashtra, West Bengal, Kerala and Gujarat
Soil hardening	21.57	Andhra Pradesh, Maharashtra, West Bengal, Kerala and Gujarat
High permeability	13.75	Rajasthan, West Bengal, Gujarat, Punjab and Tamil Nadu
Subsurface hard pan	11.31	Maharashtra, Punjab, Bihar, Rajasthan, West Bengal and Tamil Nadu
Surface crusting	10.25	Haryana, Punjab, West Bengal, Odisha and Gujarat
Temporary water logging	6.24	Madhya Pradesh, Maharashtra, Punjab, Gujarat, Kerala and Odisha
Total	89.52	

IN-SITU FIELD TESTING

Field assessment methods help to obviate the need of taking samples. Nonetheless some of the issues that confront the investigator are more or less similar to soil sampling. Three important issues are:

- Where to conduct the tests?
- When to conduct the tests?
- How many tests to conduct?

Any decision on test sites must ensure that the sites are representative of the whole area to which test results are proposed to be applied. Some strategies to select the test sites are described in the following sections.

Random Selection

The most common method of site selection is to randomly select the representative sites by going around the whole area. This strategy is most appropriate when fields are more or less uniform having minimal spatial variability in its properties. Depending upon the kind of tests and number of sites, 2-3 replications can be used at a particular site to derive average values.

Directed Random Testing

This strategy is most appropriate for areas having more than one soil type and/or areas with rolling landforms. In such cases, it is difficult to identify a single dominant area that might represent the whole project area. The easiest way in such a case is to sub-divide the whole area on the basis of visual differences in soil types. The other option is to sub-divide the area into management zones based on some unique characteristic of the soil. For instance, if there are noticeable differences in any of the soil physical property, management zones might comprise below average, average and above-average areas in relation to that property. The tests are conducted zone wise in replicates to determine zone wise averages of the soil physical properties. These average values form the basis to manage fields falling in that particular zone.

Benchmark Testing

Benchmark testing involves selecting a small (say 30 m by 30 m) site that represents the whole project area. This site is used to test one or several soil properties. The results so obtained are used as a guide for the entire project area. If a single site does not adequately represent the area, multiple benchmark sites can be selected. The strategy in that case is known as directed benchmark testing.

Grid Testing

Grid testing is the most intense and expensive sampling strategy (Fig. 1.1). The project area in this case is systematically sub-divided into small areas or blocks called grids. The grid size may range from 10 m × 10 m to more than few hectares depending upon the size of the project area and kind of tests to be conducted. Tests are conducted at some central location within each grid. Although the accuracy increases, the cost and manpower required to conduct tests increase with decreasing grid size. To strike a balance, the usual practice is to establish a small grid size based on the soil property, which requires intensive testing or sampling. The tests for other soil physical properties are conducted on the same grid but with low intensity (Fig. 1.1).

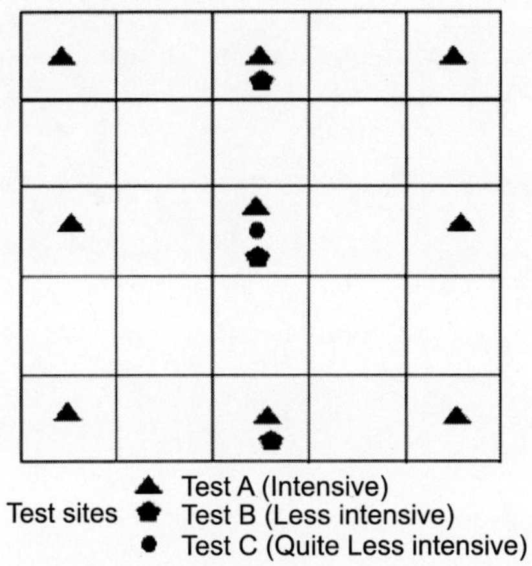

Fig. 1.1. *Grid testing with adjustment in the intensity of soil tests*

A Caution on Averaging

Since, a statement has been made on the averaging of the values of the parameters, a caution in this regard is appropriate at this stage. It is now known that many soil physical properties especially related to water movement such as infiltration rate, hydraulic conductivity, drainable pore space and many others are log-normally distributed as against the normal distribution commonly assumed in arithmetic averaging. Therefore, it will be most appropriate to use the geometric mean for such soil physical properties as being log-normally distributed. The geometric mean of a data set is the nth root of the product of n values from as many tests.

$$\text{Geometric mean} = (a_1 \times a_2 \ldots. \times a_n)^{1/n}$$

Here a_1, a_2 .... a_n are the values of a given soil property from n number of tests.

SOIL SAMPLING

A sample is a small quantity of material, which in every respect represents a much larger quantity of material. The intention of collecting a sample is not to select the best, or the worst examples of the material but a material that best represents its average conditions, which is most likely to be encountered during project implementation. In fact the accuracy of the test results will depend upon how accurately the sample represents the whole soil matrix in space and time. Sampling is therefore done on a completely random basis with no personal preferences or disliking. In addition, the samples must be processed and analyzed accurately to obtain the most reliable information. A true soil sample can prove to be a storehouse of valuable information having huge implications in project design and its performance or the utility of the soil as an engineering material.

The properties of any soil system, even on a small scale, vary in space and time. Therefore, sampling must be carefully planned to take into account the local variability likely to be encountered. Although some soil properties might require a different kind of sampling approach for their measurement, yet the general procedure described herein consists of the following steps.

- First of all, mark the horizontal boundaries of the soil to be sampled
- Take samples from depths with the greatest significance for a particular soil property
- Use an appropriate sample design to ensure that the soil samples represent the system in its entirety
- Take adequate number of samples to account for local variability
- Use proper soil sampling tools for collection and appropriate equipment and techniques for handling and pre-analysis processing of samples

Following four kinds of samples, in order of their decreasing efficacy and accuracy, are used in soil testing.

- Undisturbed samples
- Composite samples
- Representative samples
- Bulk or grab samples

Use of undisturbed samples is the best option after *in-situ* field testing. The group of last three together is categorized as disturbed samples. The bulk or grab samples scraped from a borehole, test pit or backhoe bucket, cuttings from a borehole, or directly collected by shovel are the least desirable.

Undisturbed Samples

Undisturbed samples are discrete high-quality samples collected from a specified depth under controlled conditions with minimum physical or chemical disturbance. It is extremely difficult to obtain a truly undisturbed sample. One of the means to take undisturbed samples is in the form of excavated blocks from which test specimens can be prepared. The method is quite promising to sample cohesive soils. In this case, soils in test pits or other locations are sampled in blocks, which are prepared by carefully cutting away the surrounding material followed by undercutting the block (Fig. 1.2 left). In profile or pit method, disturbed or undisturbed samples can be drawn looking at the changes in moisture condition, colour, consistency, soil type, and structure etc (Fig. 1.2 right). Other methods involve the use of core samplers (Fig. 1.3) or tube samplers such as split-spoon sampler, Shelby tube sampler and direct-push samplers.

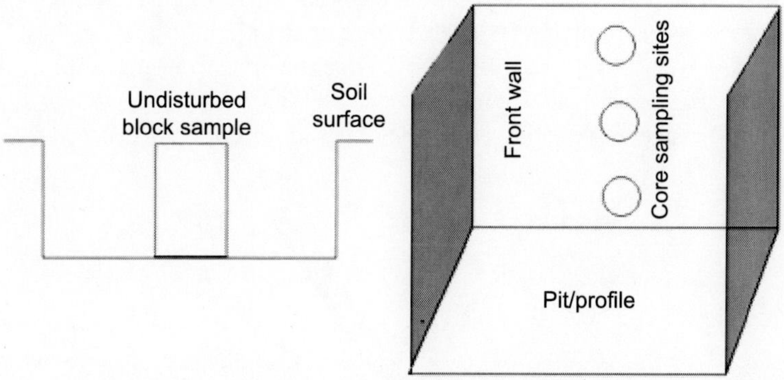

Fig. 1.2. *A schematic view of an undisturbed block sample (left) and profile sampling (right)*

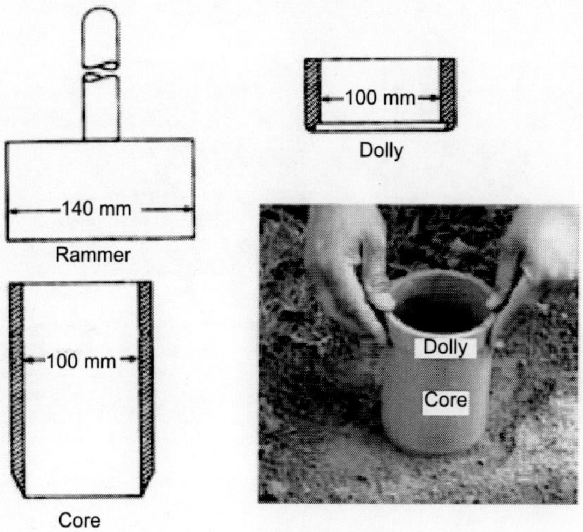

Fig. 1.3. *Core sampler for drawing undisturbed soil samples*

Disturbed Samples

Disturbed samples can be drawn from stockpiles of material and from material excavated for any kind of construction. Small disturbed samples can also be collected while carrying out other field activities. Since these methods have often failed to provide representative samples, the most common means of taking disturbed samples is to dig test pits, trenches or similar excavations, or using auger holes and boreholes. Commonly used sampling tools for this purpose are the pick and shovel, scoop, mason spatula or other appropriate hand tools. Soil sampling for agricultural purposes involving shallow depths can be accomplished using a *khurpa* or mason spatula. A V-shape groove is made to the desired depth (Fig. 1.4). Two slices from both sides of the groove are cut. Central portion of each slice (discarding the soil at two ends) represents the sample of the desired depth. Hand augers are ideal for collecting disturbed and undisturbed samples from depths up to 5 m. Commonly used augers for disturbed sampling are bucket auger, screw auger, and the Dutch auger (Fig. 1.5). Soil samples with hand auger are collected by turning auger barrel attached to an extension rod into the ground until the barrel is filled with soil. The auger is withdrawn from the borehole, and the soil is removed for examination or sampling (Fig. 1.5 right). Ample care needs to be exercised to prevent rolling off coarse material from the sides of the tool. If that happens, it may leave behind too fine a sample that might impact most soil physical properties. Power augers or drilling equipment is used for depths exceeding 5 m.

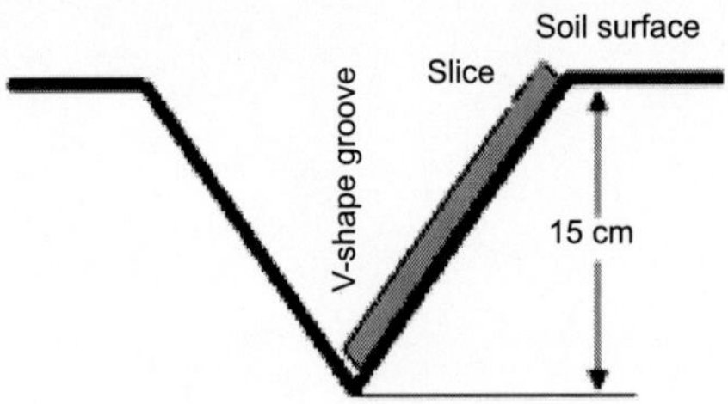

Fig. 1.4. *Sampling procedure using a khurpa, or a mason spatula*

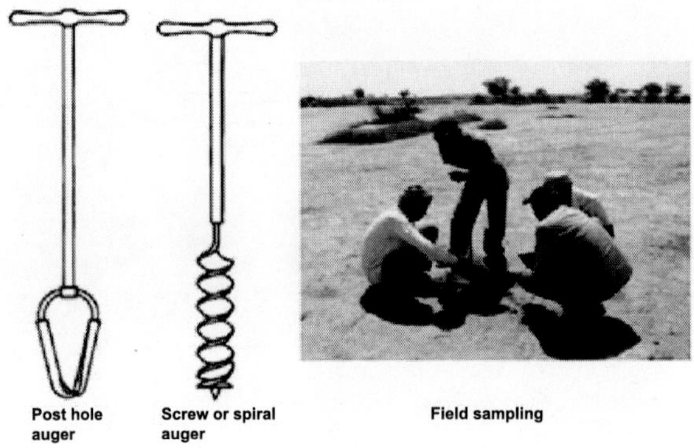

Fig. 1.5. *A schematic view of selected augers and their field application for soil sampling*

Sampling Strategies

Depending on the manpower, time available and kind of properties being investigated, sampling can be completely random, selective or benchmarking (based on judgment), or directed or stratified random (divided into sub-sites in which samples are taken randomly), directed benchmarking, regular transect (often used when some trend in a particular direction is expected in the values of some property), topography based sampling or on a grid basis as described in a previous section on *in-situ* field testing. The adoption of a startegy under given conditions depends upon few important factors such as likely field variability, time, funds available, desired sampling intensity and the likely information to be derived and its application. Selection of a particular method as dictated by these factors is illustrated in Fig. 1.6. It should be possible to take a decision on the best startegy using information provided in this figure.

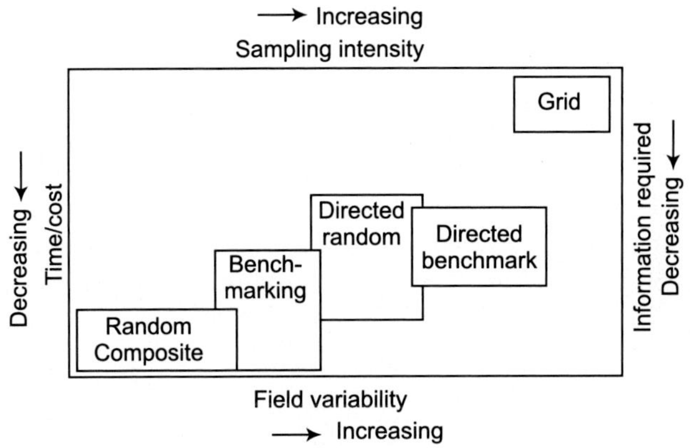

Fig. 1.6. *Selection criteria of a sampling strategy*

Other Issues of Soil Sampling

Sampling, as already stated is quite crucial to any plan of research and soil management program. To have a representative soil sample, besides the sampling strategy and equipment discussed in previous sections, following issues need to be kept in view.

- Time of sampling
- Sampling depth
- Amount and number of sample
- Sampling frequency
- Appropriate handling of samples

Time of Sampling

Soil properties vary considerably over time, showing short-term as well as long-term variations. Clearly, the analysis of samples taken at one time may yield quite different results than the results of samples taken at another time. It calls for sampling through time or alternatively, recording of the conditions when samples are taken and measurements are made. The ideal time to draw samples for assessing soil physical properties is when the fields are not cropped.

Sampling Depth

In agriculture, the soil sampling depth is governed by the depth of the root zone of the proposed cropping pattern. The following is the commonly used depth criteria in agriculture.

- For field crops (cereals, vegetables and others seasonal crops) a sampling depth of 0-15 cm (plough layer) should be sufficient

- For deep-rooted crop like sugarcane or under dry farming conditions, 3 samples from different depths to a depth of 60 cm may be enough. Normally samples are drawn from 0 - 15 cm (0 - 6 in), 15 - 30 cm (6 - 12 in) and 30 - 60 cm (12 to 24 in) depths
- For horticultural crops, sampling may be done at depths of 0-15, 15-30, 30-60, 60-90, 90-120, 120-150 and 150-180 cm
- In saline-alkali soils, salt crust (visible or suspected) should be sampled separately while drawing samples from below depths

Quantity and Number of Samples

Quantity of sample required to conduct physical tests is usually more compared to the quantity of sample required for testing chemical properties. Moreover, physical tests are replicated to arrive at a reasonably good average value. A uniform field in all respects can be treated as a single sampling unit such that samples from such units can be composited. Separate sets of composite samples need to be collected from areas differing in some characteristics. Quartering process, commonly used during soil sampling for reducing the quantity of sample, may not be required. It is because relatively more quantity of soil sample is required for physical tests.

Sampling Frequency

Unlike chemical properties, which may change seasonally/annually, physical properties may not show vast changes in a year or two except for some temporary changes. Thus, except for some specific research tests of physical properties, the sampling can be delayed for 2-3 years. If necessary, only limited *in-situ* tests may be conducted in between on few representative fields. The resulting recommendations can be used to manage un-sampled areas. The cost and time at the disposal will determine the number of such tests to arrive at the best management solutions.

Caution: The samples should not be drawn immediately after rains, irrigation, application of fertilizer and burning of crop residues. Similarly, the sample should not be drawn at sites near to or along an irrigation canal, recently fertilized plots, marshy tracts, farm ways, buildings, wells, a collector or lateral drain, dyke *(bund)*, farmyard manure pits, shady trees, threshing floor and roads. When crops are grown in rows, samples should be taken in between the rows. Soil samples should be taken in zigzag pattern or randomly.

Packing and Labeling

It is always useful to exercise the following precautions before packing and transporting soil samples to a soil-testing laboratory.

- If initial processing is done at the site, then air-dry the samples in trays in shade. Remove any visible stones/gravel from disturbed soil samples
- Air-dried samples should be packed in polyethylene / hard paper bags
- Insert the polyethylene bag in a cloth bag and put an appropriate labeling tag for identification
- Cloth bags should be labeled with a copying pencil/ink
- Seal the bag properly, and use wooden/hard paper board cartoon for transport/ mailing.

Each sample should be accompanied by a detailed Performa. It helps the analyst to make sound recommendations on the basis of soil analysis. The label should include appropriate information from the following exhaustive list.

- Name of the project
- Name of the sampler
- Location within project
- Date and time of sampling
- Sample number
- Depth of sample
- Description of the soil layer
- Test pit- borehole- auger hole number
- Type of sample
- Sampling method used
- Stakeholder's name
- Crop(s) grown
- Irrigated/un-irrigated
- Number and type of container(s)
- How samples are being sent?
- Additional information, e.g. how the soil was processed before sampling

PROCESSING OF DISTURBED SAMPLES

The samples must be promptly transported and processed to prevent any changes that might affect analysis results. Immediately after the samples are received in the laboratory; these should be tallied with the accompanying information list. The relevant information should be entered in a register and each sample be given a laboratory number. The samples should be processed in the laboratory, if these have not been processed at the site. Samples are processed as per the following processes in the listed sequence.

- Drying
- Grinding
- Sieving

The soil samples should be emptied in wooden or enameled trays. While the large chunks of soil are broken, stones and large objects are handpicked and removed. Samples are then spread out to dry in the air at room temperature. Alternatively, the trays are placed in racks in a hot-air cabinet, the temperature not exceeding 35°C with humidity remaining in the range of 30 to 60%. Take ample care to maintain the identity of each sample. It is utmost necessary to ensure that the samples are not contaminated in any manner.

Air-dried samples are ground with a wooden pestle and mortar so that clods and large aggregates are crushed and mixed. Take care that pebbles, concretions, and stones are not broken during grinding. Also ensure that individual soil particles are not crushed during the grinding process. For most routine soil tests, the entire sample should be passed through 2-mm stainless steel or plastic sieve (Jackson, 1958). The coarse portion on the sieve is returned to the mortar for further grinding. Repeat sieving and grinding till all aggregate particles are fine enough to pass the 2-mm sieve. For specific type of analysis (e.g. organic C) grind the soil further to pass it through 0.2 to 0.5-mm sieves. Mix each soil sample to homogenize before weighing or scooping a sub-sample for analysis. Prepared samples should be stored in cardboard containers for further analysis. If samples need to be preserved for a longer duration, it may be better to store them in glass or polythene jars.

SAMPLING WATER

Physical properties of soils are intricately related to the quality of water to be used for irrigation or even to determine the soil physical properties (Annexure I). Therefore, water samples must be tested for proper diagnostic analysis of soils. The following procedure is used to collect representative water samples.

Selection of Containers

As far as possible, use plastic bottles for sampling. If plastic bottles are not available, glass bottles can be used. Avoid metal containers for collection of water samples. Bottles should be thoroughly cleaned and rinsed at least 2-3 times with water of the same source of which the sample is being collected.

Time of Sampling

For gravity canal, take the sample of the water from the channel after irrigation has been in progress for about 20-30 min. For groundwater, take the sample after running the tube well for about 30 min.

Quantity of Sample

Most routine appraisal tests require about 100 mL of water. To be on the safer side and to allow specialized tests, if need be, as much as 500 mL of sample should be collected and sent to the laboratory.

Packing

Water container should be properly screwed and if possible wax sealed. Containers should be properly labeled for identification. Each sample should be accompanied by appropriate description from the list prescribed for soil samples.

Chapter **2**

Soil Profile and Soil Colour

Soil, an important natural national resource, comprises mineral matter, organic matter, water, air and living organisms. The proportions of these constituents vary from one soil to another. It is because the soil profiles develop over time due to weathering of minerals and deposition of organic matter. Therefore, great variations in depth, colour, composition and even soil properties can be observed in soil profiles. To identify and decipher these variations in components and characteristics of the soil profiles, one resorts to conduct systematic soil surveys. The objectives of the soil survey are to arrive at a representative image of the various types of soils and of the soil horizons present at the site (USDA, 1962). The information generated in these surveys helps to decide as to how the soil might be used and/or predicting how the soil might react to its intended use. Nonetheless, the information on soil profile descriptions can prove to be valuable only if we adopt a common terminology to describe the soil profiles.

Colour is often used to name soils as it reflects some of the soil's qualities. We often hear of black soils, red soils, and mottle-coloured soils etc. These terms signal a snapshot of different soils properties and even help to interpret them in a useful manner. The colour of a soil is influenced by four factors namely mineral matter derived from the constituents of the parent material, organic matter, the nature and abundance of iron and moisture content. The colour of the surface soils is often governed by the amount of organic matter. But, it may also be dependent upon the parent material, the age of the soil, and even the land slope (Lynn and Person, 2000). A careful observation of colour helps to identify problems of water logging or leaching. Although colour is widely discussed and used soil characteristics in pedology, but the causes of colour differences and its significance are matters of intensive discussions.

SOIL PROFILE

A soil profile, a historical record of all the soil-forming processes, is the unit of study in all pedological investigations. It is a vertical section of the soil that extends through all its layers going up to the parent rock, unless some obstacles are encountered or soils are too deep. It is an essential pre-requisite to classify soils and forms the basis of practical studies of soils. A study of the soil profile is important from crop husbandry point of view, since it reveals the surface and the subsurface characteristics and qualities, namely depth, texture, structure, drainage conditions and soil-moisture relationships, which directly affect plant growth and soil-moisture-plant relationships.

Apparatus: Tools to dig a pit, Trowel to take soil samples, Plastic sheet for placing soil taken out during the digging operation, Water bottle (spray type), Meter stick or tape measure, Dilute HCl (10%), pH kit, Camera-if available, Gloves, Site map, Soil sieves, Hammer, Vinegar, Munsell colour charts (See colour section of the book), plastic ruler, Plastic vials, rubbing alcohol to store soil invertebrates, Field notebook, Pen or pencil, Coloured pens or pencils.

Process

1. Select a representative field site. Identify a location where it should be possible to dig a soil pit. The pit location should be far away from concrete floors, pavements or roads. It is always better to look for less compacted and less disturbed areas. The presence of grasses, trees and other types of vegetation may help to take a knowledgeable decision on identifying the exact site for opening a pit. It is always better to move away from sites that are suspected to have buried public utilities.
2. Describe all characteristics of the site in the notebook that might help in interpreting the results. It may include features like land use whether urban, agriculture, forest, recreation site etc., vegetation types, their heights, canopy cover percentage, and buildings or other features such as distance to fences, roads, power poles, etc., surrounding landforms, and presence of any water bodies. It may be useful to assess the approximate slope of the site and its exposure to the sun.
3. Dig a pit using appropriate digging tools. The pit should have straight sides of 0.80 m × 1.50 m and may extend up to 2 m depth in deep soils, and in others up to the parent material. If ground water is encountered before reaching 2 m depth, one can stop further digging at this stage. Nonetheless, it is a good practice to sample the soil below the bottom of the pit going as down as possible with available tools.
4. The pit should be wide enough so that it is possible to have a glance at all the soil layers (horizons) down to the bottom of the pit. Normally the pit is as wide as it is deep.
5. Place the excavated soil on the plastic sheet in sequence so that each pile corresponds with depth- wise soil layers. Follow the same sequence during back-filling the pit, the piles from the deepest horizons being returned to the bottom of the pit. The pile made first is returned to surface horizon. It is used to cap the pit.
6. Select one of the well-exposed sides of the freshly opened pit. Beginning from the top, proceed downward until reaching the bottom of the pit. Examine carefully and identify different soil layers (horizons O, A, B, C, and R) in the profile. Remember, the layers will vary in thickness. Make a neat drawing of the soil profile (separate for each pit, if more than one pit is opened) and note down the presence of soil horizons along with their depth from the soil surface. The morphological characteristics, such as colour, texture, and structure, etc. are used to distinguish the soil horizons, as is described later in this chapter.
7. Use a meter stick or tape to measure the thickness of the soil horizon. Two measurements, one up to the top and other up to the bottom of each horizon, should suffice to determine the thickness. Both measurements should correspond to 0 cm, the top of the profile. Make a sketch of the horizons types and their depths in field notebook. The characteristic thicknesses of the boundaries are described as: Abrupt (a): boundary less than 2.5 cm thick, Clear (c): boundary 2.5-5 cm thick, gradual (g): boundary 5-12 cm thick and diffuse (d): boundary more than 12 cm thick.
8. A photograph that clearly reveals the horizons of the profile should be taken and kept in records. Besides, it is always useful to take few photograph(s) of the surrounding landscape. It helps to interpret the observation and draw realistic inferences.

9. Examine each horizon for moisture, colour, smell, texture, structure and mottles. In can be achieved by examining the soil of exposed profile or the soil of various horizons piled on the plastic sheet. Note down these characteristics in the notebook.
10. Refill the soil pit in such a manner so that the site is returned to its original condition i.e. as it was before the digging of the pit.

MASTER HORIZONS AND SUB-HORIZONS

A soil profile is composed of a series of horizons or layers, which are stacked one above the other. It mainly consist of six broad horizons. These are named as O, A, E, B, C and R. It is not necessary that a soil profile must contain all the six horizons (For example in Fig. 2.1 E horizon is absent). It is mainly because the soil profiles differ from one location to another and one point to another at the same location. For example, one may not observe the organic horizon (O) in many mineral soils but it is quite common in forested soils. Combined A and B horizons are known as Solum. R horizon, also known as bedrock, is the parent material. Solum together with parent material is referred to as regolith. Any horizon is identified either by a single letter or a combination of letters. For example horizon A may be sub-divided and designated as A1, A2 etc. It is achieved by adding Arabic numerals to letters of the horizon, such as O, A etc.

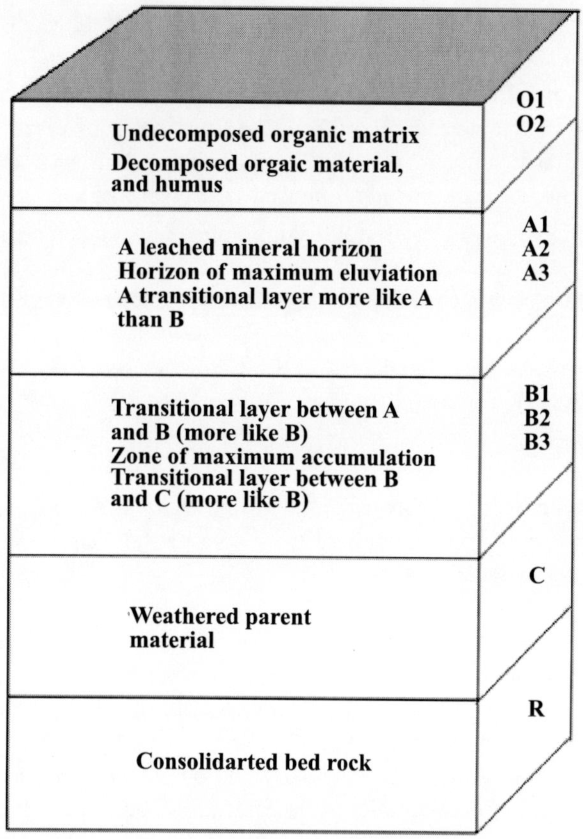

Fig. 2.1. *A schematic view of soil horizons*

O Horizon

O horizon, known as organic horizon, is formed in the upper part of the soils and comprises fresh or partly decomposed organic material in significant amounts. For soils having more than 50% clay, the organic matter content in the organic horizon should be more than 30%. For mineral fraction having less clay content, organic matter content should be more than 20%. These horizons are mostly absent in grassland, and cultivated soils but are common features of forested soils. Sometimes, this horizon is categorized in 2 layers namely O1 and O2 depending upon whether the organic material is present as wholly or partially decayed plant and animal material (Fig. 2.1).

O1: In this horizon, the organic material is present in the partially decayed form. It is possible to recognize original forms of the plant and animal residues through the naked eye.

O2: In this horizon, the organic material is present in fully decayed form. It is not possible to recognize original forms of the plant and animal residues through the naked eye.

A Horizon

The top or the mineral surface layer on which organic matter accumulates is known as the A horizon. The clay, iron, and other materials present in this layer are lost through the process of leaching, the process known as eluviation. Materials such as sand, being resistant to weathering, tend to remain in the horizon. This horizon is responsible to provide the best environment for the plant roots to grow, and sustain the growth of micro-organisms and other living elements. A horizon can be sub-divided into three layers i.e. A1, A2 and A3.

A1: It is the top most mineral horizon formed adjacent to the surface. Humified organic matter associated with mineral fraction accumulates in this layer. Compared to lower horizons, it is darker in colour mainly because of the organic matter content.

A2: Maximum eluviation of clay, iron and aluminium oxides and organic matter occur from this horizon. While these constituents are lost, there is greater accumulation of quartz, sand and silt sized particles, which resist leaching. It is lighter in colour compared to horizons above and below it.

A3: A transitional layer between A and B horizons, exhibits properties which are nearer to A1 or A2 horizons than the B horizon.

E Horizon

It is a characteristic of sandy, forested soils of high rainfall regions. It is highly leached zone from which clay, chemicals, and organic matter gets eluviated. As a result, this layer exhibits a very light colour.

B Horizon

It may be incorporated at least in part in the plough layer but may be considerably below the plough layer in soils with deep A horizon having sub-horizons. The B horizon is also known as the zone of illuviation, as the dissolved or suspended soil materials eluviated from other horizons accumulate here. Clearly, the dominant features of this horizon are accumulation of clay, iron, aluminium or humus alone or in combination. The horizon show dark or strong red colour than the horizons above or below it mainly because of the coating of sesquioxides. One should possibly look out for the presence of following three layers.

B1: It is a transitional layer between A and B, being nearer to A than B.

B2: The clay, iron and aluminium oxide either eluviated from upper horizons or formed *in-situ* accumulate in this zone. It is in fact the zone of maximum accumulation. The organic matter content in this zone is generally higher. Its colour is darker than the A2 horizon.

B3: It is a transition horizon lying between B and C horizons. Its characteristics are closer to the overlying B2 than the C horizon.

C Horizon

This horizon excludes the bedrock from which A and B horizons have been formed. It may contain weathered parent material. This horizon is unaffected by pedogenic processes, and is devoid of any major biological activity. This horizon may accumulate carbonates or sulphates of calcium and magnesium.

R Horizon

It is the underlying consolidated bedrock from which Solum is formed. It may or may not look like the parent rock.

As the soils age, these may develop additional horizons than the basic master horizons discussed in the foregoing sections. Position wise, these layers are located between the master horizons and have intermediate properties. More often such transition layers are identified by two master letters, the dominant one being written first. Thus, an AB layer lying between A and B horizons will resemble both but is more like A than B. If some parts of the horizon have properties similar to the two kinds of master horizons, then the two capital letters are separated by a virgule (/). As an example, E/B, B/E, or B/C, means that the layer has similar properties to E and B in the first, B and E in the second and B and C in the third case. Their precedence is decided by the volume, the first symbol being that of the horizon that makes up the higher volume. Note E/B and B/E in the example cited previously.

Sub-Horizons

Sub-ordinate distinctions within master horizons are designated as sub-horizons. Lowercase letters, having specific meaning as explained in Annexure II, are used as suffixes to designate specific kinds of master horizons and layers (See Brady and Weil, 1984 or any other books such as Soil Science Division Staff, 2017). As an example, sub-ordinate distinction, p meaning plowed is used with the master horizon A as Ap to indicate disturbed surface horizon. Similarly, Btg stands for master horizon B with sub-ordinate distinction t and g signifying clay accumulation (t) and reduced conditions (g). The normal convention is that letter suffixes directly follow the capital letter followed by numerals. Not more than three suffixes are used except in rare cases.

EXAMINATION OF HORIZONS

Let us now have a look at the ways to describe different soil characteristics while examining a soil profile.

Surface Topography

Surface topography may be described as: Smooth (s): the boundary is almost a plane surface, Wavy (w): the pockets are wider compared to their depth, irregular (i): the pockets are deeper compared to their width and broken (b): the horizon boundary being discontinuous.

Moisture

The feel method is commonly used to describe soil moisture. In this method, soil is held between the fingers to observe qualitatively whether the soil is dry, moist, or saturated as per the following criteria.

Description	Criteria
Dry	Soil appears dry, looks dusty with a clear absence of moisture
Moist	Soil is damp with slight moisture; just enough that soil may retain moulded form
Wet	Soil is wet; and water becomes visible, if squeezed
Saturated	Water is visible, it drains freely from the soil sample

Colour and Smell

Describe soil colour (See next section of this chapter) and smell qualitatively (subjectively). If the soil is dry, it may be worthwhile to wet the soil by spraying water. Both colour and smell are best observed when the soil is wet.

Structure

The general structure comprising consistency, presence or absence of rocks, roots and animals in each horizon of the profile should be described (See section on structure in chapter 5 of this book).

Consistency

The consistency of the soil is described as per the following criteria.

Loose	Falls apart in the hand
Friable	Soil peds break apart easily between fingers
Firm	Soil peds break apart if pressure is applied but leave dents on fingers
Extremely firm	Peds can only be crushed with a hammer

Rocks and Animals

Describe the presence of rocks (mineral fragments $\geq$ 2-mm). Actual size of rocks can be measured with the help of a ruler. The presence of roots is qualitatively described as none, few or many. An alternative option is to measure the actual number of roots exposed using a ruler. Similarly, describe the numbers of worms, insects, snails, rodents, etc. and/or their traces in the form of channels, burrows, or nests as none, few or many of each kind of animal or animal traces.

Texture

Soil texture is given by the relative proportion of different fine-earth fractions (< 2-mm size) of a soil sample. It will suffice to use the "feel" method to assess soil texture during profile investigation. The results of this investigation should be updated after laboratory determinations (Refer Chapter 5).

Soil Internal Drainage

Soil with slow drainage and temporary water logging is characterized by redoximorphic mottles, which are spots or blotches of greyish (reducing) and/or yellowish-reddish (oxidizing) colour. Higher the abundances of greyish mottles, longer is the water logging period during the year. In addition, reducing conditions of the soil should also be characterized by observing black iron/manganese concretions.

Mottles

If mottles are present, their dominant colour is determined by using the Munsell notation. These are described with regard to their colour contrast (faint, distinct and prominent), abundance (few, common and many) and size (fine, medium and coarse) as per the following guidelines (Khakural and Sharma, 1984).

Faint: Indistinct mottles that can only be recognized following the close examination of the soil. Colour hues and chromas are closely related in both the matrix and mottles.

Distinct: The mottles can be readily seen although may not be quite striking. The value, hue and chroma of the matrix and those of the mottles can be easily distinguished. The variations could be several units in chroma and one or two in value and hue.

Prominent: Mottles are present, and is an outstanding feature of the horizon. The hue, value and chroma of the matrix and mottles can be several units apart.

Abundance is described as few, common or many.

Few: Less than 2% of the exposed surface occupied by the mottles.

Common: Mottles occupy 2 to 20% of the exposed surface.

Many: More than 20% of the exposed surface is occupied by the mottles.

Size is given by the approximate diameter of individual mottles. The diameters measured along the widest dimension are described as fine, medium and coarse.

Fine: Mottles are designated as fine, if their size is less than 5 mm.

Medium: Medium mottles vary from 5 to 15 mm in size.

Coarse: Coarse mottles are more than 15 mm in size.

Other Characteristics

Other characteristics such as cutans, root distribution, reaction to dilute HCl, presence of concretion, hard pan etc. may be noted.

POTENTIAL ROOTING DEPTH

The potential rooting depth determines the ability of the soil to provide a suitable rooting medium for roots to grow vertically. Thus, it is the depth of soil that plant roots can potentially exploit before encountering a barrier that may hinder their root growth. Available-water-holding capacity of the soil, as well as availability of macro and micro-nutrients is governed by the rooting depth, increasing with increasing rooting depth. Rooting depth studies can be carried out along with the soil profile study. The shallowest root limiting layer measured from the soil surface controls the rooting depth. Some of the common limiting layers/barriers that restrict the soil depth are as follows.

- Bedrock
- Natural hardpans such as soil cemented by iron, lime, gypsum, silica, etc.
- Densely compacted material (tillage pan)
- Strongly contrasting textures
- Water table
- High ESP or salt layer

Procedure

Depth of the soil is easily determined during the examination of the soil profile. While the profile is being dug observe the following.

1. The presence of roots and old root channels, worm channels, cracks and fissures down to which roots can extend
2. Note whether there is an over-thickening of roots (a result of a high penetration resistance), and whether the roots are being forced to grow horizontally, known as right-angle syndrome
3. Note the firmness and tightness of the soil, whether the soil is grey and strongly gleyed owing to prolonged water logging
4. Presence of natural or manmade hardpans such as an iron, siliceous or calcitic pan or human-induced tillage or plough pan
5. An abrupt transition from a coarse (sandy/gravelly) layer to a fine (heavy) material.

A rough estimate of the potential rooting depth is made by noting the above characteristics. Using a measuring tape, ruler or stick graduated in cm, measure the location (depth and thickness) of the barrier below the soil surface. Record these depths and prepare a sketch of the soil profile, annotated with depth and principal soil features. Further laboratory tests on soil ESP and salt content may help to modify the field evaluated soil depth. Following guidelines may be used to characterize the soil with respect to its depth.

Potential rooting depth (cm)	Remark
>150	Very deep
90-150	Deep
50-90	Moderately deep
25-50	Shallow
<25	Very shallow

PROFILE SAMPLING

Disturbed and/or undisturbed soil samples are drawn for *in-situ* field testing or for laboratory investigations. Follow the procedure as per the following steps to collect soil samples for analysis.

- Carefully clean one face of the whole soil profile
- Sample each identified horizon. While doing so, begin with the lowest horizon and proceed upward covering all the identified horizons. Each sample for chemical analysis should weigh about 1kg except in gravelly soils where the sample should be large enough to contain at least 100 gm of fine earth particles (< 2 mm diameter)
- Since the uppermost horizon might have been disturbed during digging, enough care must be exercised to collect the representative sample. Most of the times, it may be useful to move to a nearby undisturbed site and dig a new shallow hole up to the desired depth. Use this hole to take a soil sample of the uppermost horizon. The sample of the upper horizon may also contain stones, and organic material such as leaves and roots. These may preferably be removed during packing of the sample
- Do not mix soil samples of different horizons

Following data sheet is used in the examination of the soil profile.

General Description of the Profile Site

Location: **Soil classification:**

Site: It's relative position in the landscape may be indicated:

Parent material: Slope class: Degree of erosion:

Natural drainage class: Natural vegetation or land use: Water table:

Horizon	Depth (cm)	Colour (moist soil)	Clay (%)	pH	Organic carbon (%)	CEC (Cmol(+)/kg)	Base saturation (%)

Other characters include mottles, HCl reaction etc.

As an illustration, some physico-chemical characteristics of the soil of Niduvalalu series, a member of the clayey-skeletal, mixed isohypothermic family of Rhodic Paleustalfs is reported in Table 2.1 (NBSS&LUP, RRS, Bengaluru).

Table 2.1. *Physico-chemical characteristics of a soil profile*

Horizon	**Depth (cm)**	**Colour (moist soil)**	**Clay (%)**	**pH**	**Organic carbon (%)**	**CEC (Cmol(+)/kg)**	**Base saturation (%)**
Ap	0-19	5YR4/4	6.98	7.28	0.49	6.49	73
Bt1	19-40	2.5YR2.5/4	20.05	7.28	0.45	10.45	96
Bt2	40-64	2.5YR2.5/4	42.88	7.93	0.53	19.25	100
Bt3	64-82	2.5YR3/4	38.89	8.15	0.35	15.70	100
Bt4	82-106	2.5YR3/4	35.83	8.14	0.20	15.07	100
Bt5	106-155	2.5YR3/4	21.93	8.11	0.16	16.17	100

Soil Colour

Munsell colour charts are commonly used to describe the soil colour (www.munsell.com). Although, soil scientists have developed advanced chip image-based soil sensors for quantitatively deciphering the soil colour, this section describes the application of Munsell colour chart as illustrated in USDA–SCS (1971).

Equipment: Soil colour charts (Munsell colour chart), Water bottle

Procedure

1. Take a lump of soil from the layer/horizon whose colour is to be described
2. Break the lump to expose a fresh face
3. Moisten (without glistening) the face by spraying some water, if dry
4. Stand with the sun over the shoulder, allowing the sunlight to shine on the colour chart and soil sample
5. Hold the soil sample behind apertures separated by the closest matching colour chips. The process should be repeated for both the dry and moist samples to determine their colour

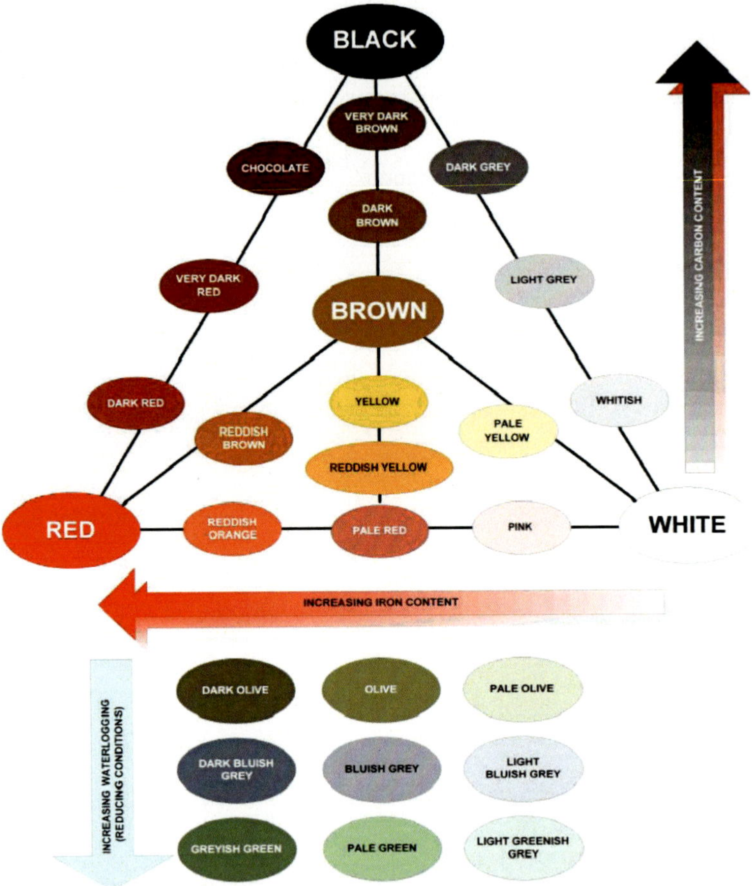

Fig. 2.2. *Soil colour triangle: relationship between soil colour and influencing factors (with permission from http://vro.agriculture.vic.gov.au/dpi/vro/vrosite.nsf/pages/soilhealth_prac_soil_colour)*

6. Use masks (a part of the Munsell colour charts) to match colour, if necessary
7. Record Munsell notation in the order and form Hue Value/Chroma, which is symbolically written as H V/C (e.g., 5R 7/2). 5R is the Hue (or colour), 7 is the Value (or lightness/darkness) and 2 is the Chroma (weak/strong). Also state (dry, moist). Colour of a moist sample is darker than dry sample. Although there is seldom change in hue, the value is darker by 1/2 to 3 steps, while chroma shows a variation of -1/2 to +2 steps. Report as 5YR 4/4, 4.5/4 moist, the latter values being for the moist sample
8. Colour characteristics such as mottle size, percentage and contrast should be observed and recorded
9. Simple names as listed in the triangle shown in Fig. 2.2 can be used in case Munsell charts are not available (Agriculture Victoria, 2019, State of Queensland, 2015).

COLOUR ATTRIBUTES: MUNSELL COLOUR CHART

Munsell soil colour chart is basically a book comprising about 250 different coloured papers, or chips. These are systematically arranged on hue cards as per the Munsell notations. An illustration of the arrangements of colour chips on a Munsell colour card is illustrated in Fig. 2.3. Hue, value, and chroma are the three elements of colour used in the Munsell colour system.

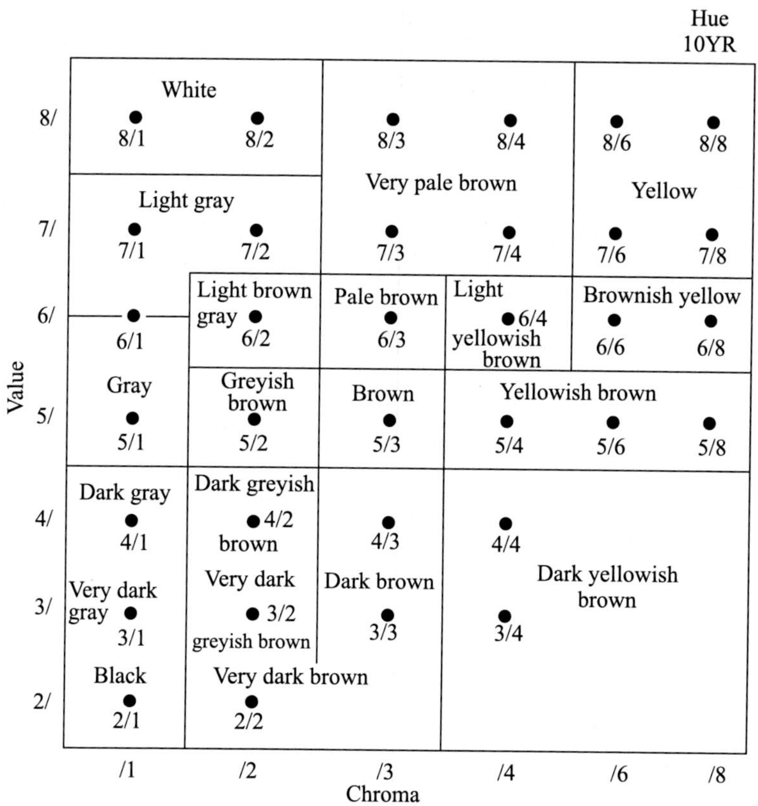

Fig. 2.3. *A sample Munsell colour chart*

Hue

Hue is an attribute of a colour that measures the chromatic composition of light to distinguish red (R) from yellow (Y), green (G), blue (B), and purple (P). Note that the alphabets in parenthesis are the letter codes for respective hues. Besides, there are five intermediate hues that represent mid-points between each pair of principal hues. Five intermediate hues namely yellow-red (YR), green-yellow (GY), blue-green (BG), purple-blue (PB), and red-purple (RP) and 5 principal hues complete the 10 major hues used to describe the system. Each hue is further divided into 4 segments of equal visual steps designated by numerical values 2.5, 5, 7.5 and 10. These values are used as prefixes to the symbol for the hue name. As an illustration, four equally spaced steps of the yellow-red (YR) hue are 2.5YR, 5YR, 7.5YR, and 10YR, respectively. Besides, standard soil chart has few separate hue cards and special charts for gley colour and for very light colours. Some of these cards are gley 1 and 2 soil charts, Munsell 5R individual soil chart, Munsell 7.5R individual soil chart, White page, 7.5R, 10YR, and 2.5Y. Cards with symbols such as N 6/ are totally achromatic having zero chroma. The extreme cases are black (N 2.5/) and white (N 8/) having no hue and no chroma. N 5/ (gray) is the notation for a neutral (achromatic) colour. The colour with notation 10YR 5/1 is also gray, because the hue at such a low chroma is hardly perceptible.

Value

The degree of lightness or darkness of a colour in relation to a neutral gray (achromatic) scale is described by values. The value extends from pure black (0) to pure white (10). The value measures the amount of light reaching the eye under standard lighting conditions. The actual amount of light reaching the eye is logarithmically related to the value. Gray, having a value notation of 5, is perceived as about halfway between black and white. Numbers between 5 and 10 indicate lighter while numbers between 5 and 0 indicate darker colours. A card of the colour chart has a series of chips arranged vertically in equal steps varying from the lightest to the darkest shades. Fig. 2.3 shows this arrangement vertically on the card for the hue 10YR. The highest value on a standard colour card is 8. Colour chips having a value of 9 are included on special colour cards, used to describe very light colours.

Chroma

Chroma represents the relative purity or strength of the spectral colour, indicating the degree of saturation of neutral gray by the spectral colour. The chroma scale for soils begins from 0, representing neutral colours to 8 representing colours with the strong expression. These colour chips are arranged horizontally, chroma increasing from left to right (Fig. 2.3).

The Munsell notation is written in the order: hue, value, chroma. A space is given between the hue letter and the value number. A virgule is used between the value and chroma. Thus, a notation hue 5YR with a value of 5 and chroma 6 is written as 5YR 5/6, expressing a yellowish-red colour. If the soil sample satisfactorily matches with some Munsell standard chip in all the three attributes, it is easy to record the colour. On the other hand, samples more often do not match with any of the chart colours. In fact, they fall somewhere in between two of them. To resolve this problem, it is necessary to compare the sample with few neighbouring Munsell colours. Following procedure is adopted.

Turn to the hue chart that displays the nearest match with the sample. Since the match is not exact, also detach a neighbouring chart, and lay them side by side to estimate the exact hue position of the sample. Since hue numbers of both the charts are known, the result is recorded in terms of hue steps. For example, if it appears that the match is in between 2.5YR and 5YR, hue of the colour is taken as (2.5+5)/2 = 3.75YR. Similar averaging is done in respect of value and chroma. A sample matching in between say 2.5YR 5/6 and 5YR 6/8 will be denoted as 3.75YR 5.5/7. Note that numbers, which are not the whole numbers, a decimal system rather than fraction is used to write the Munsell colour notation.

Notes

- Protective envelopes of the charts must be removed before using them for comparison. It improves the accuracy of measurement as the colour perception is not affected by the special effects of glossy surface
- While making colour comparisons, the observer should view the sample and the standards at an angle of 90°, with light falling upon them at an angle of about 45°
- The illumination quality is quite important in assessing the colour. It is recommended that all observations should be made in daylight. In special circumstances it can be done under adequate artificial daylight
- The appropriate masks should be used for fair comparison and accurate matching of colours. The mask holes expose equal areas of sample and standard at the same time do as to make a fair comparison

INTERPRETATION FOR SOIL MANAGEMENT

The soil colour offers a snapshot of several soil properties. It helps to identify associated management problems for their resolution (Table 2.2).

Table 2.2. *Soil colour, soil types and management implications*

Soil colour	Soil types and characteristics	Typical management implications
Black	These soils are often associated with high levels of organic matter (peats)	Water logging or drainage problems Low pH High denitrification
Black	Vertisols (cracking clay soils)	Workability and tillage problems
White/pale/ bleached	These soils are often referred to as bleached or 'washed out'. The iron and manganese particles have been leached out due to high amounts of rainfall or drainage	Leaching of nutrients (A cause of deficiencies) Low plant-available water
Red	It indicates good drainage. The iron in the soil is oxidized readily because of presence of high amounts of oxygen. As a result, a rusty colour develops. The colour is even darker, if organic matter is also present	High phosphorus fixation Low plant available water

Yellow to yellow-brown	These soils are poorly drained than the red soils. The iron compounds in these soils are in a hydrated form and therefore, do not produce the 'rusty' colour	Moderate phosphorus fixation Low plant-available water Compaction
Brown	Soils associated with moderate organic matter and iron oxides	Low to moderate phosphorus fixation Low to moderate plant available water
Gleyed/ gray/green	These soils are associated with very poor drainage or water logging. The lack of air creates conditions for iron and manganese to form compounds that give these soils their colour	Water logging or drainage problems High denitrification risk Methane emission

Chapter **3**

Soil Densities and Related Properties

Soil, one of the most important elements of crop production, is composed of solid mineral particles of different sizes that range from 2 mm to less than 0.002 mm in diameter. These particles are also bound together into aggregates by organic matter, mineral oxides, and charged clay particles. The gaps between the particles and between the aggregates constitute the total pore space, which is occupied by water or air. The water to air ratio in the pores of the soil volume varies greatly in space and time, but the sum of the two always remains the same. Any increase or decrease in one component causes a corresponding decrease or increase in the other. The soil pores act as water reservoir and supply water needed by the plants. The exchange of water and air in the pores with that of the environment controls the flow of plant nutrients and heat. The relative proportion of the solid mineral particles of various sizes and that of solids and pores together control many soil properties. The density of soils is generally measured by its bulk density and particle density. Whereas bulk density refers to the specific gravity of a volume of soil including its pore space, particle density measures the specific gravity of the soil excluding pore space. These two together control the total pore space, which is the sum of the macro and micro-pore space. This chapter briefly looks at these properties and describes the methods of their measurement.

BULK DENSITY

Bulk density (BD) of the soil is defined as the ratio of the mass (M) of oven-dried soil to its bulk volume, which includes the volume of the soil particles and the voids (pore space) between the particles. BD mostly depends on soil texture, yet is a dynamic property as it can be altered by cultivation, compression by animals and machinery, weather conditions, and organic matter content. It generally increases with depth in the soil profile and normally varies in the range of 1.0 to 1.7 g/cm^3 or Mg/m^3. The higher the number, more compacted is the soil and more difficult for roots to penetrate. Typically, sands pack more closely and BD values range from 1.4 to 1.9 Mg/m^3. Clays, which tend to bridge and cannot pack as tightly as sands, have values in the range of 0.9 to 1.4 Mg/m^3.

Bulk density of the soils is determined using a number of methods but the most common and simple method is to cut out a cylindrical core of soil of known volume. BD of the soil is then calculated from the oven-dry mass of the soil in the core. This section deals besides the core method, clod and sand pouring methods of measuring bulk density. The latter is commonly used in civil engineering but can find application in agriculture especially in soil conservation activities.

Core Method

Apparatus: Core sampler with metal cores or only metal cores, Aluminum boxes, Oven, Balance, Knife/spatula, Vernier caliper.

Core Sampler

Core sampler with cylindrical cores of varying lengths and diameter are available. A cylindrical core cutter is a seamless steel/brass tube having 7.5 to 10 cm diameter and 15-20 cm long. The core sampler with cores in place is pressed into the soil mass so that it is filled with the soil without much disturbance to the core contents.

Procedure

1. Measure the length (L) and diameter (2 times the radius) of the core using a vernier. The volume of the core is then given as $\pi r^2 L$
2. Place the core inside the core sampler. For taking samples, the assembled core cutter is positioned over a clean levelled surface and pressed inside the soil by rotating the handle or by dropping a hammer over the central portion of the upper end of the thick rod until the edge of the collar comes to rest over the soil surface
3. Loosen the grip of core cutter from soil around it by moving the stem forward and backward. Pull out the core cutter slowly from the soil and place it horizontally on the soil surface. Unscrew the collar and carefully take out the rings and core containing the soil
4. Trim off the extra soil from both ends of the core with a sharp knife or spatula
5. If soil cores are to be weighed in the laboratory, place soil filled cores into strong plastic bags that can be sealed tightly with minimal air space. If weighing is done at the site, transfer the soil core to a pre-weighed aluminum can and take the weight. Put the same cans in the oven at 105°C for 24-48 hr till constant weight is achieved. Note that core contents alone can also be transferred to the cans
6. Take the dry weight of the soil with can
7. Calculate the weight of the dry soil after subtracting the weight of the empty can.

Simple locally manufactured soil cores can also be used to measure bulk density (Fig. 1.5 and Fig. 3.1). These cores are driven into the soil using a dolly and hammer, or hammered directly placing a wooden piece at the top of the core. Once the core is inside the soil, soil around it is carefully removed to take out the core. Trim off the extra soil from both ends of the core with a sharp knife or spatula. Further procedure is same as explained in the core cutter method.

Fig. 3.1. *A pictorial view of core sampling*

Observation Sheet

Soil sample identification number:

1	Soil depth at which sample is drawn	D (cm)
2	Inner diameter of the core	d (cm)
3	Length of the core	L (cm)
4	Sample volume	$V = (\pi \times d^2 \times L)/4$ (cm^3)
5	Weight of can	W_1 (g)
6	Weight of moist soil + can	W_2 (g)
7	Weight of dry soil + can	W_3 g (g)
8	Weight of moist sample	$W_2 - W_1$ (g)
9	Weight of dry sample	$W_3 - W_1$ (g)
10	Wet bulk density	$(W_2 - W_1)/V$ (g/cm^3)
11	Dry bulk density, ρ_b	$(W_3 - W_1)/V$ (g/cm^3)

Reporting Results

Bulk density (Mg/m^3) is reported to the nearest 0.01 Mg/m^3. Note that numerical value of the bulk density in Mg/m^3 and g/cm^3 will be same.

Notes:

1. If the soil contains significant amount of gravel, grind the oven-dry soil mass in wooden pestle and mortar. Pass the sample through a 2-mm sieve. Weigh, and record the weight of the rock fragments retained on the sieve. Determine the volume of the rock fragments based on their weight and an assumed particle density of 2.65 g/cm^3, unless better information is available. Take off the weight and volume of rock fragments from the total weight and volume of the sample before calculating the bulk density
2. If for some reason or the other, the core is partially filled, then adjust the core volume in the calculations
3. As the volume of 1 g of water is equal to 1 cm^3, it is possible to calculate the percent of water by volume. In that case one can use both the % weight and % volume water to calculate bulk density as a check.
4. While retrieving the soil cores, exercise due care to preserve the natural soil structure.

Limitations: This method cannot be used to determine the bulk density of sandy soil or wet soil as the cores will not be able to retain the soil. The technique may also pose few problems in gravelly soils.

Clod Method (Blake, 1965)

The basic principle of this method is to determine the volume of the soil clod by immersing it in water (Archimedes' principle). The difference in its weight in the air and water represents the volume of the soil clod (assuming 1 g of water as 1 cm^3 of water). It is why, this method is also known as water displacement method. The clod in this method is coated with paraffin before it is immersed in water. It prevents the water to penetrate into the clod.

Apparatus/Equipment: Balance, Fine thread, Melted paraffin (wax), A cylindrical container fitted with an overflow outlet tube in the upper half, A rubber tube with clip attached to the outlet tube for measuring water outflow

Procedure

1. Collect soil clods of about 3 cm diameter from a known depth of the soil profile. Carefully cut any visible roots with a sharp scissor. Also trim out the sharp edges, if any
2. Tie the thread to the clod and hold it with the thread. Take its weight in the air (W_a)
3. Momentarily dip the clod in the melted paraffin, and hang it to dry for 30 min. The temperature of paraffin should be in the range of 60 to 70°C i.e. few degrees above its melting point. At this temperature, paraffin quickly solidifies once the clod is taken out of the paraffin pot. Thus, it is less likely to penetrate the pores of the clod as might happen at higher temperatures. Weigh the clod, thread and paraffin together in the air ($W_a + P_a$)
4. Suspend the paraffin-coated clod in water and weigh it in water (W_{pw}). If such a balance is not available, immerse the clod completely in a graduated cylinder half full with water, and detect the change in water volume in the cylinder (V). The volume V is then used to calculate, W_{pw}
5. Record the water temperature
6. To correct the BD so determined for initial moisture content of the soil, break open the clod and draw a soil sample. Determine its moisture content and calculate the oven-dry weight of the soil clod.

Calculation

BD = Weight of the clod/volume of the clod

BD = Weight of the clod/(volume of the (clod+ paraffin) – Volume of the paraffin)

$$\text{Bulk density} = O_{ds}/[V - (P_a/\rho_p)] \qquad ...(3.1)$$

Such that O_{ds} = Oven-dry weight of soil clod = $W_a/(1 + m)$ (g)

m = Soil moisture content of the soil clod (fraction) determine in step 6

P_a = Weight of paraffin coating in air (g)

ρ_p = Density of paraffin (g/cm^3), 0.88 to 0.93 g/cm^3

In terms of weight and taking the density of water at the measured temperature into account

$$\text{Bulk density} = \rho_w \times O_{ds}/[W_a + P_a - W_{pw} - (\rho_w \times P_a/\rho_p)]$$

Here ρ_w is the density of water at the temperature of determination (g/cm^3), W_a is the net weight of soil clod in air (g) and W_{pw} is the net weight of soil sample plus paraffin in water (g).

Notes:

1. The BD determined by this method may be slightly higher than other methods (Tisdall, 1951). It is because inter-clod spaces are not included in this method. The other reason could be that the soil volume is the air-dry volume, which is likely to be slightly less than the volume of a field-moist sample used in other methods.
2. If bubbles appear on the paraffin or the weight of the clod in water increases over time, it is an indication that water is penetrating the clod. Discard the sample in such events.

Comment

If paraffin is not available, rubber, Saran, wax mixtures, or some kind of oils can be used (Blake and Hartge, 1986).

Sand Replacement Method

Apparatus: Sand pouring cylinder, Pouring cone separated by a valve or shutter, Calibrating container, Tray with central circular hole, Glass plate about 450 mm/600 mm square and 10 mm thick, Chisel and scoop for excavating hole, Balance accurate to 1 g, Uniformly graded dry clean sand, preferably passing a 1 mm IS sieve and retained on a 600-micron IS sieve.

Calibration: Bulk Density of Sand

Fill the pouring cylinder with sand up to a height of about 2 cm below the top and take its initial weight (W_1). Place the cylinder over a plane surface, open the shutter and allow the sand to run out and fill the cone. When no further movement of sand takes place in the cylinder, close the shutter. Collect the sand in the cone and weigh (W_2). Refill the cylinder with sand taken out of the cone.

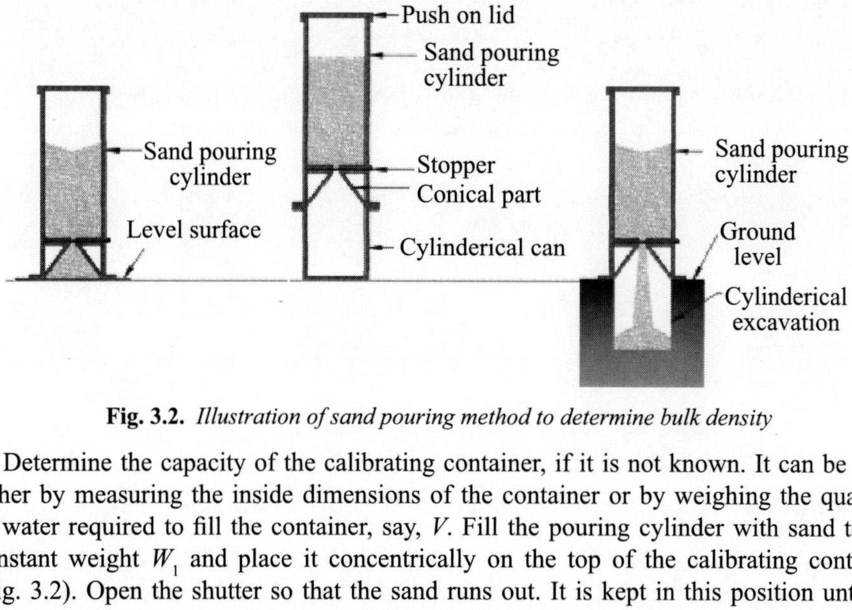

Fig. 3.2. *Illustration of sand pouring method to determine bulk density*

Determine the capacity of the calibrating container, if it is not known. It can be done either by measuring the inside dimensions of the container or by weighing the quantity of water required to fill the container, say, V. Fill the pouring cylinder with sand to the constant weight W_1 and place it concentrically on the top of the calibrating container (Fig. 3.2). Open the shutter so that the sand runs out. It is kept in this position until no further movement of sand takes place. Close the shutter and weigh the cylinder (W_3). The loss in weight is both due to sand going into the calibrating container and the cone. The weight of sand required to fill the calibrating container (W_c) is given by:

$$W_c = W_1 - W_3 - W_2 \qquad ...(3.2)$$

Here W_1 is the weight of the cylinder + sand before pouring into calibrating containers (g), W_2 is the weight of sand in the cone, and W_3 is the weight of cylinder + sand after pouring into calibrating containers (g). The bulk density of sand in g/cm³ is given by:

$$\text{BD sand} = W_c/V \qquad ...(3.3)$$

Measurement of Soil Density

1. Fill the cylinder with sand to the initial constant weight, W_1
2. Weigh the empty tray
3. Clean and level off about 50 cm² of the site where it is proposed to conduct the test. Place the tray on the prepared surface. Excavate the test hole to the desired depth within the circular hole of the tray and carefully collect all excavated material into the tray
4. Weigh the tray containing the excavated soil. Keep a representative sample (at least 100 g) to determine its moisture content
5. After removing the tray, carefully place the sand pouring cylinder, directly over the excavated hole to cover it concentrically. Open the shutter. When sand stops running out, close the shutter, remove the cylinder and weigh, W_4.

Calculations

The weight of sand required to fill the excavated hole, W_h is given by:

$$W_h = W_1 - W_4 - W_2 \qquad ...(3.4)$$

Here W_1 is the weight of the cylinder before pouring sand into the hole (g), and W_4 is the weight of the cylinder after pouring the sand into hole (g). *In-situ* volume of excavated soil is calculated as:

$$V_h = W_h/BD \text{ sand} \qquad ...(3.5)$$

Wet unit weight of the soil, $\rho_{wet} = W/V_h$ ...(3.6)

In- situ dry bulk density of soil, $\rho_{dry} = \rho_{wet}/(1 + m)$ (g/cm³) ...(3.7)

Here W is the weight of excavated soil (g), and m is the moisture content of the soil in fraction.

Note: As far as possible, try to keep the initial weight same throughout the tests i.e. equal to the weight used during calibration.

Precaution

The sand should be dry and clean. Its bulk density may considerably vary due to impurities (Singh, 1967). Closely graded sand should be chosen to get the best results. Since the dry density of soil may have large spatial variations, it is desirable to conduct number of tests to arrive at a representative average value.

Observation Sheet

Calibration of the apparatus

1.	Initial weight of cylinder + sand	W_1 (g)
2.	Weight of sand in pouring cone	W_2 (g)
3.	Capacity of calibrating container	V (mL)
4.	Weight of cylinder+ sand after pouring in the calibrating container	W_3 (g)
5.	Weight of sand to fill container	$W_c = W_1 - W_3 - W_2$ (g)
6.	Bulk density of sand	W_c/V (g/cm^3)

Density of Soil

1.	Weight of tray + excavated soil	W_{ts} (g)
2.	Weight of empty tray	W_{te} (g)
3.	Weight of excavated soil (W) = (1) – (2)	$W = W_{ts} - W_{te}$ (g)
4.	Initial weight of cylinder + sand	W_1 (g)
5.	Weight of cylinder + sand after pouring sand in the hole	W_4 (g)
6.	Weight of sand in hole + cone	$W_1 - W_4$ (g)
7.	Weight of sand in pouring cone	W_2 (g)
8.	Weight of sand in hole (W_h)	$W_h = W_1 - W_4 - W_2$ (g)
9.	Volume of hole (V)	$V_h = W_h/BD$ sand (cm^3)
10.	Bulk density of soil (wet)	W/V_h (g/cm^3)
11.	Water content	m (fraction)
12.	Dry density of soil	$(W/V_h)(1/(1+m))$ (g/cm^3)

Guidelines

USDA (1978) guidelines on soil bulk density are used to assess whether the soil is densified enough to affect root growth.

Soil texture	Ideal bulk densities for plant growth (g/cm^3)	Bulk density that affect root growth (g/cm^3)	Bulk density that restrict root growth (g/cm^3)
Sands, loamy sands	<1.60	1.69	>1.80
Sandy loams, loams	<1.40	1.63	>1.80
Sandy clay loams, clay loams	<1.40	1.60	>1.75
Silts, silt loams	<1.40	1.60	>1.75
Silt loams, silty clay loams	<1.40	1.55	>1.65
Sandy clays, silty clays, clay loams	<1.10	1.49	>1.58
Clay (>45% clay)	<1.10	1.39	>1.47

Practical Applications

Weight of a Furrow Slice

The weight of a furrow slice is the oven-dry weight of soil per hectare to a depth of 15 cm. This weight is often used to calculate the amount of fertilizer and amendments needed to meet the soil requirement. It is customary to consider that an average furrow slice of one-hectare area of a medium-textured soil having a bulk density of around 1.33 Mg/m^3 weighs about 2000 Mg or 2×10^6 kg.

$$\text{Weight} = \text{Bulk density } (Mg/m^3) \times \text{Furrow slice depth (m)} \times \text{Area } (m^2)$$
$$= 1.33 \times 0.15 \times 10^4 = 1995 \text{ or} \approx 2000 \text{ Mg or } 2 \times 10^6 \text{ kg}$$

It is possible to calculate the actual weight as per the BD of the soil by replacing 1.33 with the actual bulk density. It is always better to calculate the actual weight rather than using the guideline value.

Soil Compaction Studies

Soil compaction is inferred from the change in bulk density of a soil before and after application of compaction force. The inference drawn holds good for the same kinds of soil. Its extension for soils that vary in mineralogy is limited. Its application is also limited in soils varying in texture, organic matter content, other variables and *in- situ* soil conditions. To overcome this problem, one approach is to determine the relative bulk density (RBD), the ratio of the bulk density to the maximum bulk density determined using compaction tests (Carter, 1990, see Chapter 12 of this book).

$$\text{Relative bulk density (RBD)} = \text{Actual bulk density/ Maximum bulk density}$$

Studies have shown that the relative bulk density is a useful index to assess soil compaction. One can expect optimal crop production, if RBD values are in the range of 0.74-0.81 regardless of soil texture and organic matter content.

Particle Density

All soils are derived from minerals. Individual soil mineral has a characteristic particle density. Particle density of most mineral soils is in the range of 2.60–2.75 Mg/m^3 (representing particle density of quartz, feldspar and colloidal silicates, which constitute the mineral fractions of soils). Particle density, for porous soils or soils with high organic matter content, may be as low as 2.0 (Table 3.1). On the other hand, particle density may exceed 3.0 in soils containing heavy substances. The value for quartz, a predominant soil mineral, is about 2.65 Mg/m^3, which is why this value is frequently used as representing all soils. Soil particle density (G_s), also designated as specific gravity of a solid substance, is the ratio of the weight of a given volume of material to the weight of an equal volume of water (at 27°C, some researchers prefer 20°C). All measurements are made or converted to the stated temperatures as discussed in the following sections. The particle density tells how much heavier (or lighter) the material is than water. It is why the use of distilled or demineralized water is used to ensure measurement accuracy. Moreover, organic matter (OM) present in the soils, as stated earlier in this paragraph, can alter particle density, being low in high OM soils. Therefore, standard procedure recommends that OM in high OM soils should be removed before proceeding to determine particle density. Many researchers, on the other hand, have a contrary view. They argue that the particle density determined after OM removal may produce significantly higher value of calculated porosity, owing to the high value of the particle density. Besides the use of particle density in calculating porosity, it is also used to determine soil's void ratio and grain-size distribution in hydrometer analysis. The most commonly used method to determine particle density is pycnometer method described by Blake (1965) and Blake (2008).

Table 3.1. *General range of particle densities of various soils*

Soil type	Range (Mg/m³)
Sand	2.63-2.67
Silt	2.65-2.70
Clay and silty clay	2.67-2.90
Organic	< 2.0

Apparatus: Pycnometer (density bottle, Fig. 3.3) of 50 or 100 mL capacity with perforated stopper, Balance accurate to 0.001 g, Drying oven, Vacuum desiccator, Thermometer, Beaker 250 mL, Filter paper, Air free distilled water, Funnel, Clean and dry cloth.

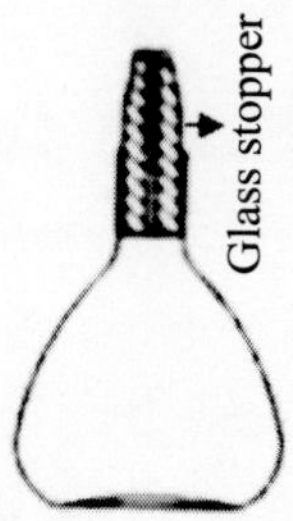

Fig. 3.3. *Diagrammatical view of a pycnometer with glass stopper*

Procedure

1. Weigh a dry, clean pycnometer with glass stopper to the nearest 0.01 g, W_1
2. Carefully transfer about 10 g (in 50 mL bottle), and 20 g (in 100 mL bottle), of oven-dry and pulverized soil sample. If air-dry soil is used, dry the contents of the bottle at the end of the test to get the weight of dry soil or determine the moisture content of a sub-sample of the soil used in the test
3. Replace the stopper and record the weight, W_2. Mass of the soil is then given as $W_s = W_2 - W_1$
4. Add air-free boiled and cooled water to the pycnometer until it is about 2/3rd full. Keep it undisturbed for few hr. For clays, it has to be at least 12 hr
5. Remove the entrapped air by gentle boiling for several minutes. During this process, gently shake the pycnometer or lightly stir the contents with a glass rod
6. Fill the pycnometer with air-free boiled and cool water to the brim and put stopper tightly so that extra water shoots out of the hole in the stopper
7. Carefully dry outside of the pycnometer by wiping with a dry cloth and weigh to the nearest 0.01 g, W_3

8. Wash the pycnometer bottle and fill the pycnometer with air-free boiled and cool water to the brim and put the stopper tightly so that extra water shoots out of the hole in the stopper. Record the weight of the pycnometer with water, W_4.

Particle density denoted as G_s at test temperature is calculated as per the following formula.

$$G_s = [W_s] / [W_4 + W_s - W_3] \quad ...(3.8)$$

The specific gravity is normally calculated and reported to the nearest 0.01g at a temperature of 27°C. If the water temperature is other than 27°C, make the following correction to the calculated value (BIS, 1980c).

$$G'_s = KG_s \quad ...(3.9)$$

Here G'_s is the corrected specific gravity at 27°C

K = [Relative density of water at room temperature]/[Relative density at 27°C] ...(3.10)

Relative densities of water and values of K are reported in Table 3.2. Some researchers report specific gravity corrected to 20°C. In that case values of K in Table 3.2 will change accordingly.

For results to be reported at 27°C, eq. (3.8) is modified as

$$G'_s = [W_s / (W_4 + W_s - W_3)] \times [\rho_{wt} / \rho_{w27}] \text{ g/cm}^3 \quad ...(3.11)$$

Here, ρ_{wt} and ρ_{w27} are the densities of water at test temperature and 27°C respectively (g/cm^3). If the reported specific gravity is to be based on water at 4°C, it is given by:

$$G'_s = [W_s] / [W_4 + W_s - W_3] \times \rho_{wt} \text{ g/cm}^3 \quad ...(3.12)$$

Since the density of water at 4°C is 1 g/cm^3.

Table 3.2. *Relative density of water and correction factor (K) at various temperatures*

Temp (°C)	Relative density (4°C as reference)	Correction factor (K)#
18.0	0.99862	1.0021
19.0	0.99843	1.0020
20.0	0.99823	1.0017
21.0	0.99802	1.0015
22.0	0.99780	1.0013
23.0	0.99757	1.0010
24.0	0.99733	1.0008
25.0	0.99708	1.0005
26.0	0.99682	1.0003
27.0	0.99655	1.0000
28.0	0.99626	0.9997
29.0	0.99598	0.9994
30.0	0.99568	0.9991
31.0	0.99537	0.9988
32.0	0.99505	0.9985

The correction factor, K, is the ratio of the relative density of water at the test temperature to the relative density of water at 27°C.

Observation Sheet

Test temperature T_t = ________ (28°C)

Sp.gr. of water at T_t = ________ (0.99626)

Sp.gr. of water at 27°C = ________ (0.99655)

Correction factor, K = ________ (0. 99626/0.99655 = 0.9997)

1.	Test No.	1
2.	Density bottle No.	5
3.	Weight of density bottle	W_1 (g)
4.	Weight of bottle + dry soil	W_2 (g)
5.	Weight of the soil	$W_s = W_2 - W_1$ (g)
6.	Weight of bottle + soil + water	W_3 (g)
7.	Weight of bottle + water	W_4 (g)
8.	Sp. gr. of soil at T_t°C	$W_s / [W_4 + W_s - W_3]$ (g/cm³)
9.	Sp. gr. of soil at 27°C	$KW_s / [W_4 + W_s - W_3]$ (g/cm³)

Precaution

Carefully remove all the entrapped air from the pycnometer. If air remains entrapped, value of particle density will be lower than the actual value. If partial vacuum is applied to remove the entrapped air, it should be applied at slow rates.

Flask Method

This method is quite similar to the density bottle method.

1. Take a flask of 500 or 250 mL capacity having a graduation mark at the 500/250 mL level with an adaptor for fitting to the vacuum line. Weigh the empty flask, W_1
2. Fill about 80-100 g soil for the test, which may be oven-dry or air-dry. Weigh the flask with soil, W_2. If soil is air-dry, the weight of soil solids ($W_d = W_2 - W_1$) is adjusted for initial moisture content by oven-drying a soil sub-sample as discussed in previous sections. Clay soils may be mechanically dispersed with a high-speed stirrer for about 15 min, before transferring to the flask for further observations
3. Add water to the flask and remove the air by boiling or by vacuum pump
4. Fill the flask up to mark and weigh the flask + soil + water, W_3
5. Fill the empty flask with water without any addition of soil, W_4
6. Measure the temperature of the water and apply a correction for the temperature as in the pycnometer method.

Observation Sheet

Capacity of flask = 250 mL

1.	Test No.	1	2
2.	Flask No.	3	4
3.	Weight of flask + dry soil	W_2 (g)	
4.	Weight of empty flask	W_1 (g)	
5.	Weight of dry soil	$W_2 - W_1$ (g)	
6.	Weight of flask + soil + water	W_3 (g)	
7.	Weight of flask + water	W_4 (g)	
8.	Specific gravity	$(W_2 - W_1)/(W_3 - W_4)$	

Note: Weights should be accurate to 0.01 g in case of 250 mL flask and to 0.05 g in case of 500 mL flask.

Soil Porosity

Soil is not a coherent solid material. It is a 3-phase system comprising solid particles (mineral grains, rock fragments) with water and air in the voids between the particles and the aggregates (Piper, 1942). A schematic diagram of the system in terms of weight and volume symbols for soil solids, water, and air respectively is shown in Fig. 3.4. In this Figure V is the total volume and W the total mass (Neglecting mass of the air). Subscripts s, w, and a respectively indicate the values of mass and volume of soil solids, water and air respectively.

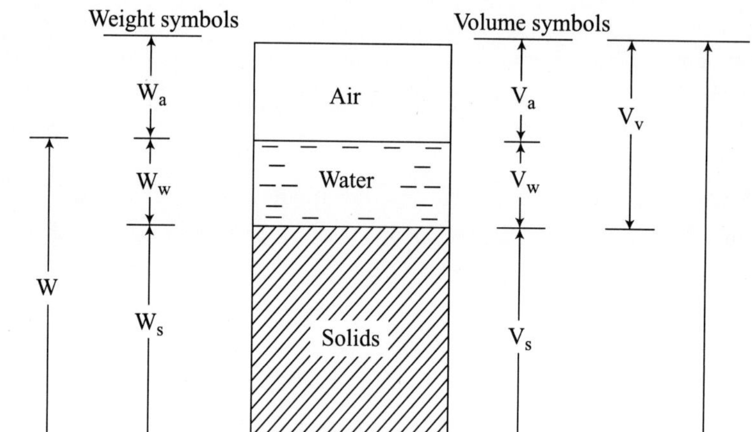

Fig. 3.4. *Schematics of soil solids and soil voids*

Total or bulk volume of the soil is written as

$$\text{Total volume, } V = V_s + V_w + V_a = V_s + V_v \quad ...(3.13)$$

Here V_v is the volume of void-space comprising volume of water, V_w, and volume of air, V_a, V_s is the volume of solids. The total porosity, simpler characterization of the soil pore system, is defined as the the volume of voids, the fraction of the soil space occupied by the air and water, divided by the total volume of the soil. More often, it is denoted by the symbol n and is given as

$n = (V_v/V)\ 100$ ...(3.14)

The porosity, n can be determined either directly or indirectly as a derived quantity. Both these approaches are discussed in following sections.

Direct Measurement

Procedure

1. Use one of the cores collected during the determination of bulk density using core method (See Bulk density section). Carefully take out the cores, so that no part of soil falls out. Put a twofold layer of cheese cloth at the bottom of the core with the help of a rubber band
2. Weigh the core with the cheese cloth and rubber band(s)
3. Set the core in a pan containing water until it is almost immersed. Leave it in the pan to soak up water for 24 hr or until it is completely saturated. Add extra water to the pan, if required
4. Remove the core from the pan, and let the excess water drain until there is no dripping of water
5. Weigh the core with soil at saturation point, W_1
6. Carefully remove the cheese cloth and replace with a square of aluminum foil. Put the core in the drying oven set at 105^0 C for 24 hr. Weigh the wet piece of cheese cloth and rubber bands; its weight needs to be subtracted from the total weight of the core + soil to get the actual weight of the water in the soil
7. Remove the core from the oven, cover top with Saran wrap so that it does not suck the moisture from the atmosphere. Weigh the dry core, W_2
8. Calculate the amount of water in the soil at saturation by subtracting the (weight of the core + dry soil) from the (weight of the core + saturated soil). Weight of water at saturation = $W_1 - W_2$ (Reduce the weight of wet cheese cloth and rubber band here)

Saturation % = 100 $(W_1 - W_2)$ / Volume of the soil core ...(3.15)

Note that $(W_1 - W_2)$ represent the volume of water at saturation assuming its density as 1 kg/m^3. Saturation percent is the total porosity in percent, which comprises macro and micro-pore space discussed in a following section of this chapter.

Indirect Measurement

The porosity of the soil can be calculated using soil densities determined as per the methods described in previous sections. Herein, we simply convert densities to mass and volumes. Referring to Fig. 3.4, porosity as per definition is given as

$n = V_v/V = (V - V_s)/V = [W_s/\rho_b - W_s/\rho_s]/(W_s/\rho_b)$ ...(3.16)

Since bulk density, ρ_b is defined as W_s/V and particle density, ρ_s as W_s/V_s. From eq. (3.16), we have

$V_v/V = n = 1 - \rho_b/\rho_s$ ...(3.17)

Multiply eq. (3.17) by 100 to calculate soil porosity in percentage.

Procedure

1. Determine the bulk density with any method described in this chapter
2. Determine particle density of soil with any method described in this chapter
3. Calculate porosity using eq. (3.17)
4. For reasonable accuracy, a value of 2.65 g/cm^3 can be used for particle density of mineral soils. In that case it is possible to dispense with step 2 of this procedure.

Guidelines for Interpretation

Total pore space (%)	Comments
< 30	Low
30-55	Medium
> 55	High

MACRO AND MICRO PORE SPACE

The total pore space of the soil comprises macro pores and micro pores. While the micro pores are mainly responsible for holding water, water in macro-pores is easily drained by gravity, pores being too large to have any significant capillary force. Nonetheless, these pores are important in holding air, an essential element of plant growth. Several reports indicate that 10% (0.10 m^3/m^3) macro-porosity is critical for soil aeration. Following characterization criteria are used to determine the relative proportions of these pores.

Total pores	The percent water content at zero cm tension
Micro-pores	The percent water content at 50 cm tension (Saha, 2008)
Macro-pores	Total pores – micro pores

Apparatus: Core sampler with rings (normally 5 cm diameter and 2 cm long), Sintered glass funnel, Vacuum flask, Filter paper, Oven, Moisture boxes, Balance

Procedure

1. Collect the soil sample with a core sampler and remove the rings with sharp knife. For disturbed samples, rings are filled with the soil sample in such a way that the bulk density of the soil is same as might be expected under field conditions
2. Set up the vacuum flask and place filter paper over the sintered plate. Place the soil ring over the filter
3. Saturate the soil overnight keeping a head of about 1 cm over the soil
4. Create zero tension on the disc of sintered glass funnel and let the system equilibrate for 8-12 hrs
5. Determine the moisture content of the soil

   Total pore space = % moisture content of the soil sample
6. Repeat the experiment with another ring
7. Apply 50 cm tension on the disc of sintered glass funnel and let the system equilibrate for 8-12 hrs
8. Determine the moisture content of the soil.

Micro-pore space = % moisture content of the soil sample

Macro-pore space = Total pore space – Micro-pore space

Notes:

- There are no codal guidelines to define macro porosity. Some reports recommend that percent water content at 60 cm tension should be used to determine macro-porosity (Kiehl, 1979). It is recommended that one should mention the tension at which macro-porosity is evaluated.
- Macro-pores can also be created by cracks, division of peds and aggregates, plant roots, and animal burrows. These pores act as preferential flow paths. This issue is discussed in a separate section in Chapter 14 of this book.

Degree of Saturation

The degree of saturation is defined as the fraction of voids filled with water. Denoted by *S*, it is normally expressed in percentage as

$$S = (V_w/V_v)\ 100$$

The degree of saturation is 0% for a fully dry, and 100% for a fully saturated soil.

Air-Filled Porosity and Air Content

It is an important criterion of soil aeration. Air-filled porosity (S_a) is the volume occupied by air (V_a) per unit volume of soil and expressed as percentage.

$$S_a = 100\ (V_a/V)$$

It is related to the degree of soil saturation as per the following relation:

$$S_a = n(1 - S)$$

A similar term is air content, A_c. It is the ratio of the volume of air to the volume of voids.

$$A_c = V_a/V_v$$

The value of air content is zero for a fully saturated soil.

Depth of Water

The depth of water, d_w in cm or mm contained in the soil depth, d_t is calculated as:

$$d_w = \theta_v \times d_t$$

Here θ_v is the volumetric moisture content.

Since $\theta_v = \theta_g \times \rho_b$

$$d_w = \theta_g \times \rho_b \times d_t$$

Here θ_g is the gravimetric moisture content.

VOID RATIO

The void ratio, denoted as *e*, is the ratio of the volume of voids to the volume of soil solids. It is expressed as a decimal (Singh, 1967).

$$e = V_v/V_s \qquad ...(3.18)$$

It indicates the volume change tendency control of soil. For loose soils with high void ratios, voids tend to minimize under loading i.e. adjacent particles contract. The opposite situation, i.e. when void ratio is relatively small (dense soils), it indicates that the volume of the soil is vulnerable to increase under loading - particles dilate. The void ratio, besides other factors also depends on the consistence and packing of the soil. It is directly affected by compaction. The void ratio of coarse-grained soils is smaller than that of fine-grained soils.

Soil Porosity and Void Ratio

The relationship between soil porosity and void ratio is established as per the following procedure.

Since $V = V_v + V_s$

$n = V_v / (V_v + V_s)$

$1/n = (V_v + V_s)/V_v = 1 + 1/e = (1 + e)/e$

$$n = e/(1 + e) \qquad ...(3.19)$$

Similarly $e = V_v/V_s = V_v/(V - V_v)$

$1/e = (V - V_v)/ V_v = (1/n) - 1 = (1 - n)/n$

$$e = n/(1 - n) \qquad ...(3.20)$$

Example 3.1. A soil sample having the volume of 75 cm^3 weighs 150.5 g at the field moisture content. When saturated, it has the mass of 153.6 g. If the oven-dry weight of the sample is 126.3 g, calculate the soil porosity, volumetric water content, dry bulk density, particle density, porosity and void ratio. Assume the water density as 0.998 g/cm^3 at the room temperature of 20°C.

Weight of the water in saturated soil = 153.6 –126.3 = 27.3 g

Volume of water at 20°C = 27.3/0.998= 27.335 cm^3

Porosity = 27.335/75 = 0.3644 or 36.44%

Bulk density = 126.3/75 = 1.684 g/cm^3

The volume of the solids = 75 – 27.335 = 47.665 cm^3

Particle density = 126.3/47.665 = 2.65 g/cm^3

Porosity = 1 – (1.684/2.65) = 0.3645 or 36.45%

(Note the value is same as determined earlier)

Void ratio, e = 0.3645/(1 – 0.3645) = 0.5735

Example 3.2. Calculate the bulk density and void ratio of the soil having saturation percentage of 40%. Assume any value required for the calculations.

Soil porosity, n = 40%

Assume the particle density = 2.65 g/cm^3

Using the formula: $n = 1 - \rho_b/\rho_s$

Bulk density = (1 – 0.4) × 2.65 = 1.59 g/cm^3

Void ratio, $e = n/(1 - n)$ = 0.4/0.6 = 0.67

Chapter 4

Sieve Analysis

The grain-size analysis is widely used in soil classification. It comprises plotting grain-size distribution curves using data from sieve/sedimentation analysis. It helps to determine soil texture, assess other soil properties through correlations, design filters applied in several engineering applications such as the construction of earth dams and assess soil suitability for roads and air field's construction, etc. Information obtained from grain-size analysis is used to predict soil water movement although permeability tests are preferred and used for this purpose. Grain-size analysis reveals that the particle size range in soils is very large. It varies from boulders having dimension of over 300 mm down to clay particles less than 0.002 mm. Within clays, some may contain particles less than 0.001 mm in size. These small size particles behave as colloids and fail to settle in water. Soils in the Indian Standard Soil Classification System (ISSCS) from an engineering point of new are classified into groups according to size. Depending upon their grain- size range, soils particles are grouped as boulder, cobble, gravel, sand, silt or clay. Symbols used for gravel, sand, silt, and clay respectively are G, S, M, and C. The terms used in this classification indicate grain-sizes and do not signify the actual soil types. Following table illustrates the ISSCS classification.

Very coarse	Boulder		> 300 mm
	Cobble		80 - 300 mm
Coarse	Gravel (G)	Coarse	20 - 80 mm
		Fine	4.75 - 20 mm
	Sand (S)	Coarse	2 - 4.75 mm
		Medium	0.425 - 2 mm
		Fine	0.075 - 0.425 mm
Fine	Silt (M)		0.002 - 0.075 mm
	Clay (C)		< 0.002 mm

Sieve analysis and sedimentation analysis tests are carried out to determine the distribution of particle sizes in a soil sample. While sieve analysis procedures and their applications are included in this chapter, sedimentation analysis is covered in the next chapter.

Dry Sieve Analysis: The process is used to quantitatively determine the particle/grain-size distribution for soil particles of sizes in the range of 4.75 mm to 0.075 mm (75 microns). In some cases, when specimens contain a high percentage of granular materials and enough fines, sample preparation process may require wet sieving. It is because fines can stick together to form clumps. In such cases, dry sieving alone may not give accurate grain-size distribution.

Wet Sieving: If the soil contains a substantial quantity, more than 5% of fine particles, wet sieve analysis is performed along with the dry sieve analysis.

Sedimentation Analysis: It is performed only for the soil fractions finer than 75 microns, when sieve test is performed. Soil passing through 2-mm sieve is used for this test, when sedimentation test is conducted without sieve analysis.

DRY SIEVE ANALYSIS

The dry sieve analysis as already stated is performed with or without wet sieving. We discuss the wet sieving technique under a separate head and then discuss the dry sieve test for which appropriate sieves are selected from the full set of sieves sizes reported in Table 4.1. IS-460 (BIS, 1985) applies to test sieves for various purposes having aperture sizes from 125 mm down to 0·032 mm (32 μm).

Table 4.1. *US standard mesh and aperture openings of the sieves*

US standard mesh	Sieve aperture (mm)	US standard mesh	Sieve aperture (mm)	US standard mesh	Sieve aperture (mm)
3.5	5.66	18	1.00	80	0.177
4	4.76	20	0.841	100	0.149
5	4.00	25	0.707	120	0.125
6	3.36	30	0.595	140	0.105
7	2.83	35	0.500	170	0.088
8	2.38	40	0.420	200	0.075
10	2.00	45	0.354	230	0.063
12	1.68	50	0.297	270	0.053
14	1.41	60	0.250	325	0.044
16	1.19	70	0.210	400	0.037

Apparatus: Stack of test sieves, Balance (with accuracy to 0.01g), Rubber pestle and mortar (for crushing the test material, if lumped or conglomerated), Sieve shaker, Oven, 2% Calgon solution, if required

Wet Sieving

- Take 500 g of oven-dry soil sample and soak it in water. For heavy clays requiring deflocculation, 2% Calgon solution is used instead of water
- The sample is stirred and left for soaking at least for 10 min
- Place the weighed sample on #200 sieve (75 micron) and "wash" with tap water. The fine particles pass through the sieve into the pan. Be careful, regulate the water to ensure that the soil does not splash out of the sieve
- The material is washed until the water coming out of the sieve appears clear
- The soil retained on 0.075-mm sieve is collected and dried in an oven. This portion alone is used in the dry sieve analysis
- The material received in pan is equal to the total mass of soil taken for wet sieve analysis minus the mass retained on the sieve. It is the weight of fines, that is the weight of the soil particles having a size 0.075 mm or less. This weight is added to

each soil fraction passing through each sieve in dry sieve analysis to make the total weight of the soil same as used in wet sieving test.

Dry Sieving

If wet sieving has not been performed, the sample is prepared as per the following steps.

1. About 500 g or more of air-dry sample free from tree roots, and pieces of bark etc is taken and is sieved through 4.75 mm IS sieve. Ensure that at least 450 g pass through the sieve. Sample above the size of 4.75 mm is sieved separately though 20 mm, and10-mm sieves and the weights retained on each are recorded
2. The sample is dried to constant weight on a hot plate or in an oven at a temperature of 110°C. When air-dry soil is used then go to the next step.
3. Weigh a small portion of the sample, oven-dry at 105°C for 24 hr, and reweigh to determine the moisture content of the soil sample. Note the moisture content of the air-dry sample and record on the data sheet. It is used to report the results on oven-dry basis
4. Prepare a stack of test sieves of various sizes say from 4-200 mesh (Table 4.2). The sieves are stacked in order, with the largest aperture size at the top, and the smallest at the bottom. The very last sieve is 0.075 mm. The sieve stack should have a lid at the top and a pan at the base to collect the fraction of soil passing 0.075-mm sieve (Fig. 4.1)
5. Weigh all the sieves and the pan separately
6. Pour the sample into the top of the stack of sieves and put the lid. Place the stack in the sieve shaker and fix the clamps, adjust the timer to 10 or 15 min, and switch on the shaker
7. Stop the sieve shaker and measure the mass of each sieve plus soil retained. Calculate the soil/material retained on each sieve

Fig. 4.1. *A pictorial view of the sieve shaker used in sieve tests*

8. If a sizable portion of soil is retained on the 0.075-mm sieve, it should be washed. It is accomplished by placing the sieve with retained soil in a pan. Pour clean water on the sieve. A spoon or glass rod may be used to gently stir the slurry. Collect the soil that passes through the sieve. Dry the soil and weigh it. The weight of soil recovered should be subtracted from the weight retained on the 0.075-mm sieve and added to the weight received in the pan
9. As already stated, sedimentation analysis can be performed with the whole soil that has passed the 2-mm sieve.

Preparation of Results

1. The mass of soil retained on each sieve is determined by subtracting the weight of the empty sieve from the mass of the sieve + retained soil. It is recorded on the data sheet as the weight of the soil retained (Column 5 of Table 4.2). The cumulative sum of the retained masses on all the sieves should approximate to the initial mass of the soil sample. A loss of more than 2% is unsatisfactory and may require retesting. If wet sieving has been done, the soil passing through 0.075-mm sieve may be added to each sieve and use the original soil weight that was taken for wet sieving to determine the percent soil passing each sieve (Not shown in Table 4.2)
2. Calculate the percent retained on each sieve by dividing the weight retained on each sieve by the original sample mass (column 6)
3. Calculate the percent passing (or percent finer) by starting with 100% and subtracting the percent retained on each sieve as a cumulative percentage
4. Plot the data with percentage passing on the y-axis and sieve size on the x-axis
5. The curve can be extended by plotting the fine sand, silt and clay fractions obtained from sedimentation analysis discussed in the next chapter.

Let us use one of the rows of Table 4.2 to illustrate its construction.

Total soil mass, $A = 830$ g

Mass retained on No. 4 sieve = $A_1 - B_1$ =155 g (Column 5)

The percent retained = Mass retained/Total mass = R_1 = (155/830) × 100 = 18.6%

The % passing = 100 – 18.6 = 81.4% (Column 8)

The process is repeated for all the sieves (Table 4.2).

Table 4.2. *Data sheet for sieve analysis*

Soil weight: A g — Moisture content of air-dry soil: %

1	2	3	4	5	6	7	8
Sieve #	**Sieve size (mm)**	**Mass of each sieve (mm)**	**Mass of each sieve + soil retained**	**Mass of soil retained (g) (Col.4-Col.3)**	**Percentage retained on the sieve, R_n = 2 (Col.5/A) × 100**	**The cumulative percent retained, ΣR_n**	**% finer, $100 - \Sigma R_n$**
4	4.75	A_1	B_1	$B_1 - A_1$	R_1	R_1	$100 - R_1$
10	2.00	A_2	B_2		R_2	$R_1 + R_2$	
16	1.18	A_3	B_3		R_3		

30	0.6	A_4	B_4		R_4		
40	0.42	A_5	B_5		R_5		
50	0.30	A_6	B_6		R_6		
100	0.15	A_7	B_7		R_7		
200	0.075	A_8	B_8		R_8		
Pan		A_9	B_9	$B_9 - A_9$	R_9	$R_1 + R_2 + \ldots R_9$	–

Interpretation

The grain-size distribution curves are typically plotted on a semi-logarithmic scale on a semi-log paper (Fig. 4.2), mainly because the soil particles size can range from 20 mm down to less than 0.002 mm. While log-sieve size is taken on x-axis, % retained or % finer is taken on y-axis. The graph so prepared is known as soil gradation curve.

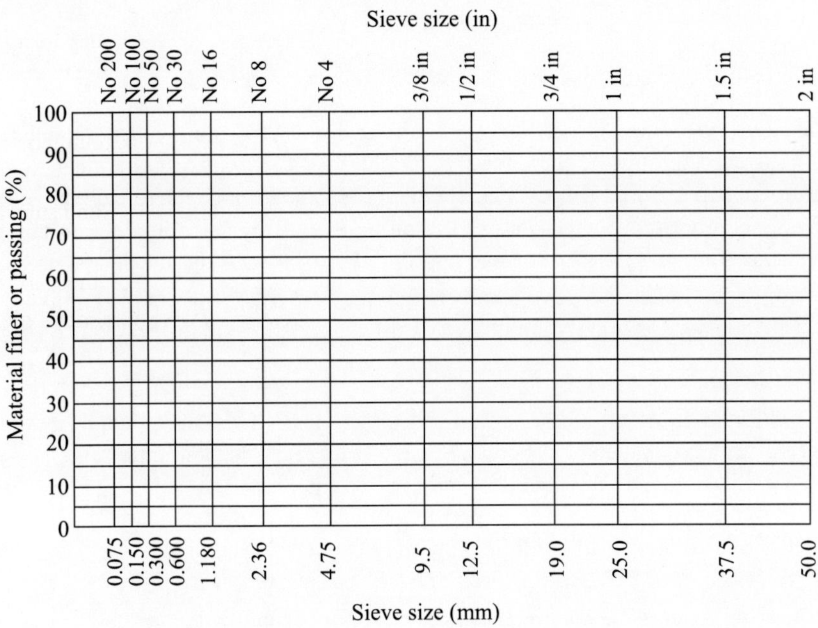

Fig. 4.2. *Semi-log paper for drawing soil gradation curve*

The shape of the cumulative particle size distribution curve is an indication of the uniformity of the soil (Fig. 4.3). A steep curve indicates a uniform soil while a flat curve reveals a well graded soil. Humps in the curve indicate a mixture of two or more uniform soils. These are but few specific instances and many more observations can be made on the basis of the shape of the curve. Three major points on the grading curve namely d_{60}, d_{30} and d_{10} are used to describe the grading characteristics of the soil.

d_{60} = size at 60% finer by weight

d_{30} = size at 30% finer by weight

d_{10} = size at 10% finer by weight

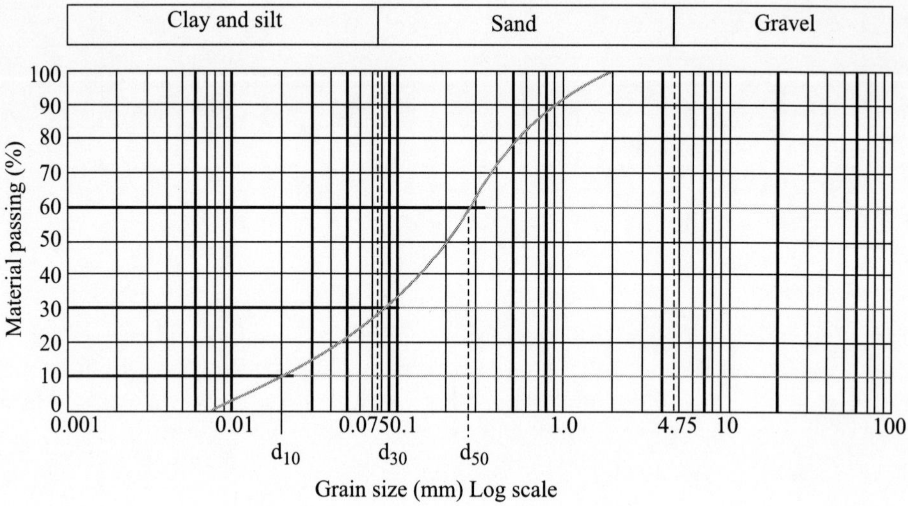

Fig. 4.3. *A typical soil gradation curve*

Value of d_{10}, meaning 10% of the grains are finer and 90% are coarser (retained) than this value, is used to indicate the effective diameters of the grains. It is a measure of the fineness of the soil material. Experimental evidences have been generated to conclude that diameter d_{10} is more representational of the properties of the soil sample than any other diameter. Therefore, it is more useful and appropriate for calculating several soil properties.

Uniformity coefficient is determined by taking the ratio of d_{60}, 60% of the grains are finer and 40% are coarser (retained) than this value, to d_{10} as defined in the previous paragraph.

Uniformity coefficient, $CU = d_{60}/d_{10}$ ...(4.1)

If CU is less than 2, it indicates a uniform material. If CU is more than 4, it is highly non-uniform material. When CU is greater than 5 (some reports mention a range of 4 to 6), it is a well graded soil (soil having a distribution of particles over a wide size range). If CU is less than 4, the soil is categorized as poorly graded or uniformly graded. The uniformity coefficient of a well graded soil will be more than about 4 for gravels. It is more than 6 for sands.

Another parameter is the coefficient of curvature or gradation denoted as C_c. It is given as:

Coefficient of curvature, $C_c = (d_{30})^2/ (d_{60} \times d_{10})$ ...(4.2)

Here d_{30} means 30% of the grains are finer and 70% are coarser (retained) than this value. The coefficient of gradation for gravels and sands remains between 1 and 3. The values of both the parameters CU and C_c will be equal to 1 for single-sized soils. As an example, the coefficient of uniformity, CU is approximately equal to 1 for desert sands.

Example 4.1. From a grain-size distribution curve, d_{10} was measured as 0.13 mm, d_{30} as 0.5 mm and d_{60} as 1.8 mm. Determine the effective diameter of the soil, coefficient of uniformity and coefficient of curvature. Comment upon the soil characteristics.

Effective diameter of the soil = d_{10} = 0.13 mm

CU = 1.8/0.13 = 13.8

$C_c = (0.5)^2/(1.8 \times 0.13) = 1.1$

Since CU is greater than 6 and C_c in the range of 1 to 3, it is a well graded soil.

Applications

Soil gradation indicates many engineering properties, namely compressibility, shear strength, and hydraulic conductivity being few of them. In drainage designs, the gradation of the *in-situ* or on site soil often controls the design and groundwater drainage of the site. A poorly graded soil, having more void spaces, has better drainage than a well graded soil having lesser void spaces. When a fill material is being selected for a project such as a highway embankment or earthen dam, a well graded soil can be compacted more than a poorly graded soil. In this section, it is proposed to illustrate the application of gradation curves for design of drainage envelopes.

Filter Design for Subsurface Drainage

Most envelope design criteria attempt to establish filter material gradation curve to suit a specific base material gradation curve (USBR, 1978). More often, a graded envelope (Natural) is designed as given in row 2 of Table 4.3. It states that

Table 4.3. *USBR criteria for envelope design*

Uniform envelope (natural)	$D_{50}/d_{50} = 5 - 10$	
Graded envelope (natural)	$D_{50}/d_{50} = 12 - 58$ $D_{15}/d_{15} = 12 - 48$	
Graded envelope (crushed rock)	$D_{50}/d_{50} = 9 - 30$ $D_{15}/d_{15} = 6 - 18$	
General	$D_{100} \leq 80$ mm	It helps to minimize segregation and bridging during envelope placement
	$D_5 \geq 0.07$ mm	It helps to prevent movement of fines
	$D_{\text{opening}} \leq 0.5D_{85}$	Size of drain perforation is adjusted to the size of the filter material used

$$D_{50}/d_{50} = 12 - 58 \qquad \text{...(4.3)}$$

$$D_{15}/d_{15} = 12 - 48 \qquad \text{...(4.4)}$$

Here D_{50}, D_{15}, d_{50} and d_{15} are 50% and 15% finer for filter and base material (soil) respectively. It means that the envelope D_{50} should be in the range of 12 to 58 times the d_{50} of base material (Soil taken at the drain depth) and envelope D_{15} in the range of 12 to 48 of the d_{15} of base material. For other kinds of gravel envelope, follow guidelines given in respective rows of Table 4.3. Besides, the general criteria listed in the last three rows of Table 4.3 should also be met with. It is common to all kinds of envelopes in row 1 to row 3 of Table 4.3.

USSCS (1988) criteria for filter and surround design gives the range of filter/envelope to that of the base material as follows.

SCS criteria for filter	$D_{15} < 7d_{85}$ $D_{15} > 4d_{15}$ $D_5 > 0.0074$ mm	Need not be smaller than 0.6 mm % passing sieve No. 200 less than 5%
SCS criteria for envelope (surround)	$D_{100} < 38.1$ mm $D_{30} > 0.25$ mm $D_5 > 0.074$ mm Minimum envelope thickness 76 mm	The whole sample should pass the sieve of 38 mm % passing sieve No. 60 less than 30% % passing sieve No. 200 less than 5%

The design of gravel envelope as per SCS criterion (USBR, 1978) is independent of the base soil curve. This criterion designs the filter utilizing only the d_{60} of the base soil. For example, if the d_{60} of the soil is 0.04 mm, select the row in which this value is listed in column 1 of Table 4.4. This value falls in the first row. Therefore, apply guidelines of row 1 of Table 4.4. The minimum and maximum values of D_m of the envelope gradation curve are chosen. The selected values are shown in Table 4.5. The envelope gradation curve should meet these guidelines so as to be an effective surround.

Table 4.4. *USBR criteria for surround design (USBR, 1978)*

Base soil for d_{60} (mm)	**Lower limits (mm)**						**Upper limits (mm)**					
	Percentage passing						**Percentage passing**					
	100	60	30	10	5	0	100	60	30	10	5	0
0.020-0.050	9.5	2.0	0.81	0.33	0.3	0.07	38.1	10.0	8.7	2.5	–	0.59
0.050-0.100	9.5	3.0	1.07	0.38	0.3	0.07	38.1	12.0	10.0	3.0	–	0.59
0.100-0.250	9.5	4.0	1.30	0.40	0.3	0.07	38.1	15.0	13.1	3.8	–	0.59
0.250-1.000	9.5	5.0	1.45	0.42	0.3	0.07	38.1	20.0	17.3	5.0	–	0.59

Table 4.5. *Minimum and maximum particle size range of the envelope gradation curve*

Diameter (% passing)	**Particle size (mm)**	
	Minimum	**Maximum**
D_{100}	9.5	38.1
D_{60}	2.0	10.0
D_{30}	0.81	8.7
D_{10}	0.33	2.5
D_5	0.3	–
D_0	0.07	0.59

Hydraulic Conductivity of Aquifers

The Hazen formula is commonly used in groundwater hydrology to estimate hydraulic conductivity of aquifers. It is a simple relationship between the hydraulic conductivity (K) and the effective grain diameter. It is given as (Freeze and Cherry, 1979):

$$K = A \times D_{10}^2 \qquad ...(4.5)$$

Here, K is hydraulic conductivity, (cm/s); D_{10} is effective grain diameter, mm, which is determined from the grain-size distribution curve and A is a constant usually taken as 1.0 (Freeze and Cherry, 1979).

SOIL AGGREGATE STABILITY

Water Stable Aggregates (Laboratory Method)

Yoder (1936) was the first researcher to have developed wet-sieving techniques to determine the aggregate stability to water. His premise has been that unstable aggregates break down easily than stable aggregates. He used a nest of six sieves secured to a holder. The nest moves up and down in a drum of water through a distance of 3.18 cm at a rate of 30 cycles/min for at least 10 min. Now the Yoder's apparatus has been modified in several ways to suit specific purposes.

Apparatus and Reagents: A wet sieve shaker, Set of sieves with sieve openings of 5.0, 2.0, 1.0, 0.5, 0.25 and 0.1 mm diameter stacked in descending order.

- **H_2O_2**
- **Dispersing agent: 1 N NaOH**

Procedure

1. Take 50 g (M_0) of air-dried soil aggregates (passing through 8 mm and retained on 5-mm sieves). Place the sample aggregates gently onto the top of the nest of sieves
2. Slowly bring the level of the water in the drum at which water just begins to wet up the aggregates. Allow the aggregates to wet for 10 min. Switch on the mechanical oscillator to move the nest of sieves up and down with a frequency of 30 cycles per minute and a stroke of 3 cm. Continue for 10 min
3. Remove the nest of sieves from the water and allow it to drain for some time. Gently separate the sieves
4. Pour and wash out the contents of each sieve into a pre-weighed beaker and let them settle for 24 h. Then carefully decant excess water without losing any soil. Dry the beakers along with the contents in an oven at 105°C for 24-48 hr. Weigh the beaker along with the dry soil aggregates and find out the weight of the dry soil aggregates in each beaker. Note down the weights as M_1, M_2, M_3, M_4, M_5 and M_6 for 5.0, 2.0, 1.0, 0.5, 0.25 and 0.1-mm sieves respectively
5. Carry out another test. Now add 30-40 mL of H_2O_2 into a beaker containing dry soil aggregates to remove organic matter, the binding agent of primary particles. Repeat the process till the organic matter is completely removed
6. Transfer the contents into the dispersion cup. Add dispersing agent (10 mL of 1N NaOH) and enough distilled water to fill the cup so that it is easier to stir the contents by a mechanical stirrer. Stir the suspension for 10 min. Wash the suspension on an identical set of sieves as used in the previous test by means of a stream of tap water. Transfer the contents of the sieves into beakers. The primary particles in each beaker are oven-dried and weighed. Note down the weights as M_7, M_8, M_9, M_{10}, M_{11} and M_{12} for 5.0, 2.0, 1.0, 0.5, 0.25 and 0.1-mm sieves respectively

7. Oven-dry and weigh the soil passing through 0.1-mm sieve (M_{13}) received in pan. To calculate the percent water stable aggregate, use the following equation.

Total aggregates (%) > 0.25 mm

$$= [(M_1 + M_2 + M_3 + M_4 + M_5)/M_0] \times 100 \quad ...(4.6)$$

Total aggregates (%) > 0.25 mm corrected for primary particles = $100\ [\sum(M_i - M_{i+6})]/M_0$

Such that i varies from 1 to 5, and M_i are the weights of aggregates retained on respective sieves from top to bottom before and after the dispersion.

$$\text{Mean weight diameter (MWD)} = \Sigma X_i \times M_i/\Sigma M_i \quad ...(4.7)$$

In eq. (4.7) and eq. (4.8) i varies from 1 to 6 such that X_i is the mean opening size e.g. 6.5, 3.5, 1.5, 0.75, 0.375, 0.175 mm for 8.0-5.0, 5.0-2.0, 2.0-1.0, 1.0-0.5, 0.5-0.25 mm size classes, respectively.

$$\text{Geometric mean weight diameter (GWD)} = \exp\ [\Sigma(M_i \times \text{Ln}\ X_i)/(\Sigma M_i)] \quad ...(4.8)$$

M_{13} is un-aggregated soil material less than 0.1 mm in diameter.

Example 4.2. Yoder's wet sieving method was used for aggregate analysis. Calculate the mean weight diameter of water stable aggregates from the data given in columns 1 and 2 of the following Table. Weight of the soil aggregates passing 8 mm sieve used in the analysis is 25 g.

Sieve size class (mm)	Weight of aggregates retained, M_i (g)	Mean diameter of each size fraction, X_i (mm)	Fraction of total weight, w_i	$X_i w_i$
1	2	3	4 = 2/SM	5 = 3 × 4
4.0	12.12	6.0	0.485	2.910
4.0-2.0	4.35	3.0	0.174	0.522
2.0-1.0	4.12	1.5	0.165	0.248
1.0-0.5	3.06	0.75	0.122	0.122
0.5-0.25	0.98	0.375	0.039	0.015
0.25-0.10	0.37	0.175	0.015	0.003
MWD				3.790 mm

Extend the Table by adding 3 more columns to the right. Note down the mean diameters in column 3 (As an example of row 2, we have (4 + 2)/2 = 3). Fill other rows of this column in the same way. Column 4 is column 3 divided by column 2 for the respective rows. Column 5 is column 3 times column 4 for the respective rows. MWD is the sum of the values of all the rows of column 5. It gives the mean diameter of the water stable aggregates as 3.79 mm.

Applications in Soil Erosion Studies

Measurement of wet Aggregates

The kinetic forces of rain are capable of disintegrating soil mass. Following method is used to assess the stability of soil particles to rain energy.

1. Soil is air-dried and placed on vertically stacked sieves of 2.0 mm, 0.25 mm and a catch pan. The soil is shaken for 15 seconds on a Tyler coarse sieve shaker to separate out aggregates of 0.25 - 2.0 mm size for analysis
2. A single layer of aggregates in the range of 0.25 - 2.0 mm size is spread on a 0.25-mm sieve
3. Sieve is placed at a distance of 500 mm (20 inches) below a rainfall simulator, which delivers individual drops of 4.0 mm diameter. The test is run for 5 min during which it delivers 12.5 mm of water (approximately 0.5 inches) as drops. A total of 0.74 J of energy impacts the sieve over the period of 5 min rainfall
4. The slaked soil material that fall through the sieve, and any stones remaining on the sieve are collected, dried and weighed

The fraction of water stable soil aggregates (WSA) is calculated using the following equation:

$$WSA = W_{stable} / W_{total} \quad ...(4.9)$$

Where: $$W_{stable} = W_{total} - (W_{slaked} + W_{stones}) \quad ...(4.10)$$

Here W is weight (g) with subscript stable denoting stable soil aggregates, total denoting aggregates tested and slaked are the aggregates slaked out of sieve and stones means the stones retained on the sieve after test. Corrections must be made for the stones.

Hazelton and Murphy (2007) rated the soils on the basis of 1-2 mm water stable aggregates as per the following guidelines. It is necessary to conduct separate tests on 1–2 mm soil aggregates to apply these guidelines.

1–2 mm water stable aggregates (%)	Rating
<10	Very low
10–20	Low
20–30	Moderate
>30	High

Measurement of Dry Aggregates

Like rain, the kinetic forces of wind also disintegrate soil mass, although effects may vary due to soil texture, moisture content and management practices. A rotary sieve, that simulates the wind action in cultivated fields, has been developed to measure size distributions of soil aggregates (Chepil, 1952).

Apparatus: Nest of sieves with openings of 25, 10, 5, 2, 1, 0.5 and 0.25 mm sizes along with a pan and a lid, Rotary sieve shaker, Balance, Spade

Procedure

1. Collect the sample with the help of a spade. The soil at this time should be reasonably dry to avoid breakdown or change in structure. Remove visible plant roots or stones in the sample

2. Keep the nest of sieves on the rotary shaker in descending order of openings. The bottom sieve is fitted in a pan
3. Place weighed soil sample (M_0) on the top sieve. Switch on the shaker for 20 min, and then remove the sieves. Collect the soil retained on each sieve
4. Weigh the amount of soil retained in each sieve and note them as M_1, M_2, M_3, M_4, M_5, M_6 and M_7.

Calculate the % distribution of dry aggregates retained on each sieve as per the following relation:

$$\text{Aggregates for a size group (\%)} > 25 \text{ mm} = (M_1/M_0) \times 100 \quad \text{...(4.11)}$$

$$\text{Aggregates for a size group (\%) } 17.5 \text{ mm } [(25+10)/2 = 17.5] = (M_2/M_0) \times 100 \quad \text{...(4.12)}$$

Similar calculations for other sizes can be made with the last size being

$$\text{Aggregates for a size group (\%) } 0.375 \text{ mm } [(0.5+0.25)/2 = 0.375] = (M_7/M_0) \times 100 \quad \text{...(4.13)}$$

Chapter 5

Soil Texture and Structure

Soil texture and soil structure are unique properties of the soil having a profound effect on their behaviour. Whereas particle size distribution is an important parameter in soil classification, it has implications for available soil water, soil aeration, and nutrient availability to plants. Soil texture, determined by the relative proportions of sand, silt and clay particles, influences fertility, drainage, water-holding capacity, aeration, tillage, and soil strength. Because texture is so important, this test is always performed as a standard physical test to estimate sand, silt, and clay fractions in soils. The three commonly used procedures to assess particle size distribution are the feel method, the pipette gravimetric method (Piper, 1942; Black, 1965, Chopra and Kanwar, 1991) and the hydrometer method (Ashworth et al., 2001).

The primary soil particles do not exist as such in natural conditions but are bonded together into larger units or aggregates, termed as secondary particles. Soil structure describes the arrangement of primary and secondary soil particles and soil pores between them. Therefore, soil structure is also defined as the arrangement of various sized soil pores in a certain structural pattern (FAO, 2006, Soil Survey Division Staff, 1993). Soil structure affects water and air movement in the soils, nutrient availability to plants, root growth, and micro-organism activity. The pore spaces created by peds (natural aggregates) are larger than those in between individual sand, silt, or clay particles. As a consequence, air and water movement is better in a structured soil allowing better root growth. Moreover, the ability of aggregates to hold water and nutrients is much superior to individual soil particles. Besides, strong aggregation decreases detachability and transportability of soil particles by water or wind resulting in reduced runoff and soil erosion (Chapter 4). Soil texture and structure are closely linked properties. Although these properties are studied together, it is difficult to change texture but easier to manipulate soil structure. The physical changes imposed by the farmer in ploughing, cultivating, draining, application of gypsum or liming, and manuring aim at bringing structural rather than textural changes.

SOIL TEXTURE

Particle-size distribution or texture class is one of the first things required during examination of the soil. Soil texture is the numerical proportion (percent by weight) of sand, silt, and clay in the fine-earth fraction (≤ 2 mm). The sand size is further grouped in 5 classes from very coarse to very fine as follows.

Soil separate	Particle size (equivalent diameter, mm)
Sand	0.05 - 2
Very coarse	1 - 2
Coarse	0.5 - 1
Medium	0.25 - 0.5
Fine	0.1 - 0.25

Very fine	0.05 - 0.1
Silt	0.002 - 0.05
Clay	< 0.002

Feel Method

Feel method is qualitative in nature. The basic principle of this method has its origin on the following characteristics of the sand, silt and clay.

Sand	Particles can be seen and are gritty in nature
Silt	Feels like talcum powder when dry and slightly plastic when wet
Clay	Feels hard when dry and very plastic and sticky when wet

More than one versions of feel methods are available in the literature. The nature of the examination in each case depends upon whether one is examining a dry or moist soil (Table 5.1) or forming a ball or ribbon (Soil Survey Division Staff, 1993).

Table 5.1. *Dry and moist examination of soil for soil texture evaluation*

Soil texture	Dry examination	Moist examination	Sand: Silt: Clay (%)
Sand	Loose and single grained, feels gritty	Very easily forms crumbled ball	85-100 : 0-15 : 0-10
Loamy sand	Silt and clay may mask sand, feel loose, gritty	Feel gritty, forms easily crumbled ball, slightly stains fingers	70-90 : 0-30 : 0-15
Sandy loam	Clods easily break, one can see and feel sand	Moderately gritty, forms ball and can stand careful handling, definitely stains fingers	43-85 : 0-50 : 0-20
Loam	Clods moderately difficult to break, slightly gritty	Neither very gritty nor very smooth, forms a ball, stains fingers	23-52 : 28-50 : 7-27
Silt loam	Clods difficult to break, when pulverized feel smooth, soft, and floury, shows fingerprints	Has smooth or slick butter like feel, stains fingers	0-50 : 50-88 : 0-27
Clay loam	Clods very difficult to break with fingers	Slight gritty feel, stains fingers, ribbons fairly well	20-45 : 15-53 : 27-40
Silty clay loam	Same as above, but very smooth		0-20 : 40-73 : 27-40
Sandy clay loam	Same as for clay loam		45-80 : 0-28 : 20-35
Clay	Clods cannot be broken with fingers without exerting extreme pressure	Quite plastic and usually sticky when wet, stains fingers. (silty clay feel smooth, sandy clay feel gritty)	0-45: 0-40 : 40-100

The method of examination by forming a ball or ribbon is briefly included from the descriptions in the Soil Survey Division Staff (1993), Schoeneberger et al. (2012) and others.

1. Place about 25 g soil in the palm. Knead the soil to break aggregates while adding little water (in drops) at a time. Stop adding water when the soil is mouldable. Shape the soil into a ball by squeezing. The soil is predominantly sand, if no ball is formed
2. If ball is formed then hold the ball between thumb and forefingers, pushing and squeezing upward to form the ball into a ribbon. If the soil fails to form a ribbon, it is likely to be loamy sand, silt loam, or coarse silt loam. If the soil forms a ribbon, test as per guideline given in step 3
3. If a ribbon 2.5 cm long is formed before it breaks, wet and rub the ribbon with the forefingers. The soil is sandy loam, if it feels gritty. It is silt, if it feels very smooth without any grittiness. If it feels smooth with only a little bit of grittiness, it is silt loam, and if neither grittiness or smoothness predominates, then it is loam
4. If soil forms a ribbon 2.5 - 5.0 cm (1-2 inches) long before breaking, wet and rub it as in step 3. If the soil feels gritty, it is sandy clay loam. If it feels smooth with no grittiness, it is silty clay loam. The soil is categorized as clay loam, if none of the grittiness or smoothness predominates
5. If the soil forms a ribbon longer than 5 cm before it breaks, wet and rub as in step 3. A sandy clay soil feels gritty, and silty clay feels smooth. If neither smoothness nor grittiness predominates, it is clay
6. Record the textural class of each sample. Mention the characteristics indicating the identified textural class

SEDIMENTATION TECHNIQUES

Stokes Law

The sedimentation method is used to measure mainly the silt and clay fractions in the soil. These fractions are separated by a process known as sedimentation. The process relies on differential settling velocities of various sizes of particles in a water column as explained by an expression derived by stokes based on drag force exerted on spherical particles of different sizes in a viscous liquid.. It states that larger particles settle through the water quickly than the smaller ones. As per Stokes law, if all other factors remain constant, then the settling velocity is proportional to the square of the radius of the particles (Stokes, 1851).

$$V = 2r^2 (\rho_s - \rho_l)\, g / 9\eta \qquad ...(5.1)$$

Here V is the velocity of fall (cm/s), r is the equivalent radius of the particle (cm), ρ_s is the density of particles (g/cm^3), ρ_l is the density of liquid (g/cm^3), g is the acceleration of gravity (cm/s^2), and η is the viscosity of liquid (g/cm-s). For soils, ρ_s is taken as 2.65 (a good approximation) and for water ρ_l ~1.0. In practice, both the density of water and its viscosity are affected by temperature. As a consequence, soil particles velocity of fall increases with increasing temperature. Therefore, a usual practice is to measure and make corrections for the temperature of the liquid.

Pipette Method

Apparatus and Reagents: Soil dispersing stirrer, Pipette (Fig 5.1 left), Stop watch, Hot plate, Drying oven, Weighing balance- accurate to 0.5 g, Steam plate or boiling water bath, Desiccator, Beakers, Measuring cylinders- 1000 mL, 70 µm sieve, Whatman No. 50 filter paper, Litmus paper, Stirrer.

- **Hydrogen peroxide:** 6% H_2O_2
- **HCL:** 2 N HCl
- **Amyl alcohol**
- **Dispersing solution:** Any of the following
- **Sodium hexametaphosphate, usually called Calgon:** Dissolve 250 g of sodium hexameta-phospahte, $(NaPO_3)_6$ in 900 mL of warm deionized water. When cool, add sufficient sodium carbonate to bring the pH to 8 or 9 and make to 1 L
- **0.5 N sodium oxalate**
- **1 N sodium hydroxide:** Dissolve 40 g of sodium hydroxide in water and dilute to 1L

Procedure

This section describes the analysis procedure without performing the sieve analysis. Therefore, air-dry sample passing the 2-mm sieve is used to conduct the test.

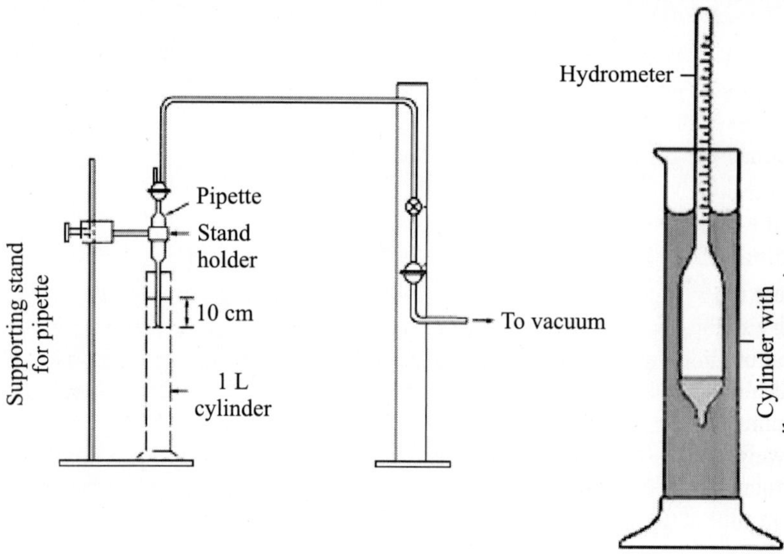

Fig. 5.1. *Soil texture analysis methods: pipette (left) and hydrometer (right)*

The Moisture Content of Air-Dry Soil

Transfer 10 g of air-dry soil into a weighed silica dish, fitted with an aluminium cover. Place the dish (uncovered) in an oven at 105°C. Dry for 12-16 hr. Put the cover and cool in a desiccator and note its weight. Multiply the loss in weight by 10, to obtain the percentage of moisture in the air-dry soil. It is used to tabulate the analysis results on oven-dry basis

$$\text{Parameter on oven-dry basis} = \text{Parameter on air-dry soil basis}/(1 + m/100) \quad ...(5.2)$$

Here m is the moisture percentage of the air-dry soil.

Removal of Organic Matter

1. Place 20 g oven-dry soil passing through a 2-mm sieve in a 400 mL beaker. Add 20 mL of 6% H_2O_2 solution (Increase the quantity of H_2O_2 to 30 to 40 mL, if the

sample contains more than 3.5% organic matter) and allow to stand overnight. Add another 10 mL of H_2O_2 solution and cover the beaker with a watch glass.

2. Place the beaker on a steam plate or over boiling water bath for 15 min. Stir when necessary to control frothing. Repeat the heat treatment with another 10 mL of H_2O_2 solution, until the colour of the soil become light enough or frothing disappears. It indicates that the organic matter has been destroyed. Boil the contents gently for another 5 min.
3. The soil may contain manganese compounds, if H_2O_2 fails to destroy the organic matter as described in the previous paragraph. In such cases, add about 40 mL of water and 0.5 g of sodium hydrogen sulphite ($NaHSO_3$) to the soil. Boil until water is half gone. Continue with step 2 to remove the organic matter.

Removal of Carbonates

1. After the organic matter is completely removed, cool the beaker and clean its sides with a rubber pestle. Add 25 mL of 2 N HCl and stir well. Add more acid at the rate of 2.5 mL for each percent of $CaCO_3$ present above 2%. Stir occasionally while the reaction proceeds. It may take an hour or so. Test after 1 hr, and make sure that an excess of acid is present. Filter the contents through a Buckner funnel using Whatman filter paper No. 50
2. Wash the soil with hot distilled water until the excess acid is fully removed. It is tested with the help of litmus paper or by adding a drop of silver nitrate solution in the filtrate
3. Repeat the same procedure to remove organic matter and carbonates using 5 g of air-dry soil. After draining as completely as possible, transfer the soil and filter paper to the original weighing dish, place in an oven at 105°C, dry for 24 hr, cool in a desiccator and weigh. The difference between the first weight (the weighing bottle containing the dried filter paper plus the weight of soil taken, 5 g in this case) and the weight after the acid treatment corresponds to the loss on peroxide and acid treatment. To express the loss in percent of the air-dry soil, simply multiply by 20. Report this value as loss on solution. Necessary corrections for reporting the results on oven-dry basis can be made since the moisture content of air-dry soil is known.

   Note: This loss is not considered while reporting the sand, silt and clay contents in the soil. Correction is only required for assessing the soil structure using the analysis results (Olmstead et al., 1930).
4. After the soil has been washed nearly free from acid and drained in step 2, transfer it from the Buckner funnel to 1000 mL long neck beaker or in a dispersing cup. After most of the soil has been removed, spread the filter paper on clock glass and clean all traces of soil from it by distilled water jet. After all the soil has been removed, roll the filter paper into a loose ball, wet it and thoroughly squeeze like a sponge. Repeat this process until no more turbid water is obtained. Also wash the Buckner funnel rinsing it with jet of distilled water.
5. When the whole soil has been transferred to the 1000 mL beaker, add 10 mL of dispersing solution as reported in reagents or 1*N* sodium hydroxide or sodium hexametaphospahte

6. Place the beaker beneath the dispersing unit, lower the propeller into position and set it into motion. Adjust the speed to cause vigorous stirring but with no splashing. Stop the stirrer after 10-15 min of stirring, raise the motor propeller and rinse the propeller blades into the beaker
7. Take a 70 mesh sieve, and hold it over the mouth of a funnel placed over a 1000 mL cylinder. Pour the suspension on to the sieve. Wash as much material as possible through the sieve using a stream of distilled water poured from the wash bottle. Transfer the coarse material left on the sieve to a weighed dish. Dry it at 105°C in an oven and find its weight. Calculate the percentage of coarse sand as follows

Weight of dish	W_1 g
Weigh of dish + dry coarse sand	W_2 g
Weight of coarse sand	$W_2 - W_1$ g
Weight of the soil sample taken	W_0 g
% coarse sand	$100(W_2 - W_1)/W_0$

If sand fractionation is required, proceed as in step 8.

8. Transfer the dried sand to a set of sieves arranged from top to bottom as follows: 1.0 mm (18-mesh), 0.5 mm (30-mesh), 0.25 mm (60-mesh), 0.105 mm (140-mesh), and 0.047 (300-mesh), and a collecting pan. Pour the sand on to the top sieve, put the cover in place, and shake the sieves on a sieve shaker for about 5 to 10 min. Weigh each sand fraction and record the weight
9. Continue from step 7. Fill the cylinder to 1000 mL mark with distilled water. Check the temperature of the suspension. Stopper the top of the cylinder and thoroughly mix the contents by inverting it several times while shaking
10. Set the cylinder on the table; and immediately record the time (hour, minute, second)
11. About 1 min before sedimentation time selected from Table 5.2, lower the tip of a closed 25 mL pipette slowly into the suspension to the proper depth (10 cm in this case) with a pre-calibrated pipette rack. At the appropriate time for silt+clay draw the sample (Fig. 5.1 left). Regulate the filling time of the pipette to about 12 s. Fill the pipette and empty into a tared 90 mL wide-mouth dish (or 100 mL beaker), and rinse the pipette into the container once

Table 5.2. *The time of sedimentation for clay and silt fractions at 10 cm depth at different temperature*

Temperature (°C)	Time for clay		Time for silt + clay	
	Hour	Min	Min	s
8	11	–	6	40
9	10	40	6	30
10	10	25	6	20
11	10	10	6	10
12	9	54	6	10
13	9	35	5	50
14	9	20	5	40

15	9	05	5	30
16	8	50	5	20
17	8	35	5	10
18	8	25	5	05
19	8	10	5	00
20	8	00	4	48
21	7	50	4	40
22	7	40	4	30
23	7	25	4	30
24	7	15	4	20
25	7	05	4	15
26	6	55	4	10
27	6	45	4	05
28	6	40	4	00
29	6	30	3	55
30	6	20	3	50
31	6	15	3	45
32	6	05	3	40
33	5	55	3	35

12. Evaporate the water by placing the dish in an oven set at 105°-110°C for at least 24 hr. Weigh the sample to 0.01 g. The % silt + clay in the sample are calculated as follows:

Weight of dish	W_1 g
Weigh of dish + silt + clay	W_2 g
Weight of silt +clay	$W_2 - W_1$ g
% silt+clay	$(W_2 - W_1) \times 1000 \times 100/(25 \times W_0) = Y$

13. Shake the contents of the cylinder for 1 min and record the temperature and time. Sample in the same way at the pre-decided depth at the settling time for clay (Table 5.2). Dry it and weigh to calculate the clay percentage as follows:

Weight of dish	W_1 g
Weigh of dish + clay	W_2 g
Weight of clay	$W_2 - W_1$ g
% clay	$(W_2 - W_1) \times 1000 \times 100/(25 \times W_0) = X$
% silt	$Y - X$

Note that during the computation of the results, the weights in steps 12 and 13 are multiplied by 40 (1000/25). Any error made at these stages will be multiplied by 40. This calls for cautious and accurate observations in these steps.

14. The weights in step 12 and 13 needs to be corrected for the amount of dispersing agent as most of it remains in solution. This correction can be made with sufficient accuracy by deducting 1.6 from the clay percentage (Piper, 1942). Alternatively, add the same amount of dispersing agent in a 1000 mL cylinder, make to mark with

distilled water, stir thoroughly, withdraw duplicate 25 mL samples, dry, and weigh. The weight so determined is subtracted from the weights. It may be around 0.012 g (Kroetsch and Wang, 2008).

Fine Sand

It can be calculated by difference or by the sedimentation method, later being preferable.

% Fine sand = 100 – (Coarse sand + Silt + Clay) ...(5.3)

Sedimentation Method

Pour away the bulk of the supernatant liquid and transfer the sediment in the cylinder to a 500 mL beaker. Make up with water to a height of 10 cm above the base, which can be marked with a pencil. Stir well and allow standing for the requisite period taken from the silt + clay column in Table 5.2 against the observed temperature. Then pour away the turbid suspension. Fill beaker again to the mark with distilled water and repeat this process till the liquid is no longer turbid. Residue left at the end of this is the fine sand, which is dried at 105°C and weighed. Calculate the percentage of fine sand as follows:

Weight of dish	W_1 g
Weight of dish + fine sand	W_2 g
Weight of fine sand	$W_2 - W_1$ g
Weight of initial soil taken	W_0 g
% of fine sand	$(W_2 - W_1) \times 100/W_0$

Cautions

If air-dry soil is taken for analysis, then moisture correction should be applied as explained in the moisture content of air-dry soil and eq. (5.2).

Expression of Results

Tabulate the analysis showing the oven-dry percentage of each fraction in the original soil as follows:

1. Coarse sand, 2. Fine sand, 3. Silt 4. Clay 5. Loss on solution, 6. Moisture (if air-dry soil is taken) 7. Total.

Note: The sum should approximate to 100. If the total turns out to be less than 98.5 or more than 102, repeat the analysis.

Hydrometer Method

A hydrometer is used to measure the density of a soil suspension at various times after settling time has been allowed for various sized particles to settle (Fig. 5.1 right). The observed densities at these times are related to the particle size distributions. Bouyoucos observed that after 40 s the hydrometer reading is not affected by sand sized particles equal to or exceeding 0.05 mm. It implies that whatever remains in the suspension is composed of silt and clay sized particles. After two hr, only the particles sizes less than 0.002 mm i.e. clay sized particles, remain in the soil suspension. More often this method is used without any pre-treatment of the sample except dispersion with dispersing agent (Day, 1965). Nonetheless, measurements can also be made on suspensions prepared by following all or any of the pre-treatments outlined in the previous section.

Apparatus and Reagents: Standard hydrometer, with Bouyoucos scale in g/L, Electric stirrer, Plunger, End-over-end shaker, Cylinders with 1000 mL mark

- **Amyl alcohol**
- **Calgont solution** (50 g/L). Contains sodium hexametaphosphate, a dispersing agent or any other dispersing agent described in the pipette method

Determination of Blank

- Pour 100 mL of CalgonT solution (or the same amount of dispersing agent as will be used in the test) in an empty cylinder and make the volume to 1000 mL with distilled water. Mix thoroughly with a plunger and let it stand until the temperature is constant
- Carefully lower the hydrometer into the solution, and determine the scale reading, RL at the upper edge of the meniscus surrounding the stem. Take the temperature
- The blank reading must be taken again each time the temperature varies by more than 2°C from 20°C during the duration of the test.

Procedure

Follow the same procedure as in case of the pipette method and transfer suspension quantitatively into a 1000 mL calibrated cylinder (hydrometer jar). Bring to volume with water. One can also begin afresh as follows:

1. Weigh 10 g air-dry sample of the soil to determine the air-dry moisture content for the reporting of results
2. Weigh 40 g air-dry soil passing 2-mm sieve into a 600-mL beaker
3. Add 100 mL of CalgonT solution (or any other dispersing solution), and 300 mL of distilled water. Cover the beaker with a watch glass, and leave overnight
4. Next day, quantitatively transfer contents of the beaker to a stirring cup, and fill the cup to about three- quarters with water

   Determination of Silt Plus Clay

5. Stir suspension at high speed for 3 min with the help of a stirrer
6. Rinse stirring paddle into the cup, and allow standing for 1 min
7. Transfer the suspension into 1L hydrometer jar, and bring to volume by adding distilled water
8. Disperse any froth by adding one drop of amyl alcohol, if required
9. Take hydrometer reading 40 s after withdrawing the paddle. Let this reading be, R_{sc}. Also measure the temperature
10. If possible, hydrometer jars should be kept in a water bath at a constant temperature of 20°C to obviate the need of temperature corrections. Otherwise, adjust the reading for temperature as follows:

    For every 1°C below 20°C, subtract 0.36 from the reading. Add 0.36 to the reading for every 1°C above 20°C.

In fact, reading is simultaneously adjusted for blank and temperature. As an example, if the blank reading is 3.5, and the 40 s reading is 22 at 21°C, then the corrected reading is given as:

Corrected reading = Reading – blank + (1 × 0.36)

$$= 22 - 3.5 + [1 \times 0.36] = 18.14 \quad ...(5.4)$$

Determination of Clay

11. Mix suspension in the hydrometer jar using paddle; withdraw the paddle, leave the suspension undisturbed
12. After 2 h, insert the hydrometer, and take hydrometer reading, say R_c
13. Correction to the reading is made as in case of silt + clay as per eq. (5.4)

Determination of Sand

14. After taking readings for clay and silt, pour suspension quantitatively through a 50 µm sieve. Wash sieve with a jet of water until water passing through the sieve is clear
12. Transfer the sand quantitatively from sieve to a 50 mL beaker of known weight
13. Allow the sand in the beaker to settle. Decant excess water. Dry beaker with sand overnight at 105°C. Cool in a desiccator, and re-weigh beaker with sand. Find out the weight of the dry sand say W_1 g.

Silt + Clay (%) = (Corrected R_{sc}) × (100/Oven-dry soil, g) ...(5.5)

Clay (%) = (Corrected R_c) × (100/Oven-dry soil, g) ...(5.6)

Sand (%) = Sand weight, W_1 × (100/ Oven-dry soil, g) ...(5.7)

Silt is determined from silt + clay by subtracting clay %.

Sum of silt + clay + sand should be 100%. The magnitude of the deviation from 100 indicates the degree of accuracy. If air-dry soil is used, a correction is applied to report the results on oven-dry basis as discussed in the previous method. An observation sheet with an example is given as follows.

Observation Sheet: Determination of soil texture by hydrometer method

1.	Field No. _____ WC-01	Date :
2.	Depth : _____ 0-15 cm	Observer name :
3.	Soil sample Identification No.	_____ AP-15
4.	Soil weight, oven-dry (g)	40
5.	Blank reading	5 at 20°C
6.	40-second hydrometer reading (g)	30.0
7.	Temperature of suspension for 40 s observation (°C)	20
8.	Corrected 40 second reading (g)	25.0
9.	2-hr hydrometer reading (g)	14.0
10.	Temperature of suspension for 2 hr observation (°C)	21.5
11.	Corrected 2-hr reading (g)	9.54
12.	Weight (g) of silt+clay (as in 8)	30 – 5 = 25.0
13.	Weight (g) of clay (as in 11)	14.0 – 5.0 + (1.5 × 0.36) = 9.54
14.	Weight (g) of silt (12 -13)	15.46
15.	Weight (g) of sand (4 – 12)	15.00

16.	% silt [(14)/(4)] × 100	38.65
17.	% clay [(13)/(4) × 100]	23.85
18.	% sand [(15/4) ×100]	37.50
19.	Textural class	?

Remarks: The simplified hydrometer method described herein is not suitable for calcareous, gypsiferous or saline soils or soils with more than 2% organic carbon. Hydrometer method to test such soils is described by Day (1965) or Hesse (1971). Readers may refer to these references for testing such soils.

Soil Texture Triangle

After the determination of sand (coarse + fine sand), silt and clay percentage, the soil texture is determined from the soil textural triangle (Fig. 5.2 (after)).

1. On the clay line, locate percent clay corresponding to the clay present in the soil sample. Draw a line parallel to the base of the textural triangle.
2. Now locate the percent silt in the soil sample on the silt line and draw a line parallel to the clay line.
3. The compartment in which these two lines intersect gives the soil texture. If the two lines intersect on the boundary of a compartment, the textural class of the adjoining compartment may be taken as the textural class.

Note: A line drawn from the percent sand in the soil sample on the sand line parallel to the silt line would also intersect the point at the same place. Hence this could also be used to arrive at the textural class instead of silt percent.

A revised triangle by USDA-NRSC is shown in Fig 5.2 (bottom). It is simpler to use than the original triangle. Clay in this diagram has been shown as clay and heavy clay.

A Comment on Soil Texture and Their Properties

Texture, one of the most stable attributes of the soil, gives an indication of the capacity of the soil to hold water and air. It is modified only slightly by farming practices such as cultivation mainly because of mixing of different layers. Textural class indicates how easy or difficult is to work with the soil. Whereas soils high in sand are easier, those high in clay are difficult to cultivate. Former soils are termed as light while latter soils are termed as heavy. Besides, soil texture governs many physico-chemical properties and processes that control production and productivity potential of the soils (Table 5.3).

Table 5.3. *Soil properties as influenced by textural classes*

Property	Sand (sand and loamy sand)	Silt dominated (silt loam, silt, clay loam and silty clay loam)	Clayey (sandy clay, silty clay or clay)
Type of soil	Light textured	Medium textured	Heavy textured
Water holding capacity	Low	Medium to high	High
Aeration	Good	Medium	Poor

Drainage	Excellent to high	Slow to medium	Poor or slow unless cracked
Soil organic matter	Low	Medium to high	High to medium
Decomposition of organic matter	Rapid	Medium	Slow
Warm-up	Rapid	Moderate	Slow
Compaction potential	Low	Medium	High
Shrink/swell potential	Very low	Low	Moderate to very high
Storage ability of plant nutrients	Poor	Medium to high	High
Sealing characteristics	Poor	Medium	Good
Erosion	Fine sands easily blown by the wind while coarse sands resist erosion by water	Easily blown by wind and susceptible to water erosion	Resist wind erosion

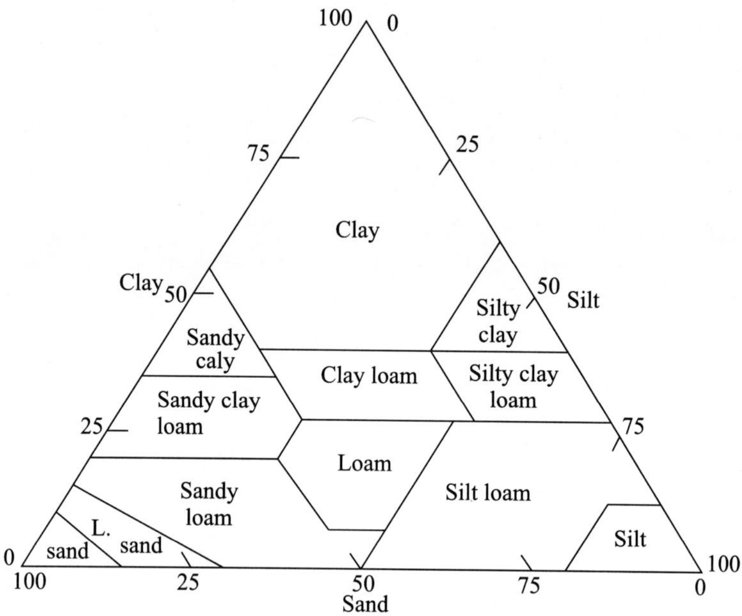

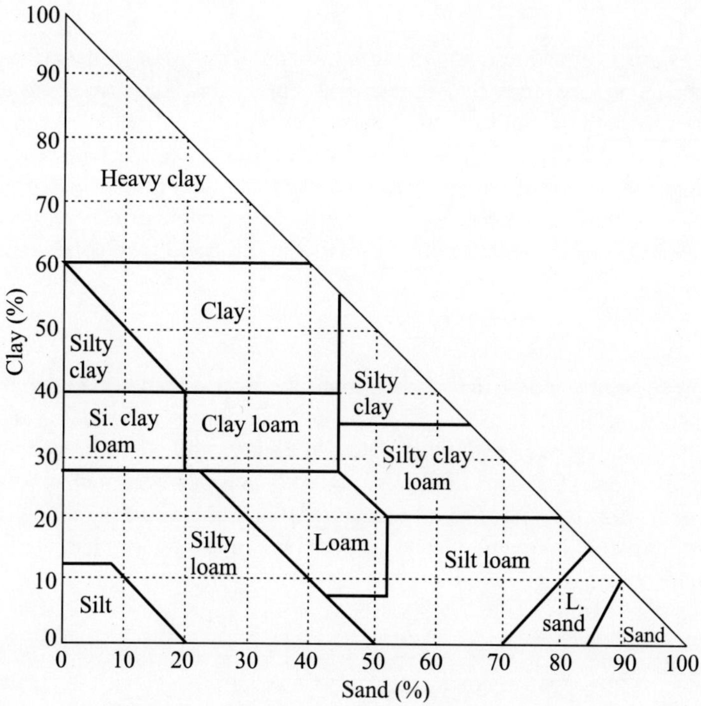

Fig. 5.2. *USDA soil textural classification chart*

SOIL STRUCTURE

Soil structure, as already stated is the way individual particles of sand, silt, and clay are assembled, and appear as larger particles called aggregates. This property of the soil is normally studied under natural field conditions. The characteristic structure of a soil can be recognized best when the soil is dry or only slightly moist. Soil structure is described in terms of the following characters.

Type: Shape or form and arrangement pattern of peds/aggregates

Class: Size of peds/aggregates

Grade: Degree of distinctness and durability of peds/aggregates

Soil structure is described in the sequence: grade - class - type. For example, we write strong coarse angular blocky or moderate thin platy or weak fine prismatic structure.

Types of Structure

Soil technicians have recognized seven types of soil structure, but mainly four principal forms of soil structure are used. Remaining types are covered in one or the other four principal forms.

Platy

The aggregates are arranged in relatively thin horizontal plates or leaflets (Fig. 5.3a). The horizontal dimensions are larger than the vertical dimensions. The structure is called platy, if the units/layers are thick, but laminar, if these are thin. Platy structure is commonly seen in the surface layers of virgin soils but can also be observed in the sub-soil. This type of structure is inherited from the parent material and its weathering, especially by the action of water or ice. It may also be formed owing to compaction of clayey soils by heavy machinery. Overlapping of plates greatly impairs the infiltration, percolation and aeration of these soils.

Prismatic

In this kind of structure, the vertical axis is more developed than horizontal axis. It gives a pillar like shape to the particles (Fig. 5.3b). The faces appear to be the casts or molds of adjoining units. The vertices are angular or sub-rounded, while the tops of the prisms are somewhat indistinct but flat. This kind of structure commonly occurs in sub-soil horizons of arid and semi- arid regions. The structure is called columnar for rounded tops but is prismatic for flat/plane, level and clear-cut tops. One of the major characteristics of this structure is poor drainage.

Blocky

The aggregates in this type of structure have been reduced to blocks, being more or less equal in the three dimensions (Fig. 5.3c). The structure is termed as angular blocky, if faces are flat and distinct and edges are sharp angular. It is sub-angular blocky, if both the faces and edges are rounded. This kind of structure is usually confined to the sub-soil. The structure has a significant influence on soil characteristics such as drainage, aeration and root penetration. Relatively large blocks usually indicate resistance to root penetration and movement of water.

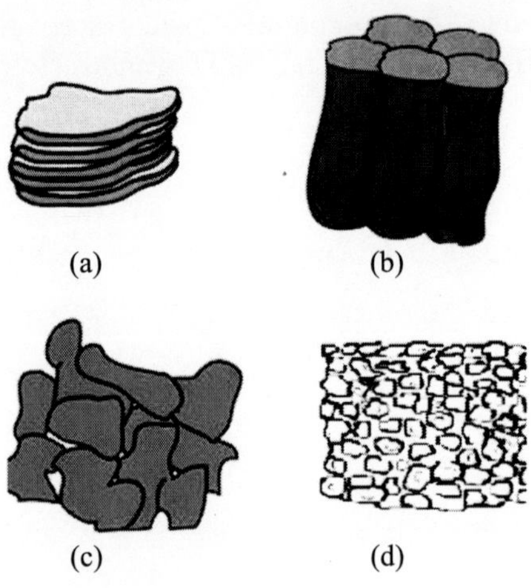

Fig. 5.3. *Principal forms of soil structures (a) Platy (b) Prismatic (c) Blocky and (d) Spheroid or Granular*

Spheroid

All rounded aggregates (peds), not exceeding 2.5 cm in diameter, are placed in this category (Fig. 5.3d). These rounded complexes are loosely arranged and readily separated. When wetted, the intervening spaces are not readily closed due to swelling as may be the case with a blocky structure. Thus, the wetting of the soil does not have much effect on infiltration, percolation and aeration in a sphere-like soil structure. The aggregates of this group, which are relatively less porous, are termed as granular. Very porous granules are termed as crumb. This kind of structure results from biotic activities in the surface horizon. It promotes infiltration, percolation, aeration and root penetration in soils.

Classes of Structure

Each primary structural type discussed in the previous section is differentiated into 5 size classes based upon the size of the individual peds (Table 5.4). The terms commonly used for the size classes are: very fine/thin, medium fine/thin, coarse/thick and very coarse/ thick (Table 5.4). The terms fine and coarse are used for all structural types except platy for which terms thin and thick are used. Plant growth in soils dominated by coarse sand sized aggregates is likely to be better than in soils with very small or very large aggregates.

Table 5.4. *Classes of different types of soil structure and their dimensions*

Class	Diameter (mm)			Thickness (mm)
	Granular (Spheroid)	Blocky	Columnar (Prismatic)	Platy
Very fine	< 1	< 5	< 10	< 1
Fine	1-2	5-10	10-20	1-2
Medium	2-5	10-20	20-50	2-5
Coarse	5-10	20-50	50-100	5-10
Very coarse	>10	>50	>100	>10

Grades of Structure

Grades indicate the degree of distinctness of the individual peds/aggregates, mainly determined by the stability of the aggregates. Soil moisture content, besides organic matter and texture, influences the grade of structure. Four commonly used terms to describe grade of structure are as follows.

Grade	Characteristics
Structure less	It is single-grain structure (non-coherent) where the individual soil particles show no tendency to cling together, as the conditions exhibited by loose sand or massive structure (coherent). In the latter, entire soil horizon appears cemented in one great mass
Weak structure	Poorly formed, indistinct formation of peds, non-durable and much un-aggregated material
Moderate structure	Moderately well-developed peds, fairly stable and distinct
Strong structure	Very well-formed peds, quite stable and distinct

Evaluation of Soil Structure

The study of soil structure is largely an art rather than science. *Per se* the most soil scientists may not be interested in soil structure yet they are greatly interested in its functions as a medium for crop growth or chemical transport. For example, it is easy to till a well-structured soil as it easily and spontaneously crumbles when relatively dry. At the same time, it does not adhere to the tools when wet. There is no single way to judge the soil structure. More often, a combination of direct and indirect methods are used to arrive at reasonably good information.

Direct Methods

Study of soil structure is normally combined with soil profile examination (Chapter 2). Soil structure is characterized by observing the shape, size and arrangement of soil aggregates macroscopically directly from the cleavages of the peds in the soil profile. For this purpose, a pit at least 60 cm deep is dug. One of its faces is made free of any spade marks. This face is examined by inserting a knife at approximately 1 cm intervals vertically. Any compacted zones where the soil offer greater resistance are observed and marked accordingly. Evidences of root development and worm activity are gathered to have a feel of how well-structured is the soil. A soil structure problem is anticipated, if the roots are shallow, their growth is restricted, and a general lack of worm activity. Other way of macroscopic evaluation is to excavate a large soil clod from the field. The clod is allowed to fall gently on a smooth surface so that it breaks into peds of different sizes and shapes. These are then examined to assess the soil structure. Thin soil sections may also be examined under a microscope to assess the shape and size of the aggregates and voids (Bullock et al. 1985).

Indirect Methods

Indirect methods involve the measurement of size distribution of stable aggregates, stability of aggregates or some soil property, which is a function of the soil structure.

Size Distribution of Stable Aggregates: Dry or wet sieving techniques discussed in Chapter 4 are used to determine different sized stable aggregates. The proportion of stable aggregates against vibrating action, which simulates the scouring effect of wind, is determined using the dry sieving technique. It provides an index for characterizing the susceptibility of soil to wind erosion. Similarly, the size distribution of water stable aggregates is determined using the wet sieving technique. Mean weight diameter of aggregates (MWD) determined through this test is a commonly used index of soil structure (Chapter 4).

$$MWD = \sum_{i=1}^{n} X_i W_i \qquad ...(5.8)$$

Here n is the number of fractions, $w_i = W_i/W$, where W_i is the weight of each size fraction, W is the total sample weight, and X_i is the mean diameter of that fraction. The soils having higher MWD of aggregates greater than 0.25 mm are said to have a good structure from crop production point of view. Variation in MWD from dry and wet sieving also is an indicator of the stability of the aggregates. The lower the difference between the two MWDs, the better is the soil structure.

Measurement of Stability of Aggregates: Several indices, namely stability index, structural coefficient and dispersion coefficient, have been developed to evaluate the degree to which soil aggregates resist dispersion. Stability index is determined by the difference in percent silt+clay determined by mechanical analysis and percent silt+clay obtained by suspension of soil sample in water. The greater is the difference, the better the soil structure. Structural coefficient is given by ((D-S)/S), where D is the percentage of particles smaller than 0.25 mm diameter as determined by mechanical analysis and S is the percentage of aggregates lesser than 0.25 mm diameter as determined by wet sieving method. A higher value of structural coefficient reveals better soil structure.

Measurement of Soil Properties: Bulk density, infiltration rate, hydraulic conductivity, aeration, available soil water and the degree of soil compaction are also used as indicators of soil structure.

DISPERSION AND EROSION RATIOS

The information generated from the mechanical analysis with some additional tests can be used to calculate dispersion ratio and erosion ratio. The dispersion ratio of soil is defined as the ratio of the percentage of silt+ clay (obtained by dispersing the soil in distilled water by end –over-end shaking in one liter cylinder) to percentage of silt + clay obtained from particle size analysis of the soil. It indicates the readiness with which silt and clay particles of soils go into suspension in water. The dispersion ratio depends upon the mechanical composition of the soil and is usually high in soils where silt content is more and clay content is less. The two properties are useful and quick way of assessing the potential for dispersion and tunneling susceptibility of soils.

Dispersion Ratio (Bryan, 1968)

Apparatus: Same as in particle size analysis

Procedure

1. Grind the soil sample with a wooden pestle and mortar
2. Take 20 g of oven-dry soil passing through 2-mm sieve in a 1 L cylinder
3. Add distilled water to make up the volume to 1 L
4. Mix the suspension thoroughly with a plunger. Note the temperature of the suspension and pipette out 25 mL suspension from a depth of 10 cm at the requisite time of silt + clay
5. Transfer the sample to pre-weighed 100 mL beaker
6. Dry the sample and weigh it
7. Calculate the per cent silt plus clay in the sample
8. Also use the total silt plus clay as determined from particle size analysis of the soil

Calculations

Mass of the soil taken	20 g
Temperature of the suspension	°C
Mass of 100 ML beaker	W_1 g
Mass of beaker + (silt + clay)	W_2 g
Mass of (silt + clay)	$(W_2 - W_1)$ g
Percent silt + clay	$\left(\frac{W_2 - W_1}{20}\right)\left(\frac{1000}{25}\right)100 = X$
% silt + clay from particle size analysis	Y
Dispersion ratio (DR, %)	100 (X/Y)

Erosion Ratio, ER (Akalan, 1967)

$$ER\ (\%) = DR(\%) \times (A/c) \quad ...(5.9)$$

Here, A is the field capacity moisture content, and c is the percentage of clay determined from particle size analysis.

Dispersion Coefficient

The dispersion coefficient is defined as the ratio of water dispersible clay to the total clay expressed as percent. This ratio is quite high in alkali soils, exceeding 90% in soils having pH more than 10.0 compared to 20-30% in normal soils. The test is similar to the particle size analysis. It can be carried out along with the test on particle size analysis (Kinney, 1979, Craze and Hamilton, 1991, Kondel, 1991).

Apparatus and Reagents: Same as in particle size analysis using a hydrometer (Pipette method may also be used)

Procedure

1. Measure the clay content of the soil as in the case of particle size analysis
2. Measure the clay content with the same method but without adding the dispersant (only add the deionized water) with limited shaking (shake only for about 10 min). Take hydrometer readings at the same time as in the previous step
3. Fill a 1000 mL cylinder with deionized water. Fill another 1000 mL cylinder with deionized water and 100 mL of dispersant (CalgonT). Take hydrometer readings at the same time as in previous steps.

$$\text{Dispersion coefficient } (\%) = 100(R_1 - R_w)/(R_p - R_d) \quad ...(5.10)$$

Here R_1 is the hydrometer reading of soil suspension for clay without dispersant and limited mechanical dispersion, R_p is the hydrometer reading of soil suspension for clay with dispersant and prolonged mechanical dispersion, R_w is the hydrometer reading of deionized water, and R_d is the hydrometer reading of deionized water plus dispersant.

In soil conservation, the dispersion coefficient is defined as the ratio, expressed as a percentage, of the soil material < 0. 005 mm diameter as determined with limited mechanical dispersion without using a dispersant to the total material < 0.005 mm diameter. Following procedure is used.

1. Carry out complete mechanical analysis with and without dispersant
2. Plot the size curves for the two cases on the same graph paper
3. Select the values of fractions of 0.005 mm from the curves
4. Calculate the ratio of these without and with dispersant and express in percentage

Following guidelines are used to characterize the soil. The dispersal index in the following guidelines is the inverse of the dispersion percentage multiplied by 100. It is used to interpret the potential for dispersion in relation to earthworks (Ritchie, 1963).

Dispersion percentage (%)	Dispersal index ratio	Rating for dispersion
<6	>16	Negligible/aggregated
6–30	3.0–16	Slight
30–50	2.0–3.0	Moderate
50–65	1.5–2.0	High
>65	<1.65	Very high

SLAKING AND DISPERSION (SCORING METHOD)

Soil aggregates collapse due to the processes of slaking and dispersion. Slaking is defined as the breakdown of a lump of soil into smaller fragments upon wetting due to clay swelling. Trapped air inside the aggregates bursts out with swelling resulting in slaking of the soil aggregates. The products of slaking can re-form larger aggregates. Dispersion, on the other hand, is the separation of soil into single particles. It is mainly governed by soil texture, clay type, soil organic matter, soil salinity and exchangeable cations. Dispersion is often seen as 'muddy' or 'milky' water. Dispersion, a more common phenomenon of alkali soils, is attributed to the swelling of clay platelets that results in collapse of clay aggregates. Dispersion into primary particles is irreversible unlike slaking. Degree of slaking and dispersion greatly influence soils' behaviour under wet conditions. Consequently, it helps in deciding the options for their management.

Dispersibility of soils was first described by Emerson (1967) using a scoring system based on 20 hr test. This section describes a 2 hr scoring test, aggregate stability in water (ASWAT proposed by Field et al. (1997), a modification of Loveday and Pyle (1973) scale. Two kinds of scores used in this test are as follows.

- Slaking score (0–4)
- Dispersion index (0–16)

Apparatus: Soil aggregates, Glass jar or dish, Distilled/rain water, Stop watch

Slaking Score

1. Take surface soil samples (only surface soil is tested for slaking) and allow them to air-dry for 1 to 5 days, depending on how dry or wet is the soil
2. Take several (at least three) 3 to 5 mm diameter crumbs of dry soil (for this test larger up to 20 mm diameter crumbs can be used) and place them in a dish or saucer of

distilled water (rainwater can be used) filled enough to completely cover the samples. Cover the dish so that water is not disturbed by the wind

3. After 5 min, assess the slaking score (0–4) as per following guidelines

Score	Observation	Management decision
0	Lump remains intact	The soil with scores of 0 and 1 are stable to wetting. No action is needed
1	Lump collapses around the edges but mainly remains intact	
2	Lump collapses into angular pieces	A typical of self-mulching clay may form a loose, granular surface layer with a thin fragile crust. If the soil does not disperse (next section on dispersion), no action is necessary from management point of view
3	Lump collapses into small (less than 2 mm diameter) rounded pieces forming a cone	A score of 3 suggests that the surface may form a crust. It is a problem in most non-swelling top soils (sands, loams and some clay loams). The problems can be overcome through reduced cultivation, stubble retention and gypsum application
4	Lump collapses into single grains, sand grains are seen	The soil is very likely to crust and hard set (sands and loams). Management actions as in score 3 may be adopted. Frequent irrigation can also help to manage the soil

Dispersion Index

The dispersion index comprises four separate dispersion scores (each consisting of a scale from 0 to 4) as follows.

- Score after 10 min (wetting)
- Score after 2 hr (wetting)
- Score after 10 min (remoulded)
- Score after 2 hr (remoulded)

Upon adding these scores, it makes a scale of 0 to16.

Dispersion on Wetting

1. Take soil samples and allow them to air-dry overnight
2. Take several (at least three) small (3 to 5 mm diameter) crumbs of dry soil and place them in a dish or saucer of distilled water (rainwater may be used) filled enough to completely cover the samples. Samples used for the slaking test can be left as such and used to assess the dispersion scores, provided 3-5 mm diameter aggregates were used in slaking tests
3. Cover the dish so that wind does not disturb the water
4. Observe and record the milky ring around the aggregates. Don't bother if the soil collapses or bubbles are formed. Assess the dispersion score (0–4) after 10 min and 2 hr as per following guidelines.

Score	Observation
0	Nil dispersion
1	Slight dispersion, recognized by slight milkiness of water adjacent to the aggregate
2	Moderate dispersion with obvious milkiness
3	Strong dispersion causing considerable milkiness, about half of the original volume of the aggregate are dispersed outwards
4	Complete dispersion, only sand grains are left in a cloud of clay around them

If the soil does not disperse on wetting (total score is 0), repeat the test using remoulded soil. It attempts to duplicate the state of the soil after cultivation. Mix some soil with rainwater (distilled water) and remould it with a knife for one min. One may also knead the soil by hand as in determination of soil texture. Make small lumps (3 to 5 mm diameter) of the remoulded soil and follow the same procedure as in the case of wetting. Remoulded samples should not be allowed to dry before being immersed in water. Scoring of dispersion after remoulding is made in the same manner as in the case of dispersion on wetting.

Adding Scores

Score of 10 min and 2 hr are added together. For soils with no dispersion in wetting, add the scores for remoulded soils. If the soil disperses in wetting test, add a score of 8 to the score of wetting test to get the total score. The guidelines on soil dispersibility are then used to assess the soil dispersion.

Dispersibility	Possible ASWAT score (Field et al., 1997)
Very high	12-16
High	10-12
High to moderate	9-10
Moderate	5-8
Slight	1-4
Negligible/aggregated	0

An advantage of the ASWAT test is that the results can be easily linked to management practices such as the need for gypsum or lime or organic mtter application and/or avoiding wet working of soils (McMullen, 2000). Severity of the problem as per ASWA scores and guidelines on management options to be adopted are reported in Table 5.5.

Table 5.5. *Response to the soil structural stability diagnosis (ASWAT scores)*

Score	Severity of dispersion	Management options
7–16	Serious dispersion	For a score in the range of 8-10, gypsum (and/or lime for pH of < 5.5) may have to be applied depending upon supporting laboratory tests. Gypsum is necessary above a score of 11 as clay disperses strongly, providing a poor environment for plant growth Apply gypsum and organic matter

2–6	Moderate dispersion	For a score of 3 to 4, cultivation under wet condition may result in dispersion and deserves caution but at 5 to 6 soil disperses strongly under wet cultivation Avoid working the soil when it is moist (also applies to the soil in remoulded category). Direct drilling and stubble retention may prove beneficial
0–1	Negligible	In general, dispersion score of <1 indicates a very stable soil that resist dispersion after tillage. At about 2, the soil is prone to compaction. Protect soil from dispersion by reducing impact of raindrops e.g. retaining stubble cover (These management options also apply to the above two categories)

SOIL STRUCTURE PROBLEMS

The intention of this section is not to elaborate numerous problems that arise from poor soil structure. The tabular information presented herein only gives a glimpse of the kinds of expected problems. Likely solutions to overcome these problems for various soil texture groups are also listed (Table 5.6)

Table 5.6. *Soil types, probable structural problems and likely solutions*

Soil/condition	Problem	Likely solution
Sandy soils	Low water and nutrient retention	The practical method to improve the structure is to add organic matter directly or through green manuring. The addition of pond sediments, clay soil or even compaction may help to increase the proportion of small sized pores. It should enhance water and nutrients retention capacity of the soil
Clayey soils	Restricted drainage and aeration due to poor aggregation	Addition of organic matter may be helpful in improving the aggregation, but the amounts required are likely to be quite high. Therefore, crop rotation and use of phosphate fertilizers may prove useful. Ridge or raised bed cultivation can reduce the aeration stress to some extent
Puddle condition	Soil aggregates are broken in areas under prolonged water logging leading to puddle conditions	Proper drainage seems to be a feasible solution of the problem
Dispersed condition	The problem of dispersion is commonly observed in alkali (sodic) soils	Addition of gypsum in combination with green manuring or incorporation of crop residues may improve soil structure besides helping to ameliorate these soils
Low aggregate stability	Commonly encountered in soils of arid and semi-arid regions	Stability of aggregates can be increased by addition of organic matter. Mulching may also be useful as it provides favourable conditions for microbes to decompose organic material

Chapter **6**

Soil Moisture Constants

Soil is a valuable resource that supports plant life. Water in the soil pores, commonly referred as soil water, is an essential component of soil-water-plant system. Soil water in a general sense is classified as gravitational water, capillary water and hygroscopic water. On the other hand, soil water constants for a particular soil define certain soil water proportions, which dictate whether the water will be available or not for the plant growth. Many management decisions concerning types of crops to grow, plant populations, irrigation scheduling, and the amount of nitrogen fertilizer to apply depend on the amount of moisture available to the crops throughout their growing season. An understanding of the soil moisture constants helps to better define as well as compare between statuses of different soils. This chapter describes some of these constants and related terms and discusses the procedures to evaluate these constants.

SOIL MOISTURE CONSTANTS AND RELATED TERMS

Saturation Capacity

The moisture content at which all soil pores are filled with water is described as saturation moisture content, which when expressed in percent is saturation moisture percentage. It is the maximum capacity of the soil to store water (Fig. 6.1). The soil moisture tension at this point is almost zero. At saturation, some water over and above the forces of cohesion and adhesion is under the force of gravity. This amount of water is known as gravitational or free water. The exchange between the soil and the atmospheric gases is very slow under this condition.

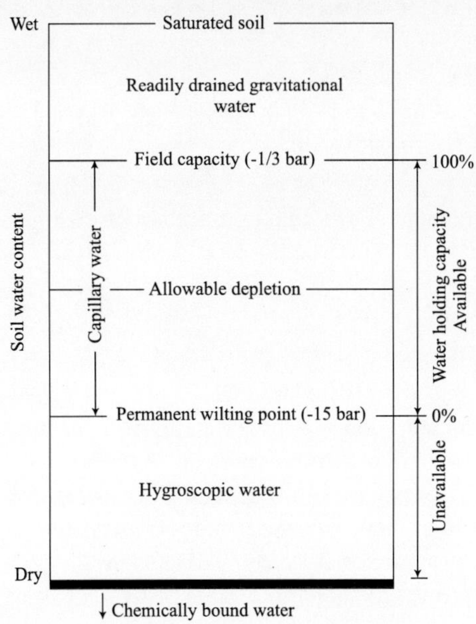

Fig. 6.1. *Definition sketch of soil water status of soils*

Field Capacity (FC)

The field capacity is the water retained by an initially saturated soil against the force of gravity. It is the amount of water remaining in the soil few days after having been saturated and free drainage has ceased. It is also known as the maximum water holding capacity of the soil. The matric potential at this soil moisture content is around –1/10 bar (for clayey soils) to -1/3 bar (for sandy soils). The water content at –1/3 bar has been found to be closely correlated to the field capacity of many soils (Fig. 6.1). At field capacity, water is retained only in the micro-pores as all the macro-pores are drained off. A term a**ir capacity (AC) is used** to specify the size of the drained macro-pore space. It coincides with the volume of the pore space filled with air at field capacity.

Permanent Wilting Point (PWP)

Plant roots are able to extract water from a soil matrix normally beginning with field capacity. As the water extraction proceeds, the moisture content of the soil goes on diminishing. At one point, the plants are unable to extract any further water and begin to wilt. Two stages of wilting points are recognized i.e. temporary wilting point and permanent wilting point. Former constant describes the soil water content at which plants wilt during the day time, but recovers during the night or when water is added to the soil. The permanent wilting point, on the other hand is the water content at which most plants wilt and fail to recover their turgor upon addition of water to the soil. The matric potential at this soil moisture condition is commonly taken as -15 bar. Currently, the term temporary wilting point has become obsolete and only permanent wilting point is in use.

Capillary Water

Soil, because of its structure and presence of organic matter, is capable to hold water against the force of gravity. The water that remains in the soil in the form of a film around the soil grains after gravitational water has been drained off is known as capillary water (Fig. 6.1).

Hygroscopic Water

Hygroscopic water is the water contained in an oven-dry soil, which is absorbed by it upon its exposure to moist air. It is held as a very thin film over the surface of the soil particles and is under tremendous negative pressure. This water is not available to plants but may be available to micro-organisms. The maximum amount of hygroscopic water absorbed by 100 g of dry soil under standard conditions of humidity and temperature is known as hygroscopic coefficient.

Moisture Equivalent

Moisture equivalent, proposed by Briggs and McLane (1910), is defined as the percentage of water, which soil after saturation will retain against a centrifugal force 1000 times the force of gravity. It is a convenient laboratory measure of soil moisture condition, and is a good indicator of field capacity moisture content of fine-textured soil materials. Studies have revealed that the field capacity of sand is higher than its moisture equivalent, and the field capacity of loams is approximately equal to their moisture equivalent. For clayey soils, the field capacity is lower than their moisture equivalent. To overcome this variation, several researchers suggested correlations to calculate field capacity using moisture equivalent.

Field capacity (%) = 0.86 × moisture equivalent (%) + 2.6 × permanent wilting point (%)

Wilting coefficient = Moisture equivalent/1.84

This concept is no longer used in soil physics, having been replaced by field capacity.

Chemically Bound Water

It is the water present in the chemical compounds in the soil particles. It is also unavailable to plants.

Water Holding Capacity

One of the foremost parameter to optimize crop production is the water holding capacity of the soil. W**ater holding capacity of a soil** is the maximum amount of water it can hold for crop use. Practically speaking, it is defined as the amount of water available between field capacity and the permanent wilting point water contents. On the other hand, most agricultural plants show signs of wilting long before the permanent wilting point is reached, typically at around -2 to -5 bars. It is why the allowable depletion comes into play (Fig. 6.1). Usually, this value is taken at a point where 50% of water holding capacity (total available water) of the soil has been depleted by the crop. Its range varies from 20 to 70% of depletion depending upon the crops and their growth stages. Soil texture, soil structure, and organic matter are the key components that determine soil water holding capacity. In terms of **soil texture**, silt and clay soils hold more water than sand, latter having quite low water holding capacity. Water holding capacity determines when to irrigate a soil depending upon the soil moisture depletion by plants.

The crux of determining the water available to plants is to determine various soil moisture constants. A number of approaches both laboratory and *in-situ* field methods are used to determine these constants, some of which are described in the following sections.

SATURATION PERCENTAGE

Apparatus: Porcelain dishes, Spatula, or mixing spoon, Vacuum filtration system

Procedure

1. Weigh the metal can with lid, W_0
2. Weigh 200-300 g air-dry soil (< 2 mm) into a porcelain dish
3. Slowly add distilled water, and mix with a spatula until the soil paste does not have free-standing water on the surface of the paste. The paste slides freely and cleanly comes off a spatula (does not apply to clay soils > 40% clay). It flows slightly when the container is tipped to 45° from horizontal. Soil surface glistens as it reflects light. It consolidates easily upon tapping after a cut is formed in the paste with the flat side of a spatula (does not apply to sandy soils >70% sand)
4. Allow the paste to stand for 1 hr, and then re-check the criteria for saturation by either adding more distilled water or the soil, as per requirement

5. Leave the paste for 6 to 16 hr or better overnight. Re-check the criteria for saturation by adding more distilled water or soil, as needed
6. Transfer a portion of the saturated soil paste into the can and weigh. Determine the weight of the moist soil and can say W_w and W_0. Find the weight of the soil, $W_1 = W_w - W_0$
7. With the lid unfitted, dry overnight in an oven at 105°C (normally for 24-48 hr)
8. Cool in a desiccator for at least 30 min, fit the lid and re-weigh, W_d
9. Find out the weight of the dry soil, $W_2 = W_d - W_0$

The saturation moisture content (%) = $100 (W_1 - W_2)/W_2$

A method to calculate the saturation moisture percentage using undisturbed soil samples is covered in the section on field capacity

Applications

A saturation extract is routinely used to determine chemical properties of soils especially the salt affected soils. The EC of the soils, cations and anions, and other chemical constituents or toxic elements such as boron are analyzed in saturation extracts.

Drawing the Saturation Extract

- Prepare the saturation paste according to the procedure described in the previous paragraph. Filter the paste with a vacuum filtration system using a Buchner funnel fitted with Whatman No. 42 filter paper. Collect the filtrate in a bottle
- If the initial filtrate is turbid, filter the extract.

FIELD CAPACITY

The following section describes a laboratory method to determine field capacity as described by Bandyopadhyay et al. (2012).

Apparatus: Soil core, Pan, Balance, Oven

Procedure

1. Use 15 cm soil cores that were used to draw undisturbed samples for the measurement of bulk density or draw fresh cores for this test. Carefully remove the caps so that soil in the core remains intact. Put a double layer of cheese cloth on the bottom of the core with a rubber band. Separately weigh the cheese cloth and rubber bands before putting them on the core, W_c
2. Weigh the core with the cheese cloth and rubber bands, W_1
3. Set the core in a pan of water until it is almost immersed; leave it in the pan to soak up water for 24 hr until it is completely saturated. Add extra water to the pan, if required
4. Remove the core from the pan, and keep it to drain off the excess water sticking to the core or at its bottom
5. Weigh the core when there is no further dripping (soil at saturation point), W_2.
6. Put the core on a grate so that gravitational water can drain out of it; cover the top of the core with some plastic wrap to prevent evaporation. Let it drain for 24-48 hr

7. Weigh the core with soil after 24 hr, W_3, 48 hr, W_4 and 72 hr, W_5
8. Carefully remove the cheese cloth and replace with an aluminium foil; put core in the drying oven for 24 -48 hr at 105°C. Weigh the wet piece of cheese cloth and rubber bands used on the core (W_6). This weight minus the dry weight W_c is subtracted from the total weight of the core + soil to determine the actual weight of the water in the soil
9. Remove the core from the oven, cover top with plastic wrap so that the core does not suck moisture from the atmosphere. Cool and weigh the dry core after removing the plastic wrap and aluminium foil, W_7
10. Calculate the amount of water in the soil at saturation and after 24, 48 and 72 hr by subtracting the weight of the core + dry soil from the weight of the core + saturated soil and after gravitational drainage, respectively
11. Weigh the empty core, W_8.

Weight of the soil + core	$W_1 - W_c = W_{sc}$
Weight of the soil + core + water (saturation)	$W_2 - W_6$
Weight of the soil + core +water (24 hrs drain)	$W_3 - W_6$
Weight of the soil + core +water (48 hrs drain)	$W_4 - W_6$
Weight of the soil + core +water (72 hrs drain)	$W_5 - W_6$
Weight of the soil (dry)+ core	W_7
Weight of the soil (dry)	$W_7 - W_8$
Moisture content at soil saturation (% dry weight)	$100 \times (W_2 - W_6 - W_7)/(W_7 - W_8)$
Moisture content at 24 hr (% dry weight)	$100 \times (W_3 - W_6 - W_7)/(W_7 - W_8)$
Moisture content at 48 hr (% dry weight)	$100 \times (W_4 - W_6 - W_7)/(W_7 - W_8)$
Moisture content at 72 hr (% dry weight)	$100 \times (W_5 - W_6 - W_7)/(W_7 - W_8)$

Moisture content at saturation in percentage is given in row 8. The constant weight after 24 to 72 hr or after 72 hr may be taken as the field capacity of the soil.

Field Method

Apparatus: Polythene sheets, Spade, Soil auger, Moisture boxes, Balance, Oven

Procedure

1. Select a level and uniform plot having dimensions of about 5 m×5 m. Remove any loose material from the surface (weeds, pebbles, etc.). Make small dykes all around the plot to retain water
2. Fill sufficient water in the plot so that soil gets fully saturated
3. Cover the plot area with a polythene sheet to check evaporation
4. Beginning after about 24 hr, take a soil sample from the centre of the plot from the desired layer, and determine soil moisture content daily as previously discussed in this chapter till the values of successive days are nearly equal. Field capacity is normally attained in the field after 24 to 72 hr of saturation depending upon the soil texture

5. The constant moisture content so determined is the field capacity moisture content. One can also plot the daily readings on a graph paper. The nearly constant lowest reading is taken as the value of the field capacity of the soil.

PERMANENT WILTING POINT

Correlation Methods

Several equations have been developed to correlate field capacity and permanent wilting point (PWP) to sand, clay and silt contents of the soils and their bulk densities. Most important relations are given in LEACHM model (Hutson and Wagenet, 1992), and are reproduced as follows.

$$SWC_{-33} = \exp\,[-3.43 + 0.419\,(Cl + Si)^{0.5} - 1.83\,(Cl + Si)^{1.5}]$$

$$SWC_{-1500} = \exp\,[-4.384 + 0.404\,(Cl + Si)^{0.5} - 9.85\,(Cl + Si)^{1.5}]$$

Bahmani and Palangi (2016) used the following equations.

$$SWC_{-33} = 0.73426 - 0.00145\,Sa - 0.29176\,BD \text{ if } Sa \geq 75$$

$$SWC_{-33} = 0.5784 + 0.00227\,Cl - 0.28438 \text{ if } Sa < 75$$

$$SWC_{-1500} = 0.02413 + 0.00373\,Cl$$

Here SWC is the soil water content at respective moisture tensions given as subscripts, Sa, Cl, and Si are sand, clay and silt contents of the soil in % respectively and BD is the bulk density in Mg/m^3.

Laboratory Methods

Equipment: Five 600 g capacity cans with lids, Sunflower seeds, Glass tube, Sealing wax, Moisture boxes, Physical balance, Drying oven, Bell jars, Water trays and Soil sampler

Procedure

1. Take five weighed cans having a drain hole at the bottom. Fill about 500 g of air-dry soil in each. Sow 4-5 seeds of sunflower (*Helianthus annuus*) in each can and allow them to germinate (Briggs and Shantz, 1912). Add sufficient water and appropriate doses of nitrogen and other nutrients
2. Moisten the soil so that its water content is suitable for plant growth and reweigh the containers. Each time the soil is watered, attempt to bring the average water content back to the starting point
3. After emergence, retain the two best seedlings while taking out the others. Allow them to pass through the two holes in the lid. Place the cans with plants in the green house. In its absence, place them in moist sawdust to avoid heating due to insolation. Grow them for about six weeks, watering as and when necessary. By this time, besides the original cotyledons the plants should have developed at least three pairs of leaves. After the third pair of leaves has developed, irrigate the soil with enough water to bring the water content to the starting point
4. Insert a glass tube in the soil for aeration and plug it with cotton wool. Seal the soil surface with wax. Also close the drainage hole at the bottom and seal it

5. At this stage, water the plants for the last time and plug all the spaces between the soil surface and the holes in the lid of the can with cotton wool to control evaporation. Allow the plants to wilt
6. As soon as both the plants show signs of wilting (loss of turgor), transfer them to a dark humid chamber. The cans with plants are covered with a bell jar and kept in a small water tray to mimic the conditions of a high humidity chamber. The bell jar should be covered with a black polythene piece. Leave the plants overnight to allow them time to extract moisture from the soil
7. If they gain turgidity, expose them to the atmosphere for a couple of hours and transfer them back to the humid chamber. Continue this process until the plants do not recover in the dark humid chamber. It may be presumed that soil at this stage is at permanent wilting point. Now remove the lid, cut the plants and weigh to assess the weight of the roots as discussed in the following notes
8. Determine the moisture percentage, which is the permanent wilting point (PWP) moisture content of the soil.

Notes:

- To ensure accuracy, conduct the test in 4-5 replications
- Correction for the oven-dry weight of the roots is necessary. It is done by assuming that the roots correspond to one-half of the fresh weight of the stem and leaves and they contain 80% water. The weight of water in the roots thus corresponds to 0.4× weight of the stem and leaves.

Permanent Wilting Point (Breazeale and McGeorge, 1946)

The method described in this section is based on the principle that the roots draw water from the soil following the path of least resistance.

Apparatus: Potted tomato plants, Cork, Paraffin-beeswax seal, Glass jackets, Soil, Culture solution

Procedure

1. Grow tomato plants in pots of soil until they are 1 to 2 months old (Fig. 6.2). Keep 1-3 seedlings per pot depending upon its size
2. Strip the lower stems of leaves to give a bare stem of about 10 cm length above the surface of the soil
3. Slip glass tubing, 5×3 cm over each plant. Fit a split cork stopper to the bottom of the tube and seal in place with a paraffin-beeswax mix to make a watertight seal
4. Pour soil sieved through 2-mm sieve into the glass tube around the stem, and then add the required amount of tap water to bring the soil to field capacity. Also seal the top in the same manner as the bottom of the tube. Continue to grow the plants in the pot as usual by adding culture solution or water and nutrients as per plant requirement
5. At the end of 2 weeks, roots should be visible on the stem enclosed in the glass jacket containing the soil

6. At the end of 7 weeks remove the jackets and find out the moisture content of jacketed soil. The moisture content at this time is the permanent wilting point moisture content of the soil
7. The other jackets can be removed after 8 weeks or so. Normally, the moisture content equilibrates within 7 weeks and one may not expect much change in water content thereafter. It is because the pull of the roots in the jacket is equal to that of the roots in the pot. Since the amount of soil in the jacket is small compared to the amount in the substrate, the roots attain equilibrium first in the smaller volume of soil.

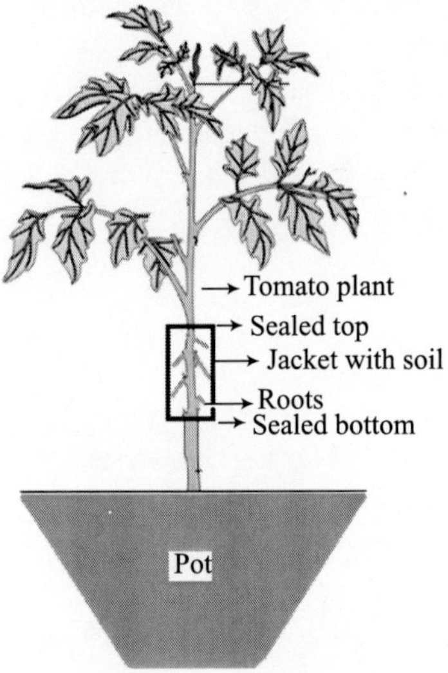

Fig. 6.2. *Experimental set-up to determine the permanent wilting point using tomato seedling*

SOIL WATER CHARACTERISTICS CURVES

The ability of soil to retain water depends on several factors, e.g., texture, organic matter (OM) content (due to its hydrophilic nature), nature of mineral colloids, arrangement of particles (soil structure) and its bulk density. The graphical representation of soil moisture tension against soil moisture content is known as water retention or soil water characteristic curve. It is used to assess the soil moisture content at a particular tension at which moisture is held in the soil. For example, volumetric moisture content at field capacity is closely related to volumetric moisture content at 33 kPa suction for normal soils. For organic and sandy soils, it is closely determined by the volumetric moisture content at 10 kPa suction. Permanent wilting point is closely replicated by the volumetric moisture content at 1500 kPa. Clearly, these curves can be used to assess the availability of soil water for plant growth. Besides, these curves are also used to determine the water that can be readily available to the plant (for irrigation purposes), the effective pore space and porosity, and specific yield (for drainage design). These curves also indicate likely changes in the soil structure caused by tillage, or mixing of soil layers. Many times, these curves are used to develop relationships between soil moisture tension and other soil physical properties, e.g. capillary conductivity, thermal conductivity, clay and OM content, although some of these are still a subject of research.

Several methods are used to develop soil moisture characteristic curves depending upon the range for which these curves are required. A list of these methods is given as follows.

Suction level	Water head (cm)	Methods used
Low suction	0-100	Hanging water column, sandbox tension table, plastic porous membrane technique
Medium suction	100-1000	Pressure plate, pressure membrane
High suction	1000 -20000	Pressure plate, pressure membrane
Very high suction	20000-100000	Vapour pressure equilibrium using vacuum desiccators

This section, except for a brief mention of few selected methods, discusses only the pressure plate apparatus, a convenient and reliable means of measuring soil water content at different pressure/suction. It enables one to develop moisture retention curve for any soil type (Richards and Fireman, 1943, Richards, 1954).

Hanging Water Column/Sandbox Tension Table Methods

Both the hanging water column and sandbox tension table methods are based on the same principle and apply suction to the soil samples in a similar manner. The only difference is between the capacities. In the former, only one sample per set is run. Thus, it may require many sets, if one wants to test several samples at the same time. In the latter, many samples can be simultaneously run. In each case, the difference in height between the suction regulator and the middle of the soil samples determines the suction applied. The samples are weighed after they have attained equilibrium at a specific tension (Barasa, 2014). The samples are oven-dried to determine the water content at the last tension. The standard sandbox tension table set with a provision for multiple testing of soil sample consists of the sandbox with control panel, suction levelling stand, water supply bottle with stand, filter cloth (140-150 micron), a number of containers, synthetic sand of grain size approximately 73 micron and various accessories. A sand/kaolin box can be used to extend the suction range. Suction heads between 0 and 200 cm can be applied, although with sand/kaolin box it may be possible to go up to 250 cm or so. The hanging water column is a simple and very accurate method for applying small tension to soil. Its use to determine drainable pore space is described in Chapter 9 of this book. Since the basic principle is being covered in Chapter 9, further discussion is not being included for the sake of brevity.

Pressure Plate Apparatus

The pressure plate apparatus can be used for disturbed samples as well as for undisturbed soil cores. Kaolin ceramic plates of this equipment are hydrophilic in nature and are the key separator between air and water. These plates allow passage of water but not air up to the air-entry value or the bubbling pressure. The smaller the pore sizes of the ceramic plate, higher the bubbling pressure or air-entry value of the plate. A given ceramic plate can be used to extract water only below its bubbling pressure. The saturated samples or cores placed on the ceramic plates of the apparatus are subjected to an air pressure above the atmospheric pressure. A pressure regulating system consists of three major components, an air compressor, an air tank and pressure regulating module. As a result, excess water is released through the microscopic pores in the ceramic plate. A number of such plates, varying from 0.5 bar to 15 bars are part of the standard equipment or can be ordered as an optional accessory. A schematic and photographic view of the pressure plate apparatus is shown in Fig. 6.3.

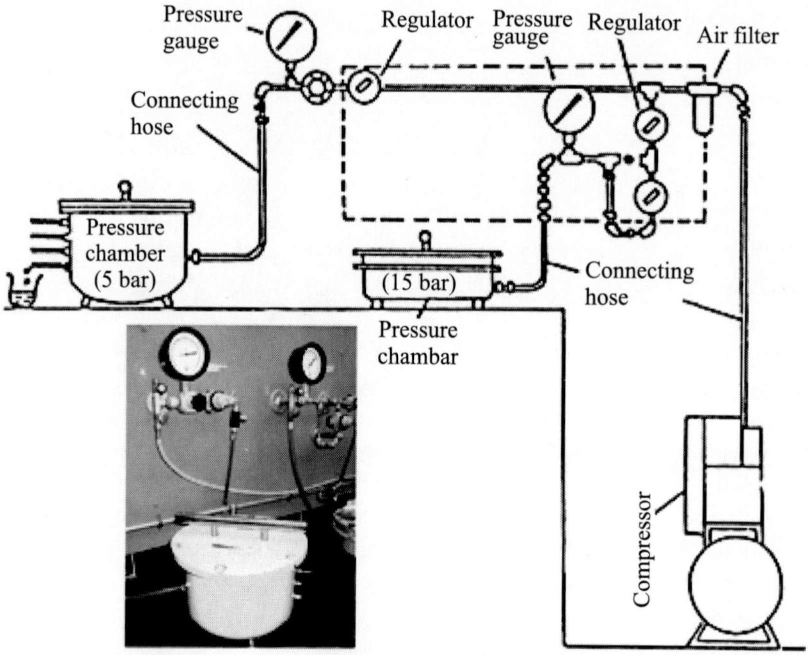

Fig. 6.3. *A schematic and photographic view of a pressure plate apparatus*

Apparatus/Equipment: Pressure plate apparatus, Brass or rubber rings of 1 cm height and 6 cm diameter, Balance, Oven, Aluminium moisture boxes, Pipette

Procedure

1. Thoroughly soak the pressure plates in deionized water overnight before their use
2. Fill rings with air-dry soil passing through 2-mm sieve using a spatula (about 20 g sample). In order to prevent segregation of particle-size, place entire soil sample into the ring, and then level, but don't pack
3. To draw undisturbed samples, use core rings straight into the wet soil in the field
4. Add water between the rings until there is an excess of water (at least 3 mm deep) on the plate. Cover plate and samples with plastic sheet. Allow samples to stand overnight
5. Remove excess water from the plate with a syringe, disposable pipette, or siphon. Place the triangular support on the bottom of the extractor vessel
6. Install plate with samples in the lower-most position in the extractor. Then install the middle and top plates with plastic disks placed in between the plates. Connect outflow tubes

7. Close the extractor, ensuring that the "O" ring is in place and all nuts are uniformly tightened. Gradually build-up the pressure in the vessel until the desired pressure is attained
8. Place beakers to collect water from the outflow tubes. Leave overnight
9. Open the pressure release regulator to release the pressure in the chamber
10. Open the chamber. Transfer the samples immediately to moisture boxes to measure their moisture content
11. Carry out the test at other pressures using appropriate ceramic plates.

Analysis

Volumetric moisture content is determined using the following relation.

Volumetric moisture content of soil, $\theta = (w/100) \times$ Bulk density

Here w is the gravimetric water content, percent. To prepare the water retention curve, water contents at 0, 10, 30, 100, 345, 1033, 3100 and 15000 cm matric suction are measured. Once the moisture content at various suctions is known, soil water retention curve is drawn with volumetric water content on the x-axis and matric suction on the *y*-axis (Fig. 6.4). The water retention or soil moisture characteristic curve is strongly affected by soil texture (Fig. 6.4). Higher the clay content, higher the water content at any particular suction and more gradual the slope of the curve.

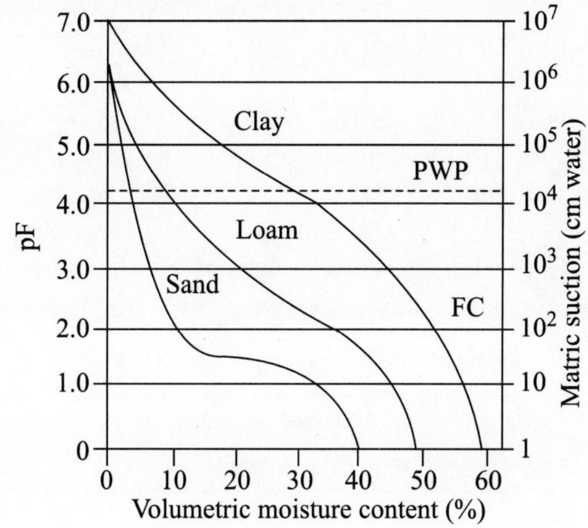

Fig. 6.4. *Schematic view of water retention curves for 3 different textured soils*

Precautions

- Increase pressure in the pressure chamber gradually until the desired level is reached. Take necessary precautions to control air leakage from the pressure plate assembly
- Fully release the pressure before opening the extractor. Don't remove the extractor lid with pressure still in the container
- Pressure plates are washed after use and dried with a coating of sand over the ceramic. This prevents fine air-borne particles from lodging in the pores of the ceramic. Brush off the dry sand before storing the plates for future use

- Use of aerated 0.005 *M* $CaSO_4$ is recommended for determining water retention curves (Dane and Hopmans, 2002), especially for fine-textured soils that contain significant amounts of swelling clays. It is because distilled or deionized water possibly can promote dispersion of clay in the soil samples. Moreover, fresh tap water is often supersaturated with air, which may affect the water content at a given pressure head (Dane and Hopmans, 2002).

Pressure Membrane Apparatus

In a pressure membrane method, soil samples are placed on a carefully supported cellophane membrane with microscopic pores for solution extraction instead of porous ceramic plate in the pressure plate apparatus. The semi-permeable cellophane membrane allows the passage of water from the sample but retains the air pressure applied to the upper surface of the membrane. The moisture content of soil in contact with the membrane is reduced by the amount that would be necessary under normal atmospheric conditions to make the pressure deficiency of the soil water equal to the excess gas pressure in the extraction chamber (Richards, 1941). The procedure mostly remains similar to the pressure plate apparatus. Depending upon the availability of the equipment, any of the two methods can be used.

A Word about pF

Tension in the past has also been expressed as the logarithmic value of tension expressed in cm of water. This value is known as pF and the graph between this value and the volumetric moisture content is known as a pF-curve. Since, this term may appear at few places in this book as well, a word on it is in order. The pF-value is obtained from soil water pressure head according to the formula: $pF = \log_{10}(-h)$

Here, h is the negative soil water pressure head (cm of water). The soil water pressure head is zero when the soil is fully saturated (no air in the pores), the permanent wilting point (PWP) is at approximately -15000 cm, i.e. pF = 4.2, and the field capacity (FC) ranges between -100 cm (pF = 2) to -500 cm (pF = 2.69) depending on the soil texture. Water can exist in soil under a tension varying from pF = 0 (no tension) to pF = 7.0 (high tension) as illustrated in Table 6.1. Most researchers now avoid using this term.

Table 6.1. *Soil moisture/moisture constants and range of tension and pF*

Moisture constant	Tension (bar)	pF
Hygroscopic moisture	31 – 10000	4.50 – 7.00
Hygroscopic coefficient	31	4.50
Permanent wilting point	15	4.20
Capillary or available water	1/3 - 31	2.54 – 4.50
Moisture equivalent	1/2 to 1	2.70 – 3.0
Field capacity	1/3	2.54
Gravitational water	0 or <1/3	<2.54

PORE-SIZE DISTRIBUTION

The relative proportions by volume of different-sized pores (pore-size distribution) can be estimated for non-expansive soils with the assumption that the soil pores are equivalent to a bundle of capillary tubes having continuous variation in diameter. From the capillary rise equation, it is possible to determine the radius of the largest equivalent cylindrical pores remaining filled with water at the applied matric suction. The equation is given as:

$$r = 2\gamma \cos u \,/\, \rho_t g h \qquad \text{...(6.1)}$$

Here r is the equivalent cylindrical pore radius, h the matric suction, cm of water, γ the surface tension of water, u the contact angle (generally taken as zero such that $\cos u = 1.0$, unless the soil is resistant to wetting), ρ_t the density of water at temperature t, and g the acceleration due to gravity. Using the moisture characteristic curve (with water content expressed on a volumetric basis) and the above equation, it is possible to calculate the volume of pores in any pore-size range.

Procedure

1. Determine the moisture characteristic curve for the soil
2. Determine the bulk density by one of the methods. It is required for converting water content from gravimetric to volumetric basis
3. Using the capillary rise equation (eq. 6.1), calculate the matric suctions corresponding to the equivalent pore radii of interest. Read the volumetric water contents from the volumetric moisture characteristic curve corresponding to these matric suctions. These values are then used to calculate the volume of pores finer than the nominated equivalent pore radii
4. The proportion of pores in a particular range can be found by the difference between the two ranges.

TOTAL WATER HOLDING CAPACITY

Use of Moisture Characteristic Curve

The actual pressure in saturated soil at equilibrium is atmospheric and the suction (positive pressure) is zero. With a slight increase in suction, water is drawn out from relatively large pores of the soil, which are unable to retain water against the applied suction. A gradual increase in suction will result in emptying the progressively smaller pores, until at high suctions, only the very narrow pores retain water. If the soil moisture characteristic curve for the soil in question is either available or can be experimentally generated, then the total water holding capacity of the soil can be determined as follows.

Water content at –1/3 bar = θ_f

Water content at –15 bar = θ_w

Total available water in the root zone depth, $d = (\theta_f - \theta_w)\, d/100$ ...(6.2)

Here θ_f and θ_w are the volumetric moisture contents in percentage and d is the root zone depth, m. Total available water is in m^3/m^2 or m, same as the units of d. Since most investigators report gravimetric moisture content, the following relation is used to convert the gravimetric moisture content, θ_g to volumetric moisture content, θ_v.

$$\theta_v = \theta_g \times \rho_b / \rho_w$$

Here ρ_b is the bulk density of the soil, kg/m^3 and ρ_w is the density of water. Since the density of water is taken as 1 kg/m^3, eq. (6.2) can be written as:

Available water = $\rho_b (\theta_{fg} - \theta_{wg}) d/100$ m ...(6.3)

In eq. (6.3) θ_{fg} and θ_{wg} are the gravimetric moisture contents in percentage.

Method Based on Laboratory/Field Evaluated Moisture Constants

Water holding capacity can be determined in the same fashion (using eq. 6.2 or eq. 6.3) from known field capacity and PWP determined from laboratory and field investigations. The value so calculated is more realistic than the value obtained from moisture characteristics curves. Following guidelines are used to qualitative classify the soils with respect to its available water.

Available water (cm/m depth of soil)	Class
<5	Very low
5-10	Low
10-15	Medium
15-20	High
>20	Very high

Keen and Rackzowski (1921) Method

Soil physical characteristics such as bulk density, porosity, particle density and water holding capacity are determined using a keen-Rackzowski box (Keen and Rackzowski, 1921, Gupta and Dakshinamoorthi, 1980, Viji and Prasanna, 2012). The box is a thick metal sheet circular box of 5.6 cm internal diameter and 1.6 cm long, having numerous perforations at its bottom. Following step-wise procedure is used.

1. The air-dry soil passing through 2-mm sieve is mixed well for use in the analysis. Measure the diameter, d and depth, L of the box with a vernier caliper. The volume of the air-dry soil is then given as: $V = \pi d^2 L/4$
2. Determine the moisture content of air-dry soil as per procedure discussed elsewhere in this chapter
3. Clean the dish of the box, dry and weigh. Place Whatman No.1 and 44 filter papers at the bottom of the box. Use the split ring supplied with the box to hold the filter paper. Weigh the box again, W_2
4. Fill the box with soil giving repeated tapings (20-30 times) on the top of the table for complete and uniform packing. The surface is levelled with the spatula. The Keen box with air-dry soil is weighed, W_3
5. The weight of the air-dry soil in the Keen box, $W_4 = W_3 - W_2$
6. Place the box in the soaking tray. Gradually fill the tray with water from the side till the water level is about 1 cm above the base of the box. Cover the tray to prevent evaporation from the soil surface. Let it remain as such for 12 hr or more to equilibrate. A continuous and shining film of water at the soil surface confirms the saturation. Carefully remove the box from the water and wipe it dry from outside and weigh quickly, W_5
7. Cut-off the expanded soil (above the dish's rim) with a spatula and transfer it to previously weighed moisture box. Weigh the moisture box with expanded out wet

soil (F_1). It is dried at 105°C until constant weight is recorded (F_2). Weigh the oven-dry expanded soil with the moisture box. The weight of the oven-dry expanded soil is calculated as $W_6 = F_1 - F_2$

8. Weigh the keen box with moist residual soil (G_1). The keen box with residual soil is dried in the oven at 105°C until constant weight is recorded (G_2). The oven-dry weight of the residual soil is given as $W_7 = G_2 - W_2$. The weight of water in the residual soil plus filter is $W_8 = G_1 - G_2$
9. Simultaneously run a blank test (only with keen box and filter paper). Weigh after the blank test (H). It helps to find the weight of the water absorbed by the filter paper alone, $W_9 = H - W_2$
10. Bulk density = Weight of the air-dry soil/volume of the soil = $(W_6 + W_7)/V$ g/cm³
11. Particle density = Weight of the oven-dry residual soil/Volume of the residual soil particles g/cm³.
    - Volume of the residual soil particles = Bulk volume of the residual soil (V) – Volume of water present in wet residual soil (Weight of water in wet residual soil – weight of water in the filter = $W_8 - W_9$)
    - Particle density = $W_7/(A - W_8 - W_9)$ g/cm³. (Since the weight of the water is equal to its volume)
12. Maximum water holding capacity = 100 [Weight of the total water in wet saturated soil/Weight of the total oven-dry soil]
    - Weight of the total water in wet saturated soil = $W_5 - W_2 - W_6 - W_7 - W_9$
    - Weight of the total oven-dry soil = $W_6 + W_7$
    - Maximum water holding capacity = $100 (W_5 - W_2 - W_6 - W_7 - W_9)/(W_6 + W_7)$
13. Volume expansion = (Volume of the expanded soil/Volume of the soil) × 100
    - Volume of the expanded soil = Total volume (Weight of oven-dry expanded soil/ Weight of the oven-dry residual soil)
    - Volume expansion = $100\ V (W_6/W_7)/V = 100\ W_6/W_7$
14. Water content in air-dry soil = $100\ W_4/(W_6 + W_7)$

Interpretation

Comments	Pore space (%)	Volume of expansion (%)
Low	<30	<10
Medium	30-35	10-30
High	>55	>30

Example 6.1. A 1200 cm³ of soil (depth =12 cm and area = 100 cm²) has a total wet weight of 1550 g including 280 g of water. Assuming the particle density as 2.65 Mg/m³, find out the water content on a mass and volume basis, bulk density, porosity, maximum water holding capacity, air-filled porosity, degree of saturation and depth of soil water.

Depth of soil cube (d) = 12 cm; Area of soil cube (A) = 100 cm², Volume of soil cube (V) = 1200 cm³, Wet mass of soil (M_t) = 1550 g, Mass of water (M_w) = 280 g and particle density = 2.65 g/cm³.

Moisture content on weight basis

$\theta_g = M_w/(M_t - M_w) = 280/(1550 - 280) = 22.04\%$

Bulk density $\rho_b = (M_t - M_w)/V = (1550 - 280)/\ 1200 = 1.06$ g/cm^3

Moisture content on volume basis

$\theta_v = \theta_g \times \rho_b = 22.04 \times 1.06 = 23.36\%$

Porosity, $n = 1 - \rho_b$/Particle density $= 1 - 1.06/2.65 = 0.6$ or 60%.

Air-filled porosity $n_a = (60 - 23.36) = 36.64\%$

Degree of saturation

Degree of saturation $= \theta_v/n = 23.36/60 = 38.9\%$

Maximum water holding capacity = Volume of the soil×soil porosity = 1200×0.6 = 720 cm^3

Or maximum water holding capacity = 720/100 = 7.2 cm of water

Note: Some researchers do not agree with this term as they consider the maximum water holding capacity of the soil after gravity drainage has taken place.

Initial depth of water in the soil profile is given as:

Depth of water = Volume of water/Area = 280/100 = 2.8 cm

Example 6.2. Soil of an irrigated field has bulk density of 1.5 g/cm^3 and field capacity and permanent wilting point of 26% and 12% respectively. What is the storage capacity of the field for root zone depth of 1.0 m? If the irrigation water is applied when the water content of the soil reaches 18%, find the water depth required to fill the root zone for a field application efficiency of 82%. Also calculate the irrigation interval, if the reference crop ET is 0.9 mm/day.

Assuming the density of water as 1.0 g/cm^3

Maximum storage capacity as per eq. (6.2)

$= \rho_b(\theta_f - \theta_w)\ d/100 = 1.5 \times (26 - 12) \times 1.0/100 = 0.21$ m or 21 cm

Water required at each irrigation, $D = \rho_b(\theta_f - \theta_i)d/100 = 1.5 \times (26 - 18) \times 1.0/100$

$= 0.12$ m or 12 cm (θ_i is the moisture content at which irrigation is initiated)

Quantity of water required to irrigate the field = D/Application efficiency

= 12/0.82 = 14.63 cm

Crop water use per day = 5.5×0.9 = 4.95 mm

Irrigation interval = 120/4.95 = 24.24 or ~ 24 days

HYGROSCOPIC COEFFICIENT

Hygroscopic water is the water adsorbed by the soil from the environment due to electro-molecular surface forces and is greatly affected by the relative humidity (RH) of the surrounding atmosphere (Brady and Weil, 2002). The tension is equal to a force of 31 atmospheres or a pF of 4.5. The water is non-mobile and can only be removed by heating to 105-110°C, and expressed as a percentage of the oven-dry weight. The hygroscopic coefficient as previously defined is the maximum amount of hygroscopic water absorbed by an oven-dry soil under standard conditions of humidity and temperature. Hygroscopic coefficient has been determined differently by different investigators because there are no standard codal provisions to quantify this property. For example some investigators take the air-dry moisture content as the hygroscopic coefficient. Others expose the oven-dry soil to an atmosphere of 100% relative humidity to determine the hygroscopic coefficient. Exposure to atmosphere having relative humidity less than 100% is also in use. A term maximum hygroscopic moisture (MHM) is sometimes used for the highest amount of moisture absorbed by the solid phase of the soil from the surrounding air nearly 96-98% saturated with water vapour at 22 ± 0.5^0. MHM depends on soil texture (particle size composition), the amount and quality of soil colloidal matter, organic matter content, and specific surface area of the solid phase. Maximum hygroscopic moisture of the soil is higher for heavier soils having high organic matter content, because of their high specific surface area and greater ability to absorb water from the atmosphere. The MHM can be as low as 0.5 to 1% of oven-dry weight in sand to as high as 30-40% in lowland peat (Petrenko and Berezhnyak, 2008). Hygroscopic moisture content of the soil can be related with several soil properties such as surface area, cation exchange capacity, liquid limit, swelling potential, permanent wilting point, and some electrical properties.

Correlation Method

Approximate values of the MHM for mineral soils, having low to moderate amounts of organic matter, can be calculated with the help of the following regression equation.

$$\text{MHM} = 0.3083\text{K} - 1.0965$$

Here K is the clay fraction (< 0.001 mm in diameter) in percent of oven-dry weight.

Moisture Content of Air-Dry Soil

Apparatus: Moisture boxes, Balance, Oven, Desiccator

Procedure

1. Weigh the metal can with a lid, W_0
2. Pour air-dry soil (< 2 mm) into the metal can. Weigh the metal can with a lid, W_w
3. Weight of the soil, $W_1 = W_w - W_0$
4. Dry in an oven, with the lid unfitted, at 105°C for 24-48 hr
5. Cool in a desiccator for at least 30 min, fit the lid and re-weigh, say W_d. Find out the weight of the dry soil, $W_2 = W_d - W_0$.

$$\text{Moisture content, M (\%)} = 100(W_1 - W_2)/W_2 \quad ...(6.4)$$

Applications: Reporting of Chemical Determinations

Air-dry samples are commonly used for analysis of physico-chemical properties of the soils because some properties might undergo irreversible changes at high temperature. On the other hand, it is customary to report the results on oven-dry basis. Therefore, the results obtained from the analysis of air-dry samples are multiplied by a moisture factor (F) calculated as per the following equation.

$$F = 1/(1 + M) \quad ...(6.5)$$

Here M is the moisture content of the air-dry soil, fraction. Note that F is always less than 1.0. The analytical results are then reported as

Analytical results on oven-dry basis = F × Analytical results on an air-dry basis

Petrenko and Berezhnyak (2008) Method

Equipment and Reagents: Glass drying containers/tray, A desiccator with a lid, Equipment needed to determine soil moisture

- **Saturated K_2SO_4 solution:** Dissolve110-130 g of salt in 1L of water with some un-dissolved crystals of salt in the solution

Procedure

1. Sufficient amount of soil is oven-dried. Clumps are broken up with the help of wooden rammer. The sample is passed through 1-mm sieve
2. A thin layer of air-dry soil is placed in a weighed drying container 5 cm in diameter and 3 cm in height. The clay and loam soils should be within 10-15 g while loamy-sand and sand soils may be within 15-20 g. Weighed amount of soil can also be spread uniformly in a tray
3. Samples are placed on a porcelain platform of a desiccator, the bottom part of which is filled with saturated K_2SO_4 solution. It creates the relative humidity of about 98% in the air under the desiccators' lid
4. After 3 to 5 days, the desiccator is opened, glass containers covered with lids, weighed and put back in the desiccator. The containers are weighed again after 3 to 5 days. The process continues until the constant weight or the difference in two adjacent weight measurements is 0.01 g or less. The equilibration of water vapour and the soil may take some 2 to 4 weeks, depending on soil texture and OM content. Shah and Singh (2006) suggested that the hygroscopic moisture at 7 days storage of the soil sample at 90% RH, referred as the optimal hygroscopic moisture, is about 1.06 times the hygroscopic moisture after 1 day of storage
5. After equilibrium is reached, soil moisture is determined by the gravimetric method.

Notes:

- The desiccator must be tightly covered with its lid, the edges of which must be properly lubricated with Vaseline
- The desiccator with drying containers and soil inside should be kept in a shaded place with minimal changes of temperature.

MHM and Other Soil Properties

The maximum adsorption capacity (MAC) = 0.6 to 0.7 times MHM

Permanent wilting point = 1.5 times MHM

Meteo-service recommends a factor of 1.34 instead of 1.5.

Specific surface area of the soil, S = 4 MHM

Here S is the specific surface area of the soil in m^2/g of oven-dry soil.

Free swelling potential FSI is given as (Shah and Singh, 2006)

FSI = 2.62 + 2.5 exp (optimum hygroscopic water content/3)

It may be noted that various coefficients used in correlation studies may differ depending upon the way the hygroscopic moisture content is defined or how it is evaluated. It is why no good relations have emerged so far.

MOISTURE EQUIVALENT

Briggs and Mclane (1907) introduced the term moisture equivalent and defined it as the moisture content of 1 cm thick saturated soil sample subjected to a centrifugal force equivalent to 1000 times that of gravity for a period of 30 min. The relative centrifugal force (RCF), which is the ratio of the acceleration in the centrifuge to the acceleration of gravity, is given as:

$$\text{RCF} = 0.0000112 \times r \times (rpm)^2$$

Here rpm is centrifuge speed in revolutions/ min and r is the radius in cm from the axis of the centrifuge to the bottom of the centrifuge vessel when in the rotating position.

Apparatus: Sieve 2 mm, Centrifuge, Soil moisture boxes, Oven, Balance, Filter paper

Procedure

1. Take soil sample passing through 2-mm sieve
2. Place the filter paper at the bottom of the rectangular steel boxes of the centrifuge and fill each of them with 30 g soil. Lightly tap to level the soil
3. Place the boxes in the water tray. Leave overnight to saturate
4. Cover the boxes to avoid evaporation and place them on a wet towel for about 15 min so that surplus water is absorbed by the towel. Put filter paper at the top and cover with respective lids
5. Arrange the boxes in the centrifuge and operate the centrifuge. Bring the speed to 2440 RPM (May vary from centrifuge to centrifuge) in about 3 min. Continue to run at this speed for about 30 min
6. Stop the machine and bring it to rest in about 3 min
7. The soil is quickly transferred to pre-weighed moisture boxes to determine the moisture content
8. Calculate the moisture content as described in previous sections of this chapter and take the average of the boxes. If different soils are used, then the results may be used to compare moisture equivalents of one soil with the other.

LEAST LIMITING WATER RANGE OF THE SOIL

The least limiting water range (LLWR) is a single effective soil water content based parameter that includes crop-limiting values of soil strength, aeration, and water supply to plant roots. The LLWR is a useful indicator of soil quality and soil physical constraints on crop production. The concept of least limiting water range (LLWR), first developed by Silva et al. (1994), characterizes a single range of soil water content beyond which the available water, soil aeration and mechanical resistance impose significant limitations to root growth. Since LLWR integrates all these three directly associated factors with the plant, it can be regarded as a useful single soil physical index to characterize soil physical environment and more specifically soil structural quality (Silva and Kay, 1997).

WHAT IS LLWR?

Like available water, LLWR is characterized by upper and lower limits of water available to plants.

Upper limit of LLWR, θ_{wet} is either the soil water content at 10% aeration porosity (θ_{ap}) or soil water content at field capacity (θ_{fc}), whichever is lower

Lower limit, θ_{dry} is either the soil water content corresponding to 2MPa soil penetration resistance (θ_{2MPa}) or soil water content at permanent wilting point (θ_{pwp}), whichever is higher.

Thus, LLWR is calculated as

$$LLWR = \theta_{wet} - \theta_{dry} \quad ...(6.6)$$

$$\theta_{wet} = \min(\theta_{fc}, \theta_{ap}) \quad ...(6.7)$$

$$\theta_{dry} = \max(\theta_{2MPa} - \theta_{pwp}) \quad ...(6.8)$$

Higher the value of LLWR better is the soil physical environment for plant growth.

Procedure

1. Determine the soil water contents at field capacity (θ_{fc}) and wilting point (θ_{pwp}). Use any of the methods described in this chapter to assess the water contents. Gravimetric soil water contents are multiplied with bulk density to get volumetric water contents
2. Determine the soil water content at saturation. Multiply by bulk density to convert it to volumetric water content
3. Soil water content at 10% aeration porosity, θ_{ap} is given as: $\theta_{ap} = \theta_{sat} - 10$

   Here θ_{sat} is soil water content at saturation in %
4. Soil water content at 2Mpa soil penetration resistance (θ_{2MPa}) is assessed by developing an empirical equation that relates penetration resistance (PR) to the bulk density, BD and the water content, θ_w. Step-wise procedure comprises the following steps (Keller et al., 2015).
   - Use an appropriate cone penetrometer with data logger. The soil strength (PR in MPa) is calculated by the procedure outlined in respective manuals of penetrometers supplied by the manufacturer of the instrument. General procedures for using 3 kinds of penetrometers are described in Chapter 12 on Engineering Properties of this book. Likely impact of PR so obtained on root growth can be understood from the guidelines given in Table 6.2 (Hazelton and Murphy, 2007).
   - Develop a regression equation by using any statistical tool of regression analysis. A linear or log-linear relationship may be obtained.

     $$PR = a + b \times \theta_w + c \times BD \quad ...(6.9)$$

     $$\text{Ln } PR = \text{Ln } d + e \times \text{Ln } \theta_w + f \times BD \quad ...(6.10)$$

     Here *a*, *b* and *c* and *d*, *e* and f are regression constants in the respective equations.
   - Incorporate PR = 2MPa in any of the regression equatiosn(eq. 6.9 or eq. 6.10), and calculate θ_w, which is equal to θ_{2MPa} on weight basis. Multiply this value with bulk density to get volumetric water content θ_{2Mpa}
5. Calculate LLWR (m^3/m^3) using eq. (6.6) through (6.8).

Magnitude of LLWR = soil water content between the upper and lower limit of LLWR. It is found to decrease with increasing bulk density (Fig. 6.5).

Table 6.2. *Effect of penetration resistance on root growth*

Penetration resistance*, (MPa)	Degree of soil consolidation	Effect on root growth
<0.50	Loose	Not affected
0.50–1.25	Medium	Root growth of some cereal plants may be restricted
1.25–2.00	Dense	Cereal root growth badly restricted
2.00–3.00	Very dense	Very few plant roots penetrate the soil
>3.00	Extremely dense	Root growth virtually ceases

*The resistance is dependent on moisture content and technique used

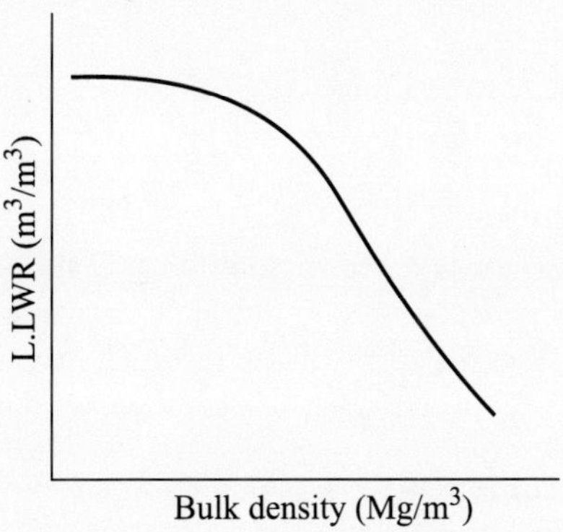

Fig. 6.5. *Effect of bulk density on LLWR*

Physico-Chemical Characteristics and Available Water Capacity

Most plants can extract water up to a maximum suction of about -15 bars, or -150 m of water before they are unable to extract any further water from the soil. Besides the physical, several chemical properties are known to influence the available water holding capacity of the soils. Soil salinity can significantly affect the water uptake by the plants as the plants have to exert additional suction against the osmotic pressure of the soil water. This section highlights the osmotic effect of salts on available water capacity through its modifying effect on various moisture constants.

The osmotic pressure of the soil solution is calculated from the equation (Richards, 1954):

$$OP = 0.36\ EC \qquad ...(6.11)$$

Here EC is the electrical conductivity of the soil solution, dS/m and OP, bars.

In terms of head of water, H_o eq. (6.11) is approximated as follows.

$$H_o = 3.6\ EC \qquad ...(6.12)$$

Here H_o is the osmotic head, *m* of water.

Groenevelt et al. (2004), used the term osmo-matric head, H_{om} and defined it as

$H_{om} = H_m + H_o\,\sigma(t, H_m, H_o)$

Here H_m is the matric head and $\sigma(t, H_m, H_o)$ is the reflection coefficient that governs the effectiveness of the osmotic suction of the soil solution. Its value changes over time (t) and is a function of both the matric and the osmotic suction of the soil solution. Besides it also depends upon the plant species (Groenevelt et al., 2004). The value of σ is zero for non-saline environment and can be a maximum of 1.0., if the plant behaves like a perfect osmometer. Taking the value of $\sigma(t, H_m, H_o)$ as 1.0, we have

$$H_{om} = H_m + H_o \qquad ...(6.13)$$

Apparently, in a saline environment H_{om} rather than H_m alone determines the water availability to plants. Eq. (6.13) can be used to calculate the matric head, H_m where H_{om} reaches a value that describes a given moisture constant. For example, for permanent wilting point we set the value H_{om} as 150 m in eq. (6.13) to write it as (Groenevelt et al., 2004):

$$H_m + H_o = 150 \qquad ...(6.14)$$

To calculate the value of H_m as a function of EC_e where H_{om} reaches a value of 150 m, we incorporate eq. (6.12) in eq. (6.14):

$$H_m = 150 - H_o = 150 - 3.6\ \mathrm{EC} \qquad ...(6.15)$$

It is assumed that soil water is an ideal solution and its contents change by perfect evapotranspiration without loss of solutes through drainage, salt-precipitation, anion exclusion or plant uptake. Then the approximate mass balance is given as:

$$\theta_s/\theta = EC/EC_e$$

Here θ_s is the moisture content at the soil saturation and EC and θ corresponds to the permanent wilting point. Thus

$$EC_{pwp} = EC_e\,(\theta_s/\theta_{pwp}) \qquad ...(6.16)$$

Incorporating eq. (6.16) in eq. (6.15), we have

$$H_m = 150 - 3.6EC_e\,(\theta_s/\theta_{pwp}) \qquad ...(6.17)$$

The value of θ_{pwp} and θ_s are determined from the water retention curve. In a saline soil value of H_m where H_{om} reaches a value of 150 m (wilting point) is a function of EC_e of the soil and it decreases with increase in EC_e. Since most plants are irrigated much before the permanent wilting point has reached, we can use eq. (6.17) to determine the matric head at which irrigation should be scheduled in a saline soil. If irrigation is to be given at a matric head of H_i in a non-saline soil, H_m at which irrigation should be scheduled in a saline soil is given by rewriting eq. (6.17) as:

$$H_m = H_i - 3.6EC_e\,(\theta_s/\theta_{hi}) \qquad ...(6.18)$$

Here θ_{hi} is the moisture content of the soil at H_i, which is read from the water retention curve.

Chapter **7**

Soil Moisture Measurement

Water is the most limiting factor for optimum crop production. Therefore, soil moisture is routinely measured in irrigation water management studies. In fact, measurements of soil water content are needed practically in all types of soil studies, e.g., determination of water holding capacity, plant available water, infiltration, pore size distribution, permeability and even with respect to soil microbial processes and biological soil remediation. Soil moisture status of the soil influences crop growth, as it controls the soil matric stress to which plants are exposed. Besides, it directly or indirectly affects nutrient availability through its influence on nutrient transformations and soil biological behavior. Soil scientists, irrigation engineers and agronomists, all are interested in the current soil moisture status of the field to plan irrigation scheduling, or soil moisture status before and/or after irrigation to calculate field water application efficiencies. They are also interested in the moisture contents of the soil at field capacity and permanent wilting points, which has been a subject of discussion in the previous chapter. Both the crop growers and researchers are interested in knowing the dynamics of soil moisture at various times of the crop growth cycle. Several methods are used to assess *in-situ* moisture content of field soils. Among them gravimetric, gypsum block, tensiometers, neutron moisture probe, time-domain refractrometry, frequency-domain refractrometry and remote sensing are commonly employed to measure/assess field soil moisture contents. This chapter describes these techniques of moisture measurements.

SOIL MOISTURE (GRAVIMETRIC METHOD)

Gravimetric method is most commonly employed technique for evaluating soil moisture. Although simple and easy, being an invasive method, it requires considerable efforts in collecting soil samples and their moisture evaluation.

Apparatus: Auger, Electric oven with thermostat, Desiccator, Aluminium cans with lids, Weighing balance

Procedure

1. Use an auger to draw soil samples at appropriate depths (Chapter 1). Transfer the samples directly from the auger to the aluminium cans
2. Weigh the aluminium cans containing moist soil samples
3. Keep the cans for about 24 hr in an oven maintained at 105°C. Weigh the oven-dry samples
4. The soil moisture content on oven-dry weight basis is calculated as follows.

Weight of the container	W_0
Weight of the container with moist soil	W_1
Weight of the moist soil	$W_1 - W_0$
Weight of the container with oven-dry soil	W_2

Weight of the oven-dry soil	$W_2 - W_0$
Weight of the water in the sample	$W_1 - W_2$
Moisture content (%)	$[(W_1 - W_2)/(W_2 - W_0)] \times 100$

Multiplication by 100 in the last row is to convert the moisture content on a percentage basis. Use the following relation to express the moisture content on a volume basis.

$$\text{Volumetric moisture content (\%)} = [(W_1 - W_2)/(W_2 - W_0)] \times 100 \times \rho_b \qquad ..(7.1)$$

Here ρ_b is the bulk density of the soil.

GYPSUM BLOCKS

The gravimetric method of measuring soil moisture poses insurmountable problems. Therefore, non-invasive methods have come into being. The application of these methods has facilitated the soil moisture measurements on a regular basis for planning irrigation schedules. Gypsum blocks have been used in soil science for the past 50 years (Bouyoucos, 1965). Currently, several types of commercial gypsum blocks are available in the market. These are cylindrical blocks of gypsum (may be square or rectangular as well) into which two electrodes are inserted (Fig. 7.1 top left). Now, fiber glass and nylon is also used to fabricate these blocks. The moisture content of these porous blocks placed in the soil remains in equilibrium with the soil moisture content. The electrical resistance between the electrodes varies, being low for high and high for low moisture contents. The measured voltage is related to moisture content or soil moisture tension, the measuring range being 0.1 to 2 bars. Nylon blocks are more sensitive and function best at tensions less than 2 bars. The depth at which gypsum blocks are placed is governed by the root depth. If roots extend below 800 mm then three blocks are installed at 300 mm, 600 mm and at the bottom of the root zone. If roots are confined to the top 800 mm then only two gypsum blocks may suffice to give a representative value, one in the middle of the fibrous root zone (usually 200 or 400 mm depth) and the other at the bottom of the root zone. To take measurements from above the ground, wires attached to these electrodes extend to the soil surface. A small transistorized meter is used to measure electrical resistance (Fig. 7.1 bottom left). Gypsum blocks are usually supplied with a calibration curve showing the relationship between measured resistance and soil moisture tension. An example of such a standard calibration curve is Fig. 7.1 right (Goodwin, 2009). It is always useful to develop the calibration curve on site or in the laboratory using the field soils. Advantages and limitations of these blocks are as follows.

Advantages

- Cheap in comparison to other systems
- Easy to install and use
- Measurements can be taken quickly and can be logged to a recorder to obtain a continuous record of soil moisture changes *in-situ*
- Direct indication of soil moisture easily helps to determine when to irrigate

Limitations

- Life of the gypsum blocks is less in some soil types, because the blocks gradually deteriorate in the soil due to their solubility

- Relatively small sphere of influence
- Blocks made of plaster of Paris (gypsum) maintain a nearly constant electrolyte concentration corresponding primarily to that of a saturated solution of $CaSO_4$. This tends to mask, or buffer, the effect of small or even moderate variations in the soil solution (such as those due to fertilization or low levels of salinity). However, its accuracy and usefulness may be questionable in saline soils or soils irrigated with saline water.

  Apparatus: Gypsum blocks, Auger, Meter, De-aerated water, Measuring tape, Glassware
- Gypsum

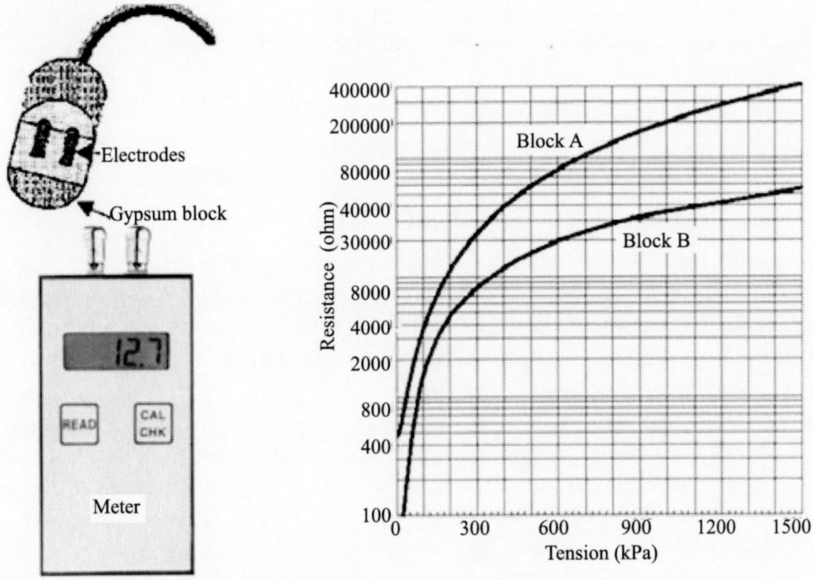

Fig. 7.1. *Gypsum block with electrodes (top left). Meter to measure resistance (bottom left) and the calibration curves for two commercially available gypsum blocks (right)*

Installation of Gypsum Blocks

1. The gypsum blocks are thoroughly soaked in water to expel all air from the blocks before installation in a representative area of the field
2. Auger a bore to a depth slightly greater than the desired monitoring depth. The hole may be augured at an angle of 45^0 to the horizontal. It helps to prevent any preferential water flow down the backfilled augur hole. Keep the soil intact in the same depth sequence so that the same sequence is mimicked while refilling the hole
3. A small quantity of a fine mixture of soil and gypsum is added to the bottom of the hole to ensure good contact between the soil and the block. A small amount of water can also be added to the hole to soften the soil at the bottom. Holding the wire leads of the block, push the block to firmly set it at the bottom of the hole. A conduit can also be used to push the block at the bottom of the hole. Carefully remove the pipe after the block has been set at the right place

4. Fill the hole with soil in steps not exceeding 100 mm at a time. Lightly tap the soil to firm-up the refilled soil
5. Repeat steps 3 and 4 until all the gypsum blocks are installed at desired depths
6. Mark the block depth either with a surface tag or knots in the wire. The number of knots in the wire may indicate the depth
7. Fasten the wire leads around a stake nearby
8. Measure the block resistance and use the calibration curve to determine the soil moisture tension

Notes:

- Gypsum blocks should be installed immediately after the plants or crop is established. It helps the plant roots to grow around the blocks. The blocks installed late in the growing season will not provide realistic readings of the matric tension experienced by the roots.
- Don't place blocks in low or high areas of the field
- Install blocks at least 6 m away from wheel tracks.

SOIL MATRIC POTENTIAL (TENSIOMETER)

A tensiometer is designed to measure the tension or suction - the force that plants' roots must exert to extract water from the soil. A tensiometer essentially consists of a PVC/acrylic tube connected with a porous ceramic cup and a pressure gauge (Cassel and Klute, 1986). Several kinds of commercial tensiometers are available in the market e.g. tensiometers with different lengths or micro-tensiometers. Now-a-days, electronic pressure transducers with recording devices are also used. These enable the connection of several tensiometers to data logger and computer. In the gauge type tensiometer, which is the subject of discussion of this section, a Bourdon (vacuum) gauge is attached few centimeters below the Stoppard end of the tube (Fig. 7.2 right). The earlier versions of Hg-H_2O type tensiometers, which were capable to read pressure values in saturated soils with positive pressure heads (Fig. 7.2 left) are not currently in use because of handling concerns and environmental issues attached to the use of mercury.

Principle

Water from the tensiometer is capable of moving in and out of the tensiometer through the porous ceramic tip (Fig 7.2). Since the water can flow in and out of the tube, the gauge reading is always in balance with the soil suction. As the soil dries, water flows out from the tensiometer. It creates a vacuum inside the tensiometer, which is indicated on the gauge. Since the vacuum corresponds to the soil suction at equilibrium, the dial gauge reading is a direct measure of the force required to take up water from the soil. If the soil dries further, additional water flows out until a high vacuum level is reached. On the contrary, a reverse process takes place when moisture is added to the soil. In this case, soil moisture moves into the tensiometer through the porous tip. The vacuum level in the gauge reestablishes at a value, which corresponds to the soil suction value at this time. The gauge reading may drop to zero, if enough water is added to the soil to become fully saturated.

Apparatus and Equipment: Tensiometer, De-aerated/distilled water, Auger, Spatula, Measuring tape, Plastic wash bottle, Distilled water

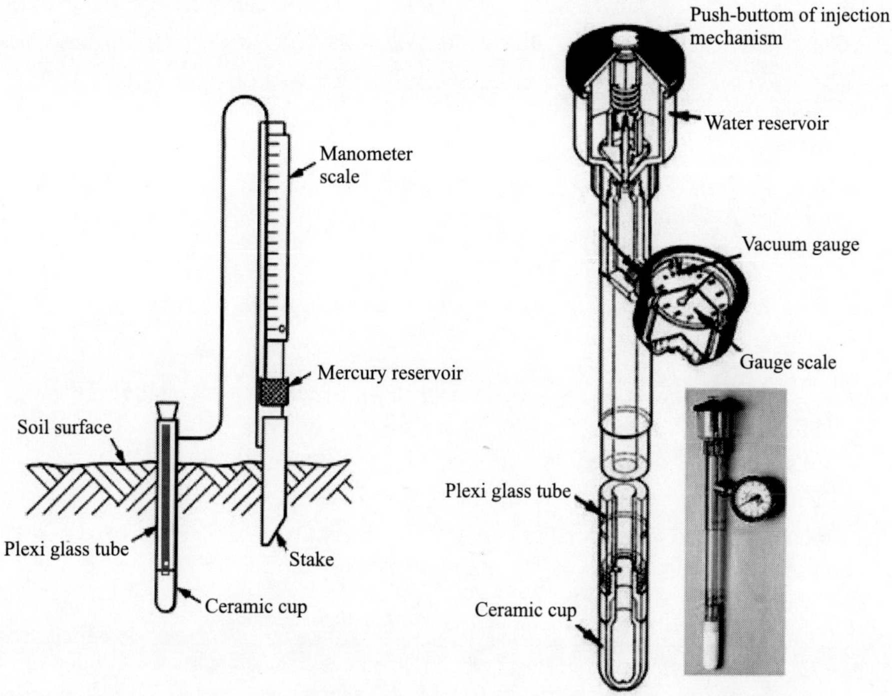

Fig. 7.2. *Hg-H_2O manometer type tensiometer (left) and a schematic and pictorial view of gauge tensiometers (right)*

Installation and Use

1. Ceramic cups of the tensiometers are kept submerged into air-free water for at least 2-3 days prior to installation to ensure that the cups are fully water-saturated. Visual tests for a leak and performance may also be conducted at this stage. One of the common tests comprises testing the sensitivity of the tensiometer cups. In this test, water-filled cups are allowed to dry in the air or near a hot plate so that the vacuum gauge shows a suction head of around 0.8 bar (800 cm of water or 80 kPa). Following this, the ceramic cups are dipped again in a beaker full of water. For a reasonably sensitive tensiometer, Bourdon gauge needle should quickly drop to zero upon immersion in water
2. To install the tensiometer, use a soil auger to make a hole. The diameter of the hole should be slightly less than the outer diameter of the tensiometer to give it a snug fit
3. Drop a handful of loose friable soil or fine sand into the hole. Unless the soil is likely to puddle easily, a small amount of water can also be poured into the hole
4. Fill the tensiometer with de-aerated water and cork it
5. Push the tensiometer into the hole by giving a firm twisting downward motion applied to the connecting PVC tube. Set the cup at the desired soil depth. It ensures a close contact between the porous cup and the soil in its vicinity

6. Backfill the hole as may be necessary so that the tensiometer is firmly held in the soil
7. Flush the tensiometer with air-free water. Allow the tensiometer to reach steady state
8. Take the observation and do the calculations as follows.

Calculations

If a tensiometer cup is placed at a depth Z cm below the gauge, then the pressure at tensiometer cup, P_a is given as

$$P_a = P_b + \rho g Z$$

Here P_b is the pressure reading of the gauge. In terms of head, we have

$$h_a = h_b + Z \qquad ...(7.2)$$

Here h_a is the reading on the dial in terms of head of water. If the dial of the gauge is calibrated from 0 to 100, which on a weight basis corresponds to 0-100 centibar (1centibar is equal to one kilopascal) or 0-1000 millibar, the 1 centibar corresponds to 10 cm.

$$h_b = -10 \times h_a + Z$$

This relation is used to convert the dial reading in centibars to pressure head or matric head. Interpret the results as per the information given in Table 7.1.

Table 7.1. *Interpretation of the tensiometer readings*

0 kPa	The surrounding soil irrespective of its texture is completely saturated with water. If the reading persists, soil aeration problem is anticipated
>0-10 kPa	Gauge readings in this range indicate a surplus of water for plant growth, which drains off within a few days. If readings in this range persist, it indicates poor drainage conditions
10-20 kPa	Gauge readings in the range indicate that there is ample moisture and air in the soil for healthy plant growth in most soil types
20-40 kPa	Available moisture and aeration in the root zone are good for plant growth. Heavy clay soils: No irrigation required Medium textured soils: No irrigation required Sandy soils: Irrigation may be given for coarse sandy soils in the 20-30 kPa range. For fine sandy soils in the 30-40 kPa range
40-60 kPa	Available moisture and aeration are good for plant growth in fine textured soils. Heavy clay soils: No irrigation required Medium textured soils: Irrigation may be given in this range. Irrigation in a fine texture soil begins at a higher reading than for a coarse textured soil Sandy soils: Too dry. Under hot windy conditions productivity may reduce
60-80 kPa	Except for heavy clay soils, readily available moisture in this range of readings for other soils is scarce. Heavy clay soils: Irrigation desirable as soil suction reaches 70-80 kPa Medium textured soils: Too dry. Hot, windy conditions can quickly force soil suction to become high and can damage plants Sandy soils: Too dry. Damage to plants has occurred before irrigation is applied

Limitations

The useful range of a tensiometer is 0 to 80 kPa (0 to 800 cm or 0 to 0.8 bar), since air-entry value of most ceramic cups is about -85 kPa (-85 centibars or -850 cm of pressure head). If the soil is drier than this range, the water column in the tensiometer breaks as the air enters through the cup

The tensiometers are usually slow to respond in soils with rapidly changing matric potential. Actual response time depends on the hydraulic conductivity of the porous cup and sensitivity of the gauge

A tensiometer measures the matric water potential only, being insensitive to osmotic potential

In gravelly soils, poor contact between the cup and the soil may affect the working of the tensiometer.

Maintenance

A tensiometer needs to be skillfully managed. The most common problem encountered during its use is the dial reading staying at zero. It can happen because of the following reasons.

- The soil remains saturated
- Air leakage through the ceramic tip or around joints and seals
- Soil tip is not in proper contact with the soil
- There is some fault in the gauge

If higher readings than the anticipated values persist, especially after irrigation, it may be possible that the tensiometer is installed at a deeper depth than the water can reach. In that case, place the tensiometer at a shallower depth within the wetted zone.

Example 7.1. A tensiometer cup is placed at 90 cm below the tensiometer gauge. If the tensiometer reading in the dial is 35 centibar, what is the suction head?

Matric head, $h = -10 \times 35 + 90 = -260$ cm $= -26$ kPa

The negative sign is usually omitted for convenience when soil water potentials are measured with tensiometers. Thus

Matric head = 26 kPa

Example 7.2. A tensiometer shows a gauge reading of -45000 Pa at 25^{0}C. The water column in the tensiometer is 40 cm. Calculate the total water potential head at 20 cm below the soil surface. Assume the soil surface as the reference.

Pressure head $= (\Psi_g/\rho_w g) + Z = [-45000/(997 \times 9.81)] + 0.40$

Pressure head $= -4.6 + 0.4 = -4.2$ m

Gravitational potential head $= Z_{soil} - Z = 0 - 0.20 = -0.20$

Neglecting the osmotic pressure head, total potential head is given as:

$H = h + Z = -4.2 - 0.2 = -4.4$ m

NEUTRON PROBE

Neutron probe or neutron moisture meter is an indirect non-destructive method to determine the soil water content. In this device, a radioactive source either a mixture of Americium-Beryllium or Radium-Beryllium is used to produce alpha particles, which in turn react with Beryllium to produce fast neutrons (^{1}n). The energies of the neutrons emitted by a source of fast neutrons vary from 1 to 15 MeV (million electron volts) with a preponderant energy range of 2-4 MeV and an average speed of about 1600 km/s. These neutrons collide with hydrogen atoms of water molecules in surrounding soil. The fast neutrons, as a result of the collision, are slowed down to about 2.7 km/s, corresponding to an energy of about 0.03 eV (Fig. 7.3 left). The instrument is equipped with a scintillator/detector filled with boron trifluoride (BF_3) gas. It detects these thermalized neutrons. Whenever thermalized neutron encounters a ^{10}B nucleus and is absorbed, an alpha particle is emitted, which creates an electrical pulse on a charged wire. A scalar or rate meter monitors the flux of slow neutrons scattered by the soil by counting numbers of pulses over a measured time interval. Count rate of thermal or slow neutrons is proportional to the density of hydrogen atoms in the vicinity of the radioactive source. A relation between count rate and volumetric water content is established for future use.

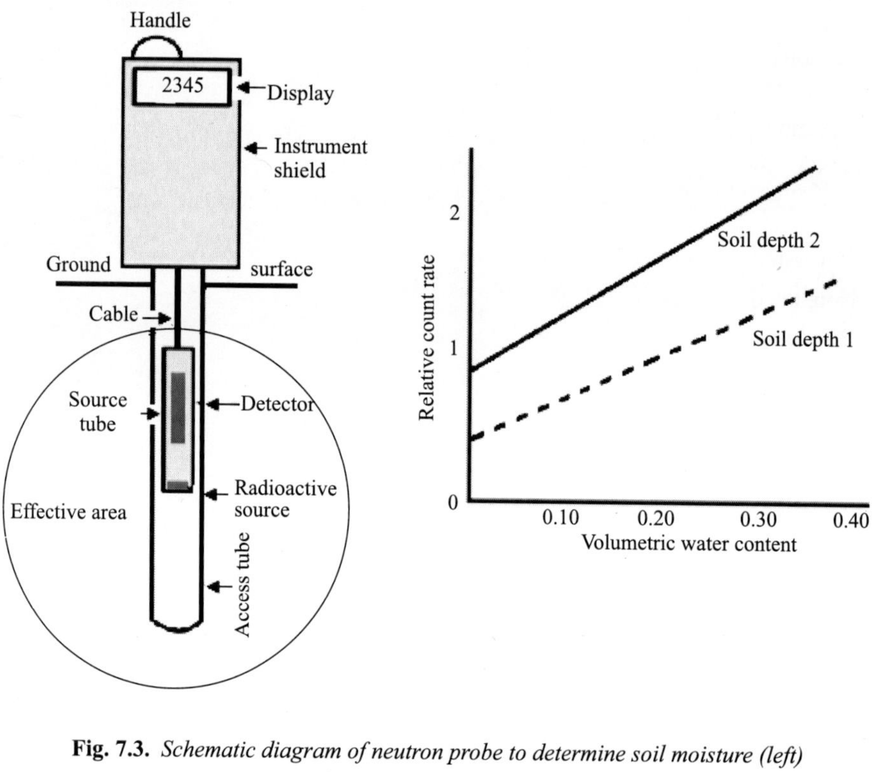

Fig. 7.3. *Schematic diagram of neutron probe to determine soil moisture (left) and a calibration curve (right)*

Calibration

To prepare a calibration curve for a particular soil depth, a series of observations are made using a neutron moisture meter. The soil samples from the same depth are drawn using soil sampler to measure the gravimetric and volumetric moisture contents. The calibration curve is prepared by plotting the count ratio, CR (observed count/standard count) and the volumetric moisture content, θ_v of soil for varying moisture contents of soil. A regression equation is developed between count ratio (CR) and the θ_v content of soil as:

$$\theta_v = a + b \times \text{CR} \qquad \text{...(7.3)}$$

A neutron moisture meter should be calibrated for each soil and soil depth especially the surface soil separately rather than using a universal calibration (Fig. 7.3 right).

Apparatus/Equipment: Neutron moisture meter, Aluminium access tube, Soil auger

Procedure

1. Drill a hole with the help of soil auger and insert the access tube into the soil. The diameter of the hole should be slightly smaller than the diameter of the access tube, so as to ensure a snug fit
2. The lower end of the access tube is plugged with a rubber stopper, especially if saturated soil conditions are likely to prevail (permanent or perched water table)
3. The length of the access tube is selected in such a way that it protrudes few cm above the soil surface. It is covered with an inverted can or a plastic container to prevent the entry of trash/rain water
4. Take initial standard count by placing the radioactive source in the shield. It helps to account for electronic drift, instability of electronics, and radioactive decay of neutron source
5. The neutron probe is inserted in the access tube and held at the desired depths using adjustable knobs
6. Note the observed count at that particular depth
7. Divide the observed count by standard count to get the count ratio
8. Refer to the calibration curve of that particular soil depth and determine volumetric water content for the observed count ratio. Regression equation like that of eq. (7.3) can also be used to calculate the moisture content.

The accuracy of the measured moisture content depends on several factors, especially on the number of calibration points used to prepare the calibration curve, and/or errors associated with gravimetric sampling.

Precautions

- Always use dent-free access tubes
- Protect the neutron source from free water
- Never touch the open probe
- Normally measurements of moisture content of the surface soil using a neutron moisture meter are not recommended so as to avoid radiation hazards. If necessary, prepare a separate calibration curve for surface soil.

Limitations

- Radioactive source of a neutron moisture meter is a major hindrance in the unrestricted use of this equipment. For the sake of safety, and to provide a convenient means of making standard readings, the probe containing the fast neutron source is normally carried inside a protective shield, a cylindrical container filled with lead. The equipment can be used without undue risk by strictly following the guidelines on its safe use
- Measurements respond to total hydrogen content and may vary with organic matter content, slightly by soil bulk density and soil chemical composition. Thus, calibration for specific site is paramount to ensure high measurement accuracies
- Measured soil volumes or sphere of influence is large (a radius of 15-25 cm), and therefore, problems may arise when measurements at narrow depth intervals are desired. It is more so as the effective measurement volume is function of water content, being larger for soil containing less water.

TIME DOMAIN REFLECTOMETRY (TDR)

It was nearly half a decade back that a TDR was first applied to measure the moisture content of soils and earth materials (Davis and Chudobiak, 1975). Since then, TDR has become a standard method of soil moisture measurements. This method dispenses with the two main limitation of neutron moisture probe. One, there is no radiation hazard associated with it and two the moisture content of the top soil can also be easily determined. TDR basically measures the dielectric constant of the soil, water having a very high dielectric constant (Table 7.2). The dielectric value of pure water is about 81, whereas that of dry soils ranges from 3-5. Clearly, the dielectric constant of the soil increases with its increasing water content. It is for the same reason, the reference liquid mostly water or any other liquid of known dielectric constant is used to check measurement repeatability and instrument drift.

Table 7.2. *Dielectric constants of common materials*

Material	Dielectric constant	Material	Dielectric constant
Water	81	Rubber	2.3–4.0
Air	1	Dry paper	3.0
Dry soil	3.0–5.0	Dry wood	2.0–4.0
Plastic	2.1–2.6		

The Equipment

A TDR probe consists of four basic components: a timing circuit, a pulse generator, a sampling receiver, and a display or recording device. The basic elements of a TDR probe are its conductive components, the parallel metallic rods (Fig. 7.4). These act as wave-guides in the soil material in which wave or signal propagates. Currently, the most common soil probe is a balanced pair transmission line consisting of two parallel rods, rod lengths varying from 0.1 to 1.0 m depending on the measurement requirements (Topp and Davis 1985, Topp and Ferré, 1998). The separation distance between the rods vary in the range of 0.01 to 0.1 m. Currently, multipronged probes are used wherein one prong or wire is centrally located and variable numbers of prongs are located circumferentially around the central wire. The trace may be displayed and analyzed manually, or data may be recorded and analyzed digitally on a PC. Most commercial instruments now offer automated analysis having all components for analysis, displaying and recording of the water contents in a single unit.

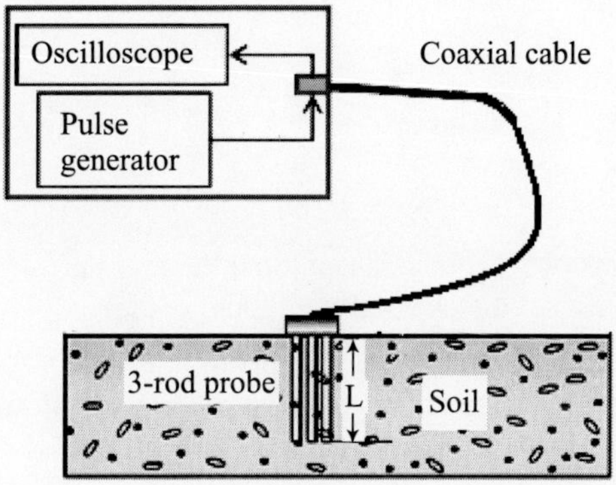

Fig. 7.4. *Schematic view of a TDR meter for soil moisture measurement*

Theory

Following theory gives some insight for manual calculations. The speed at which electromagnetic (EM) wave moves along the conductors and through the soil, consisting of air, water and soil particles, is reduced as the dielectric constant of the moist soil is higher than dry soil. The velocity (v) of an EM wave through a transmission line in a non-magnetic medium is given by (Topp and Davis, 1985)

$$v = C / K^{0.5} \qquad ...(7.4)$$

Here C is the velocity of light (3×10^8 m/s) and K is the dielectric constant of the non-magnetic medium, such as soil. Velocity is also given as:

$$v = 2L/t \qquad ...(7.5)$$

Here L is the length of the probe and t is the time period between the sending and receiving signal (The measured value). For a fixed line length, the time interval relates inversely to the propagation velocity of the signal in the soil. The time interval increases with soil wetness. This measurement is independent of soil texture and salt content. Equating eq. (7.4) and eq. (7.5), we have

$$2L/t = C/K^{0.5} \qquad ...(7.6)$$

$$K = (Ct/2L)^2 \qquad ...(7.7)$$

Travel time (t) or dielectric constant (K) is plotted versus θ_v. Regression equations between K and θ_v or between θ_v and K are developed to assess water content from measured K values. The calibration curve is of the following type.

$$K = 3.03 + 9.3\theta_v + 146.0\theta_v^2 - 76.7\theta_v^3 \qquad ...(7.8)$$

Such an equation appeared satisfactory for most mineral soils but failed to work for soils such as clayey (water close to electrically-charged surfaces has different dielectric constant) and organic soils (low bulk density). The other form of the equation that correlates volumetric moisture content, θ_v with K is proposed by Topp *et al.* (1980).

$$\theta_v = (-0.053) + (0.03K) - (0.00055\ K^2) + (4.3 \times 10^{-6}\ K^3) \qquad ...(7.9)$$

A simpler version of the equation is (Topp et al., 2008)

$$\theta_v = 0.115\sqrt{K} - 0.176 \qquad ...(7.10)$$

Apparatus/Equipment: TDR meter

Procedure

1. Insert the soil probe into the soil. Ensure that there are no air gaps around the probe that may impact measurements resulting in low values
2. Connect probe to TDR instrument using coaxial cable and initiate signal transmission and recovery of the TDR waveform. For acceptable signal-to-noise ratio, the length of cable connecting probe and the instrument is best limited to 25 m
3. For many instruments, calculations of volumetric water content are made within the instrument and the value is directly read from the meter screen. The simple calculation sequence given in the previous section applies to those instruments where travel time measurement is made explicit
4. For profile-water measurements, use series of probes of varying lengths, vertically installed from the surface. The differencing the water contents of two adjacent depths gives the water content of the non-overlapping depth interval (Fig. 7.5). Alternatively, the number of horizontally oriented probes can be installed, one at each measuring depth (Fig. 7.5). Such an arrangement provides a more precise profile of water content since it is not strongly influenced by lateral spatial variations
5. In many instruments, provision to calibrate the equipment for site-specific situations is in-built.

Most commercially available instruments are easy to operate. Turn on the TDR soil moisture meter and set the latitude and longitude by using GPS (GPS system is in-built in most TDRs). Once it is done, put the rod into the soil and press READ button, it will show the volumetric water content percentage in soil. Depending upon the measurement depth chose rod length by pressing the mode button. Save the readings on TDR data logger. Note that manufacturers are constantly upgrading the equipment with new features. Therefore, readers are advised to follow the manufacturer's instructions.

Accuracy

Topp et al. (1980) claimed that the volume wetness of the soils can be determined with an accuracy of 2% and a precision of 1%. Potential sources of errors are:

- Air gaps between the electrodes and the soil
- Temperature variations
- Effect of soil solutions on the intensity of the transmitted signals

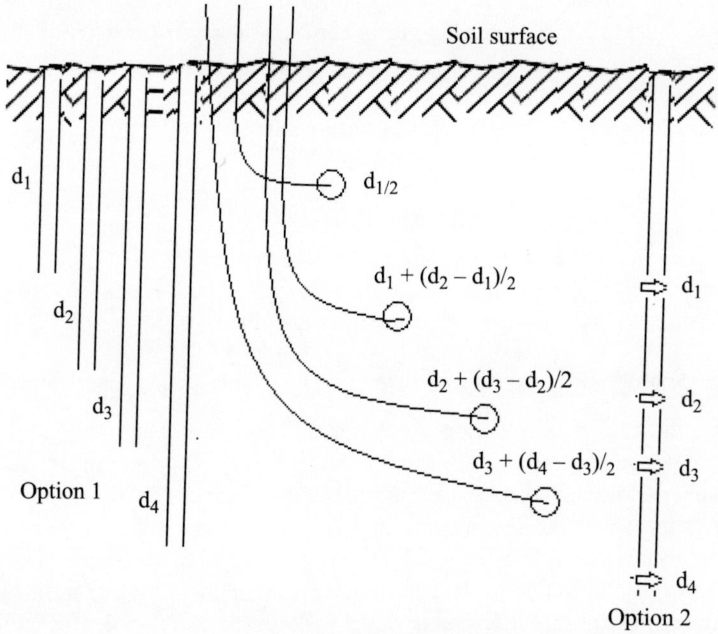

Fig. 7.5. *Two layouts of probes to measure depth-wise water content in the soil profile*

FREQUENCY DOMAIN REFLECTOMETRY

Both the time-domain reflectometry (TDR) and frequency-domain reflectometry (FDR) measure the apparent dielectric constant of the soil. Whereas TDR determines soil moisture by measuring the travel time of an electromagnetic pulse launched along a parallel metallic probe buried in the soil, FDR records the frequency of the reflected electromagnetic waves transmitted along probes. It measures variations in the dielectric properties of the soil through the capacitance.

Construction and Theory

The probe consists of a water-proof synthetic body, which contains an oscillator, a specially designed internal transmission line, and measuring circuitry. The oscillator produces electromagnetic waves that propagate in a sinusoidal waveform (Jackson, 1990). This allows the capacitor to interact with the soil outside of the tube and measure amplitude and phase at a particular frequency. The sensing head comprises a group of four rods, the center one being the signal rod. The remaining outer rods form an electrical shield around the signal rod. The sensor is connected to the soil moisture meter and to a data logger. The data logger displays the measurements and stores them for retrieval. The instrument measures the volumetric moisture content of the soil, θ_v as a percentage. Soil measurements are referenced to air and are calibrated with dielectric blocks and/or liquids of known dielectric properties (Jackson, 1990). The relation between the volumetric water content and the apparent dielectric constant is given by the following equation.

$$\theta_v = \sqrt{(\varepsilon - A)}/B$$

Here ε is the apparent dielectric constant, and A and B are soil dependent constants. The sensor is calibrated to establish a relationship between the square root of dielectric constant and volumetric water content. Indirectly, it predicts the soil dependent constants in the aforementioned equation. The sensor can measure the volumetric moisture contents in the range of 5 – 55% with an accuracy of 5% with a standard calibration. It may show large variability on field scale unless more soil specific calibrations are performed (Dirksen, 1999). It is because FDR works on frequencies in the range of 40-70 MHz whereas frequencies used in TDR are in GHz. The instrument is relatively cheaper than a TDR. It is quite accurate having excellent performance under laboratory conditions and for point measurements. Detailed information on this is being withheld because users must follow the manufacturer's instructions while using the instrument.

REMOTE SENSING AND SOIL MOISTURE MEASUREMENTS

Great variations in space and time are observed in soil moisture. Regional scale soil moisture measurements required by researchers for modelling purposes are difficult to measure using instruments discussed so far (Seneviratne et al., 2010). Ground Penetrating Radar (GPR) is one of the instruments that can cover large areas in a short time. It is quite accurate but requires relatively large investments. The other approach is space based remote sensing of soil moisture. It is capable of accommodating this need by providing surface soil moisture observations on a global scale every one to two days under a variety of conditions. In general, remote sensing measures properties of the Earth's surface by analyzing the interactions between the ground and electromagnetic radiation. Since the emissive and scattering characteristics of the soil surface depend on several variables such as soil moisture, surface temperature, surface roughness, and vegetation, the effect of soil moisture is separated from other variables. Active and passive microwave remote sensing provides a unique capability to obtain soil moisture. It has been observed that non-soil-moisture effects increase at higher frequencies, while effects of moisture become pronounced at low-frequencies. Therefore, observations at low frequencies are desired for observing the variations in soil moisture. Algorithms are then used to correlate changes in radiation or thermal signals to decipher soil moisture.

Over the last decades many soil moisture datasets have been developed from various space-borne instruments (Owe et al., 2001, Naemi et al., 2009). Although remote sensing has proven to be a valuable tool to measure soil moisture on a global scale, it is not possible to completely rule out the use of the instruments at the local scale for the following reasons.

- *In-situ* soil moisture measurements are imperative for the calibration and validation of satellite-based soil moisture data. It is now emerging that regular soil moisture determination should begin at meteorological stations to provide ground-truth data to improve accuracy of the algorithms used in the analysis of remotely sensed data
- Local measurements are necessary to provide information about moisture in deeper soil layers, which cannot be observed from satellites or aircraft
- Satellite data may not provide frequent values as may be required for many research purposes.

Chapter **8**

Infiltration Rate

Soil Science Society of America (1956) defined infiltration as the downward entry of water into soil. Infiltration rate (infiltration capacity) has been defined as the maximum rate at which a soil will absorb water impounded on the surface as a shallow depth when adequate precautions are taken regarding border, or fringe, effects. As per this definition, downward flow into and through the soil should be non-divergent. Defined as the volume of water passing into the soil per unit area per unit time, infiltration rate has the dimensions of velocity, L/T or mm/hr or cm/hr. Soil surface conditions, such as surface sealing or the soil crust or even an impermeable layer that may completely or partially hinder infiltration, predominantly determine whether infiltration will even occur. On the other hand, prevalence of cracks and open pores may increase infiltration rate. Nevertheless, several factors related to subsurface soil conditions too influence the soil infiltration. Amongst them, soil profile characteristics specifically texture, structure, and size of soil pores of different soil layers, and the existence of soil layers near to the soil surface that conducts water in a different manner than the surface layer govern the infiltration rate. The initial water content of the profile also impacts infiltration especially of short duration, being smaller for higher moisture contents. Other factors that may influence infiltration rate are: bulk density, organic matter content, chemical and physical nature of the water, the head of the applied water, the depth to groundwater, the time of application of water, biological activity, the temperature of the water and its sediment content, the air entrapped in the soil, the atmospheric pressure, and the type of equipment and method of measurement. Infiltration rate is a quite useful parameter in designing irrigation systems, hydrological investigations especially in describing rainfall-runoff relationships and in soil science. Being soil surface characteristics, field tests are performed to determine the infiltration rate. Although several methods are used to determine the infiltration characteristics, frequently used methods are:

1. Basin infiltrometer
2. Cylinder (ring) infiltrometer

This chapter discusses both the methods but only the latter is described in a detailed manner. The latter method is widely used because of its simplicity, ease of use, and precision.

BASIN METHOD

Basically, the method is similar to the cylinder infiltrometer discussed in the following section. The only difference is that in this case water is ponded over a relatively large area say few m by few m (5 m × 3 m).

Apparatus: Gauge for measuring water elevation, watch, Shovel

Procedure

1. Select the test site. It should be as uniform as possible. Remove the surface vegetation etc., if any
2. Enclose the area by constructing dykes or ridges on its four sides. Pond water on the surface. The flow can be assumed non-divergent, as the area covered is relatively large
3. Measure the rate of subsidence of the water surface at the center or at an appropriate representative site using a staff gauge, hook gauge, or a water-stage recorder
4. The infiltration rate is the depth of water that has entered the soil in a particular time. Recoup the water as and when necessary. For higher accuracy, reduce the likely evaporation from the measured infiltration during the test periods. Express infiltration rate in cm/hr.

Calculations

A curve showing the depth of water that has entered the soil as a function of time is plotted from the water elevation-and time readings. Average or instantaneous values of the infiltration rate besides the basic infiltration rate can be determined from this curve. The analysis procedure and fitting of relevant curves to the data sets is similar to that described in the following section. This method is commonly used in many Asian countries to measure the irrigation water requirement of rice fields (IRRI, 1987).

CYLINDER (RING) INFILTROMETER

Apparatus: Infiltrometer rings, Driving cap, Hammer, A hook or point gauge, Rule or tape, Shovel, Auger, Water containers, Graduated cylinder or a graduated Marriott tube, Bucket, Piece of burlap or plastic sheet, Stop watch, Recording notebook

Description

Infiltrometer cylinders are made of smooth steel or hard-alloy aluminium sheet rigid enough to drive into the ground, but thin enough to enter the soil with minimum disturbance. Two mm thick (14 gauge) steel will usually suffice. The rings are 30-60 cm diameters about 50 cm high. The larger the diameter of the cylinder and deeper the cylinder penetrates into the soil, the lesser will be the edge effects and greater the measurement accuracy. On the contrary, larger sizes of the rings may cause several other problems. Therefore, one needs to strike a balance between the size of the ring and handling difficulties. The bottom edge of the rings is beveled from outside to the inside so that it is easy to drive them into the soil. A driving disk of a diameter slightly larger than the infiltrometer rings with centering pins on its edge and a hammer is used to drive the rings into the soil. If rings are reinforced around their upper edge, then one can dispense with the use of driving disc. In that case rings can directly be hammered into the soil. To make the flow non-divergent, two concentric rings are used in infiltration tests (Fig. 8.1). The outer ring is at least 30 cm larger in diameter than the inner ring but has the same height. It is similar to the inner cylinder in all other aspects. Alternatively, a buffer ring of earth can be made surrounding the inner cylinder. When a guard ring or a buffer ring of earth is used, then the method is called double ring infiltrometer (Johnson, 1991). In both cases, water is ponded in both the rings or in a dike area and the inner ring. If accuracy demands then the water level in the cylinder should be maintained at a constant level. In that case, a Marriott siphon or other regulated supply reservoir with attached water level gauge should be used (Fig. 8.2 left). In normal cases, a simple point gauge or hook gauge is used to measure the changes in water level in the inner cylinder (Fig. 8.2 right).

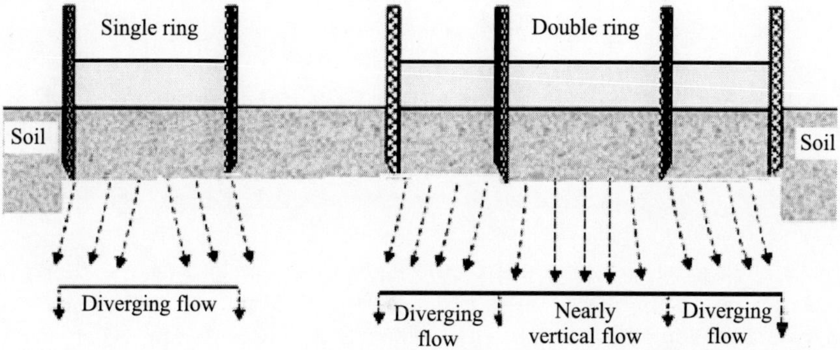

Fig. 8.1. *Flow behaviour in a single (left) and a double ring infiltrometer (right)*

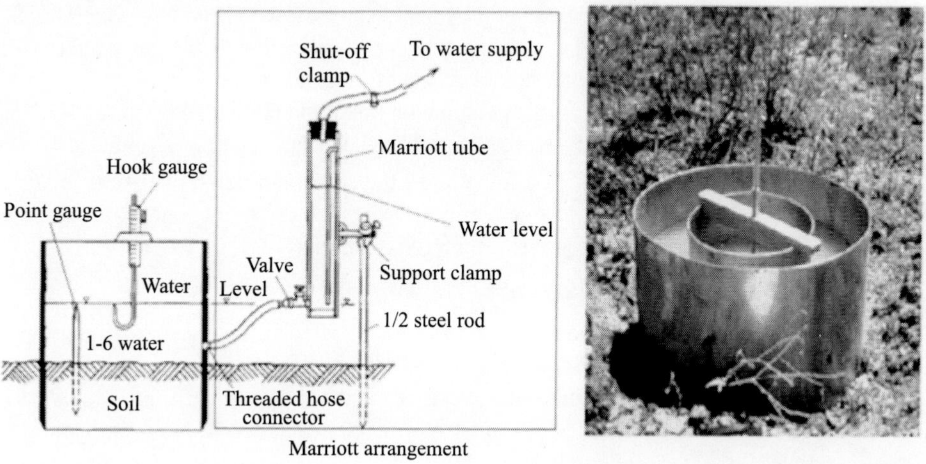

Fig. 8.2. *Set-up of the infiltration rate test: Marriott arrangement (left) and point gauge (right)*

Procedure

1. Select a representative site and examine carefully for signs of unusual surface disturbance, such as stones, cracks etc, which might give non-representative values. Remove any vegetation or loose material before installing the cylinders
2. The two concentric rings (or cylinders) are gradually hammered into the soil to a depth of at least 15-20 cm causing the least disturbance to the soil surface. The cylinders are installed as vertically as possible (Fig. 8.2). The outside cylinder may be driven to 10 -15 cm as it need not be driven as deep as the inner cylinder

3. Fill the buffer pond with water to a depth roughly equal to the depth likely to be maintained in the inner ring. Roughly equal depths should also be maintained throughout the period of observations
4. Place the plastic sheet/burlap piece on the soil surface within the inner cylinder. The plastic/burlap should be in contact with the soil at the bottom of the infiltrometer and extend up the walls of the infiltrometer at least to 15 cm
5. Place the hook/point gauge board on the cylinder and set the point of the gauge at about the likely water level (Fig. 8.2 right)
6. Fill the cylinder infiltrometer with water to a depth of about 10 cm without disturbing the soil. It should be higher in the case of light textured soils. Gently but quickly remove the plastic/burlap sheet. Set the gauge on the water surface as quickly as possible. This time and depth may be taken as the initial reading. Begin recording the depth and time immediately thereafter
7. Make hook gauge readings at periodic intervals and record the depth and times. Intervals between observations may be short (2 to 5 min) at the beginning of the test. After 20 min, the intervals may be increased to 10 to 20 min. After about 2 hr, measurements may be taken at 30-to 50 min intervals
8. Carefully add sufficient volume of water to return the water surface to the approximate initial level whenever the water level has dropped 4 to 5 cm in the inner ring. It is not required, if constant head arrangement is made. The known volume of water added, divided by the cross-sectional area of the cylinder, gives the depth of water added to the cylinder. To mimic the constant head, many practitioners add water to a pre-decided level each time they take the reading. The volume of water added divided by the area of the inner cylinder is the depth of water that has infiltrated during the time period
9. The measurements should be continued until the change in depth over time is fairly constant. The time required may generally vary from 3 to 6 hr, depending on the pre-existing moisture conditions and soil hydraulic conductivity.

Observation and Analysis Sheet

Data are recorded as per the observation sheet (Table 8.1). Clock time is recorded in column 2 and is used to calculate the time interval and cumulative time (Column 3 and 4). Depth of water is filled in column 5 to find out the intake of water during the time interval. An additional column can be made to keep account of the water added in between. Columns 6 and 7 are then filled using calculations as described in these columns. Depending on the purpose of the measurement, express infiltration rate in cm/hr or mm/hr in column 8.

Table 8.1. *Observation sheet for collection and partial analysis of data*

S. No.	Clock time	Time interval (min)	Cumulative time (min)	Depth of water infiltrated (mm)	Intake (mm)*	Cumulative infiltration (mm)	Infiltration rate (mm/hr)
1	2	3	4	5	6	7	8
1	T_1	0	0	A_1	0	0	0
2	T_2	$T_2 - T_1$	Row (1 +2) from column 3	A_2	$A_2 - A_1$	Row (1+2) from column 6	(Column 6/ Column 3) × 60

3	T_3	$T_3 - T_2$	Row (1+2+3)	A_3	$A_3 - A_2$	Row (1+2+3)	
4							
5							
n	T_n	$T_n - T_{n-1}$	Row (1+2+3...n)	A_n	$A_n - A_{n-1}$	Row (1+2+3....n) from column 6	(Column 6/ Column 3) × 60

*Any addition of water to recoup level has to be taken into account while making calculations in column 6

Plot the data of column 4 on X-axis and column 8 on Y-axis on a normal arithmetic paper. It gives an asymptotically decreasing infiltration rate curve. The constant rate towards the end of the curve is the basic infiltration rate (mm/h). On a normal arithmetic paper, plot the data of column 4 on X-axis and column 7 on Y-axis. It gives asymptotically increasing cumulative infiltration curve. It can be used to measure the total water infiltrated within a given time. This curve can also be used to calculate the average or instantaneous values of the infiltration rate. The same data platted on a log-log paper will appear as straight line since the data are well represented by Kostiakov model discussed in one of the section of this chapter.

Field Test

The observed data of a field test, shown in Table 8.2, are used to explain the analysis procedure and application of data for fitting mathematical models. On a plain arithmetic graph, the plots of the data (Table 8.2), as already said appear asymptotically decreasing or increasing curves (Fig. 8.3). The infiltration rate curve has the following two characteristics.

Table 8.2. *Instantaneous and cumulative infiltration rate of the soils at Sampla*

Elapsed time (min)	Intake during the elapsed time (cm)	Infiltration rate (cm/hr) (col. 2 /time interval between current and previous reading) × 60	Cumulative intake (cm)
1	2	3	4
5	0.333	4.0	0.333
10	0.29	3.5	0.625
15	0.2	2.4	0.824
20	0.175	2.1	1.000
25	0.125	1.5	1.125
35	0.216	1.3	1.342
45	0.2	1.2	1.542
55	0.2	1.2	1.742
85	0.6	1.2	2.342
115	0.6	1.2	2.942
145	0.6	1.2	3.542
205	1.3	1.3	4.842

- The infiltration rate decreases with the elapse of time. The asymptotically decreasing curve has a tendency to approach a final minimum value known as the basic infiltration rate
- The rate of decrease of the infiltration rate also decreases with time

Besides, the initial moisture content has a great bearing on the infiltration rate curve of a soil. The infiltration rate curve of a wet soil always lies below that of an initially dry soil. The two curves meet at a time when the infiltration rate approaches the basic infiltration rate. For the data reported in Table 8.2 and plotted in Fig. 8.3, the basic infiltration rate is 1.2 cm/hr. Basic infiltration rates of some soil texture groups are given in Table 8.3 for reference.

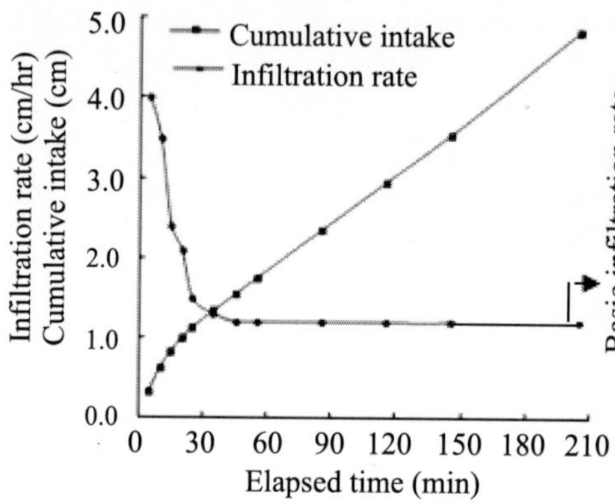

Fig. 8.3. *Instantaneous rate and cumulative infiltration curves on a normal paper*

Table 8.3. *Basic infiltration rate of some soil textural groups*

Soil group	Basic infiltration rate (mm/hr)	Interpretation
Sand	< 30	Rapid
Sandy loam	20-30	Moderate to rapid
Loam	10-20	Moderate
Clay loam	5-10	Slow
Clay	1-5	Very slow to slow

SINGLE RING INFILTROMETER

In some situations, one is forced to conduct infiltration rate studies using a single ring. In that case, the procedural details of conducting the test are similar to a double ring but without the outer ring (Fig. 8.4, Rice et al., 2014). The water level in the ring is measured and recouped at short intervals, say after the depletion of water by y cm (Fig. 8.4). At the conclusion of the test, the area around the ring is partially dug using a spade. The digging is carried out in such a way, that it is possible to identify the wetted soil to determine the wetting depth, L and distance (x) of lateral divergence (Fig. 8.4). Measured values of infiltration are corrected for divergence to determine the actual infiltration rate. For this correction, measured infiltration, I_n is multiplied by the ratio of cylinder area to wetted area as per the following equation (Bouwer, 1966, Bouwer et al., 1999).

$$I_w = I_n \, [r^2 / (r + x)^2] \qquad ...(8.1)$$

Here I_w is the corrected infiltration rate, I_n is the actual infiltration rate from the single ring infiltrometer, r is the radius of the infiltrometer and x is distance covered by wetting front from the cylinder walls (Fig. 8.4). A modified version of the single ring infiltrometer, known as percolation meter, is used to measure infiltration rate in rice fields (IRRI, 1987). There are no chances of lateral flow in flooded rice fields.

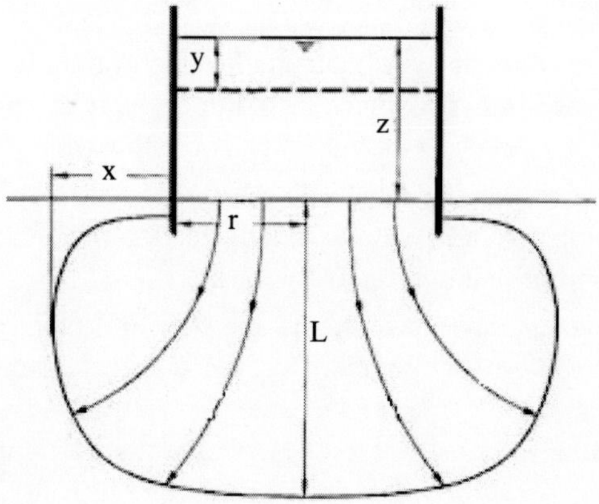

Fig. 8.4. *Flow pattern in a single ring infiltrometer showing parameters of eq. (8.1) and (8.2)*

INFILTRATION AND VERTICAL HYDRAULIC CONDUCTIVITY

The infiltration rate of water, I in the soil is governed by Darcy's law such that

$$I = K\,L/(L + z - P) \qquad ...(8.2)$$

Here I is the infiltration rate of the inner cylinder in a double ring infiltrometer, cm/hr, z is the water depth at the soil surface, m, L is the depth to the wetting front below the soil surface, K is the hydraulic conductivity, cm/hr, and P is the soil water pressure at the wetting front inside the transmission zone. As the time elapses, the moisture content in the transmission zone is equal to that of the saturated soil, such that K tends to K_{sat} and P tends to zero in medium to heavy textured soils. Moreover, z tends to be quite small in comparison to L such that

$$I = K_{sat}\,L/L = K_{sat} \qquad ...(8.3)$$

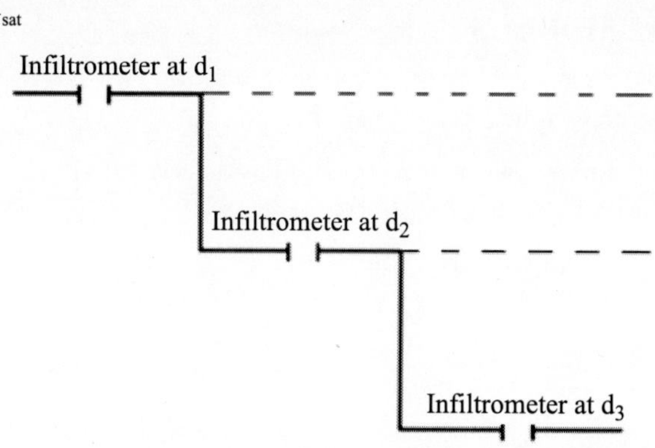

Fig. 8.5. *A schematic view of measuring vertical hydraulic conductivity at different depths*

For single ring infiltrometer, I in the equation is I_w. Although the theory is not quite elegant (especially for coarse textured soils), the basic infiltration rate has been taken as the vertical hydraulic conductivity (eq. 8.3). It is why; infiltration rate measured at different depths is used as a measure of variation in vertical hydraulic conductivity of different soil layers (Fig. 8.5).

Notes:

- In many cases, it may not be possible to neglect the evaporation rate for the duration of the test. In that case, either take account of this factor in calculations, or evaporation may be minimized by covering the water surface in the cylinder with a thin film of oil
- There is no single method that may be recommended to suit all field conditions. Experience and judgment are required in the selection of the method, conducting test and evaluating infiltration rates
- Cylinder infiltrometer may not represent the true infiltration rates because of few problems associated with the method. One, the soil gets disturbed while driving the rings into the soil. Secondly, the ponded water does not duplicate real field conditions. Although infiltration rates are not too sensitive to the water pressure at the surface, or equivalently the depth of water on the surface, still it may have some effect, if variations are large. The infiltration rate measured with an infiltrometer with 10-20 cm of ponded water will not be equivalent to that of a border with 5 cm water depth, or that of a sprinkler where no water is ponded. Thirdly, air entrapped under a level basin may not be replicated in the ring infiltration test, as air may escape laterally from around the infiltrometer ring. Fourthly, under real field conditions, water moves across the soil surface rearranging surface particles. Such erosion and aggregation do not occur in ponded infiltration tests resulting in sealing of the soil surface that may tend to reduce the infiltration rate
- When abnormally high or low infiltration rates are observed, dig out the infiltrometer to examine the soil for possible causes of such a phenomenon. If an obvious cause of high or low rates is observed, it is better to repeat the test at a nearby location
- The water used to measure the infiltration rate should be of the same quality as is likely to be used for irrigation.

MATHEMATICAL MODELS

Amongst the various models used to monitor the infiltration process, following three have been frequently used and studied in a detailed manner.

- Kostiakov model in irrigation engineering (Kostiakov, 1932)
- Philip model in soil science (Philip, 1957)
- Horton model in hydrology (Horton, 1939)

Kostiakov Model

It is an empirical model that expresses the cumulative infiltration (I_c) as a function of time (t) as follows:

$$I_c = ct^d \qquad \text{...(8.4)}$$

The infiltration rate, I_r is obtained by differentiating eq. (8.1). It is written in the same form as eq. (8.4) as follows:

$$I_r = at^b \quad \text{...(8.5)}$$

Here I_c is the cumulative infiltration, cm, I_r is the infiltration rate, dI_c/dt, cm/hr, t is the time, min, c and d are fitting constants, such that $a = c\ d$ and $b = d - 1$. Since d is always less than 1.0 ($0 < d < 1$), b is always negative. Whereas constant, a describes the instant infiltration rate at $t = 1$, the exponent b describes the decay rate of the infiltration rate. Looking at the characteristics of eq. (8.5), it is noticed that the initial value of the infiltration rate at time $t = 0$ is infinite. Besides, the infiltration rate with increasing time approaches zero instead of a constant (non-zero) commonly observed under the field conditions. In fact, it is the major weakness of this model that it does not predict a final constant infiltration rate. In spite of this weakness, it has performed well under many field applications.

If we take the log of both sides of eq. (8.4), we can write the equation as

$$\text{Log } I_c = \log c + d \log t \quad \text{...(8.6)}$$

The plot of this equation with the use of field data as reported in Table 8.2, shown in Fig. 8.6 is a straight line. The value of c is given by the intercept at $t = 1$ whereas the slope of the line gives the value of d. It may be noted that the straight line meets the y-axis at 0.13 such that the value of $c = 0.13$. The value of d is given by the slope of the straight line; in this case slope of the line being 0.66 gives the value of d as 0.66. The value of a and b can be determined either by plotting the infiltration rate curve on a similar paper or by the relations: $b = d - 1$ and $a = cd$. Thus, $b = -0.34$ and $a = 0.086$, which for eq. (8.5) will become 5.16 ($0.086 \times 60 = 5.16$). The multiplication is required because time is taken in minutes while the infiltration rate is in cm/hr.

The average infiltration rate, I_{av} at a particular elapsed time can be given as

$$I_{av} = I_c/t = ct^{d-1} \quad \text{...(8.7)}$$

Philip Model

Philip (1957) observed that time dependence of cumulative infiltration for all practical purposes can be described by the following relation:

$$I_c = St^{1/2} + At \quad \text{...(8.8)}$$

$$I_r = 0.5St^{-1/2} + A \quad \text{...(8.9)}$$

Here S is the sorptivity, cm/hr$^{0.5}$, and A the transmissivity of water, cm/hr. Other parameters are same as described for eqs. (8.4) and (8.5). The sorptivity term (with dimensions of $L/T^{1/2}$) reflects the soil's ability to absorb water by matric forces during the initial stages of the infiltration process. During this period, the gravity effects are negligible and the contribution of the term At in the equation is negligible. The transmissivity term (A) having dimensions, L/T reflects the soil's ability to transmit water under the influence of gravity. According to Dunin (1976), transmissivity is related to the soil's saturated hydraulic conductivity (K) and Philip (1969) observed that for long intervals, the equation for the cumulative infiltration can be approximated by

$$I_c = St^{1/2} + Kt \quad \text{...(8.10)}$$

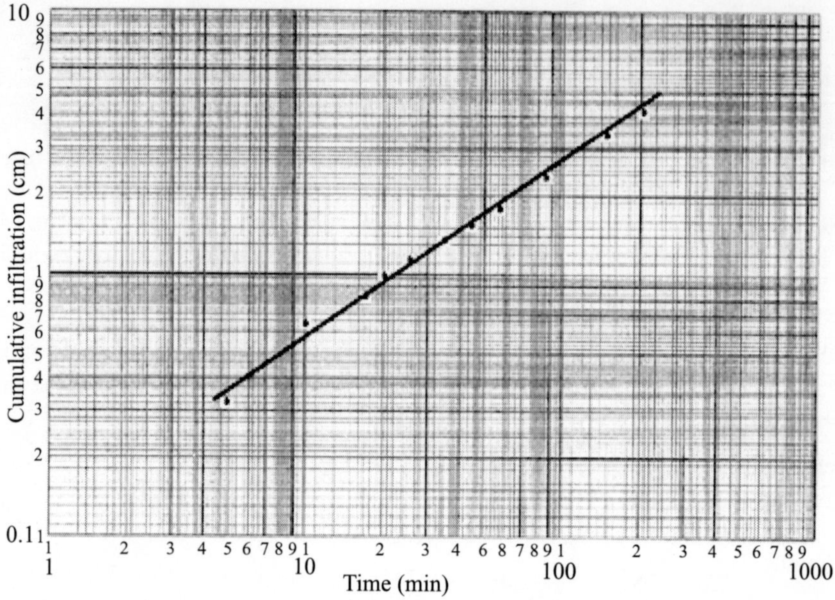

Fig. 8.6. *Plot of cumulative infiltration on log-log scale to fit eq. (8.4)*

Here K is soil's saturated hydraulic conductivity. Later some workers have observed that this may not be a realistic assumption. To describe the data given in Table 8.2, we can use eq. (8.9) and plot infiltration rate on y-axis to $1/t^{0.5}$ on the x-axis. The slope of the fitted linear curve represents the value of S/2 and the intercept gives the value of A. Students can do this as an laboratory exercise.

Horton Model

The Horton's model is widely used in hydrology (Horton, 1939). In developing this model, Horton recognized that infiltration capacity (I_c) decreases with time and approaches a constant minimum rate (f_c) (Horton, 1939). The decrease in infiltration is primarily attributed to factors operating at the soil surface rather than flow processes within the soil. Horton's equation is written as:

$$I_c = f_c + (f_o - f_c)e^{-kt} \qquad ...(8.11)$$

Here I_c is the infiltration capacity or potential infiltration rate, cm/hr, f_c is the final constant infiltration rate, cm/hr, f_o is the infiltration capacity at $t = 0$, cm/hr, t is the time, hr, k is the Horton's decay coefficient, which depends upon soil characteristics and vegetation cover and t is the time after the start of infiltration. The parameters, f_o, k, and f_c must be evaluated from measured infiltration data.

Subtracting f_c from both sides of eq. (8.11) and taking the Ln of each side, eq. (8.11) can be written as:

$$\text{Ln}(I_c - f_c) = \text{Ln}\,(f_o - f_c) + -kt$$

It is an equation of a straight line. Clearly, the plot of t versus $\text{Ln}(I_c - f_c)$ on semi-log paper with t on a normal scale will be a straight line. The base value of parameters k is equal to a negative slope and the initial infiltration capacity (f_o) is calculated using the $f_o - f_c = e$ intercept. Students may also take up this as an exercise for their laboratory study.

Chapter 9

Hydraulic Conductivity

Hydraulic conductivity is one of the most important physical features of the soils that have wide applications in soil and water science. Hydraulic conductivity of a material is defined as the ability of the fluid to pass through the pores and fractured rocks. Darcy's law quantitatively describes one-dimensional water flow in saturated soils. Simply stated, it relates the flux, J to the hydraulic gradient, i and is written as:

$$J = -Ki \qquad ...(9.1)$$

Here K is the hydraulic conductivity, a proportionality constant that defines the relationship of flux to the hydraulic gradient. The units of K are governed by the units assigned to J and i. Since hydraulic gradient always decreases in the direction of water flow, minus sign is appended to keep K positive and maintain directional integrity. If the water flux is expressed as:

$$J = V/At \qquad ...(9.2)$$

Then equating eq. (9.1) and eq. (9.2), we have

$$V/At = -Ki \qquad ...(9.3)$$

Such that V is the volume of water flow, A is the cross-sectional area, and t is the time. For simplicity, we can neglect the negative sign and write

$$K = Vi/At \qquad ...(9.4)$$

Thus, hydraulic conductivity is defined as the rate of flow of water through a porous medium under laminar conditions through a unit cross-sectional area perpendicular to the direction of flow for a unit hydraulic gradient and standard temperature conditions. V/t can also be written as discharge, Q, with units of L^3/T. Saturated hydraulic conductivity is affected by properties of the both the soil and the fluid. Thus, soil pore geometry as well as fluid viscosity and density affect the hydraulic conductivity. Whereas soil hydraulic conductivity, K is normally expressed in units of mm/hr or cm/hr, for groundwater aquifers it is expressed in m/day.

A number of methods to measure saturated hydraulic conductivity are in vogue (Fig. 9.1). Since, it is not possible to include all the methods, only important ones in each group are included in the chapter.

CORRELATION METHODS

To assess a representative value of the hydraulic conductivity is rather difficult because of wide spatial variability in the soil properties. In view of such difficulties, it has been common to correlate the hydraulic conductivity to other soil properties that are known or can be easily determined. Some such parameters are:

- Soil texture

- Grain size distribution
- Pore size distribution
- Soil mapping unit

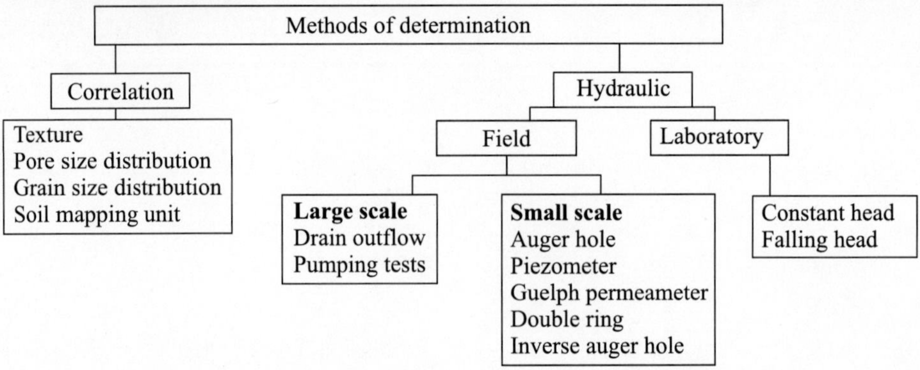

Fig. 9.1. *Listing of methods for determining hydraulic conductivity*

Correlation methods derive or use pre-determined relations between hydraulic conductivity and these parameters. Many empirical formulae are available that relate hydraulic conductivity and soil texture or grain-size distribution characteristics. It is mainly because the information on textural properties and grain size distribution of soils or rock is easily obtained (See Chapters 4 and 5). Therefore, these properties provide a potential alternative for estimating hydraulic conductivity of soils.

Soil Texture and Hydraulic Conductivity

Although, soil texture and hydraulic conductivity relationship may appear arbitrary; such tables have been quite successfully used in drainage design in the USA and the Netherlands (Table 9.1). In place testing however, continues to be the most desirable option since the conductivity may vary by a factor of 10 or more for soils of the same texture group (Table 9.1).

Grain Size Distribution and Hydraulic Conductivity

The hydraulic conductivity of granular, porous soils depends on the grain-size distribution. Therefore, relations between hydraulic conductivity and grain-size distribution of good permeable soils (e.g. sandy or light loamy soils) have been derived by various workers and tested by many researchers (Odong, 2007, Musavi-Jahromi and Shiravand, 2012). Each method is derived from the general equation given as:

$$K = (g/\upsilon) \times C \times f(n) \times (d_e)^2 \qquad \text{...(9.5)}$$

Here K is the hydraulic conductivity, g is the acceleration due to gravity; υ is kinematic viscosity; C is sorting coefficient; $f(n)$ is porosity function; and d_e is effective grain diameter. The values of C, $f(n)$ and d depend on the method used in the grain-size analysis. Only few equations among many are included to illustrate the kind of parameters required in the use of these equations.

Table 9.1. *Relationship between hydraulic conductivity and soil texture*

Soil texture	Hydraulic conductivity (cm/h)
Clay	0.0025 – 0.0380
Silty clay	0.0381 – 0635
Silty clay loam	0.0635 – 0.1016
Clay loam	0.1016 – 0.1778
Sandy clay loam	0.1778 – 0.3810
Silt loam	0.3810 – 0.8890
Loam	0.8890 – 1.7780
Very fine sandy loam	1.778 – 3.048
Fine sandy loam	3.048 – 4.318
Sandy loam	4.318 – 6.350
Loamy fine sand	6.350 – 8.890
Loamy sand	8.890 – 11.430
Fine sand	11.430 – 15.240
Sand	15.240 – 20.320
Coarse sand	20.320 – 30.480

Kozeny-Carman Equation

This equation can be used for most soil textures except for soils having effective grain size above 3 mm or for clay textured soils.

$$K = (g/\upsilon) \times 8.3 \times 10^{-3} [n^3/(1-n)^2] \times (d_{10})^2 \qquad ...(9.6)$$

Here n is the porosity and d_{10}, the effective diameter (See Chapter 4)

Hazen Equation

It can be used for soil textures varying from fine sand to gravel provided uniformity coefficient of the soil is less than five (Cu < 5) and effective grain size between 0.1 mm and 3 mm. The accuracy of this formula is less than that of the Kozeny-Carman formula:

$$K = (g/\upsilon) \times (6 \times 10^{-4}) [1 + 10(n - 0.26)](d_{10})^2 \qquad ...(9.7)$$

Breyer Equation

Use of Breyer equation is recommended for materials having heterogeneous distribution and poorly sorted grains with a uniformity coefficient between 1 and 20 (1<Cu<20) and an effective grain size between 0.06 mm and 0.6 mm.

$$K = (g/\upsilon) (6 \times 10^{-4}) \times \log (500/Cu)(d_{10})^2 \qquad ...(9.8)$$

Here Cu is the uniformity coefficient.

USBR Equation

The US Bureau of Reclamation (USBR) equation is applicable to medium grain sand with a uniformity coefficient of less than 5 (Cu < 5). This equation uses an effective grain size of d_{20} and does not depend on porosity. It is less accurate than other formulae.

$$K = (g/\upsilon)\,(4.8 \times 10^{-4})(d_{20})^{0.3} \times (d_{20})^2 \qquad ...(9.9)$$

The value of g in the above equations is 9.81 m/s^2 and that of υ for water at a temperature of 20°C is 0.0874 m^2/day or 1.01×10^{-6} m^2/s resulting in the value of g/υ for water as 9.71×10^6 s/m.

LABORATORY METHODS

Locally prepared set-ups as well as commercially available instruments are used to measure hydraulic conductivity of disturbed and undisturbed soil samples in the laboratory. Guidelines .or standardized measurement procedures published by Bureau of Indian Standards should be kept in view while using local set-ups (BIS, 1986, 1987). Two commonly used procedures to measure hydraulic conductivity in the laboratory are the constant head and variable head methods.

Apparatus/Equipment: Constant or /variable head permeameter(s) (Fig. 9.2), Core sampling kit for undisturbed soil sampling, Permeability stand with funnels, Perforated brass plates, Water reservoir with Marriott arrangement, stop watch, Graduated cylinder, Knife, Muslin cloth, Filter paper, Rubber rings or tape

Constant Head Method

The constant-head test is one of the most commonly used methods to determine saturated hydraulic conductivity of coarse-grained soils. The test is based on the direct application of Darcy's law, (eq. 9.4) to a soil liquid configuration. It represents a one-dimensional steady flow of a percolating liquid through a saturated soil column of uniform cross-sectional area (Fig. 9.2 left). In this method, a constant head is maintained at the soil surface of a saturated sample and water is allowed to flow through the sample till the measured outflow attains a constant value.

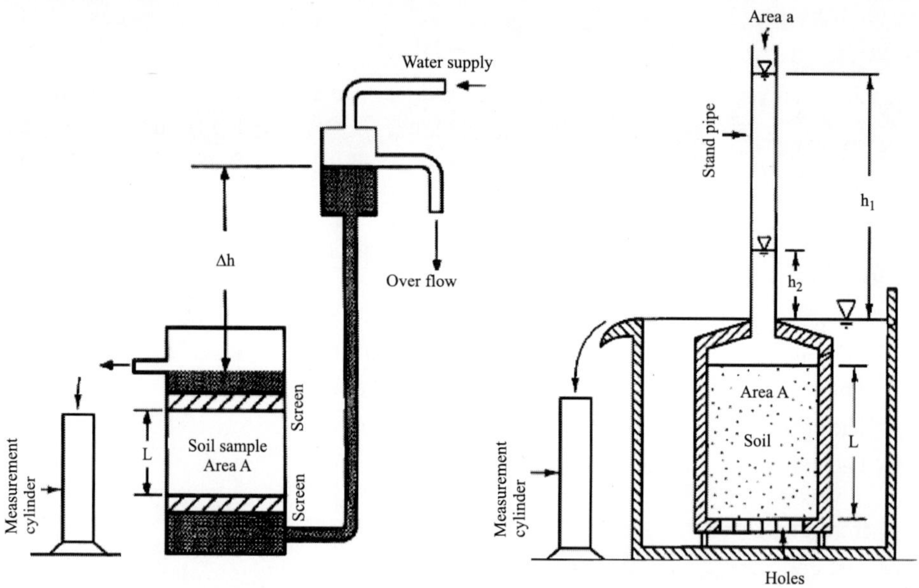

Fig. 9.2. *Schematic diagrams of the basic set-ups of constant head (left) and variable head permeameter (right)*

Procedure: Disturbed Samples

1. Measure the internal dimensions of the mould. Weigh the mould with dummy plate to the nearest gram. Apply little grease on the inside of the mould. Clamp the mould between the base plate and the extension collar and place the assembly on a solid base
2. Take about 2.5 kg of soil from a thoroughly mixed sample. Compact the soil at the required dry density using a suitable compacting device. If the soil profile is layered, ensure to take separate samples for each layer and fill in the same fashion. The idea is to create real field situation of the soil profile
3. Remove the collar and the base plate. Trim the excess soil so that it is level with the top of the mould. Clean the outside of the mould and the dummy plate. Find the mass of the soil in the mould and calculate its bulk density
4. The mould with the sample is placed on a porous plate covered with filter. The plate is now placed over the permeameter. Connect the specimen to the top inlet of the constant head reservoir
5. The bottom outlet is opened and water is allowed to flow until a steady state is established. Measure the steady state flow for a given time interval say t min
6. Measure the length of the soil column, L and head difference in levels between the constant head reservoir and the outlet at the base (Δh). In some set-ups, two manometers separated by a distance, L are provided (Fig. 9.2 left). In that case, the difference in readings of these manometers is taken as Δh in eq. (9.5).
7. Take 3 or more readings for the same interval and average out the values. The 3 values should be nearly same, being the flow under steady state. If there are large differences in the measured values, allow more time to establish steady state. Readings should be taken once the steady state has reached.

Undisturbed Soil Samples

Clear the soil surface of vegetation, leaves or any other trash. Push the core gently into the soil and obtain an undisturbed soil core. Trim the excess soil on both ends. Put lids on both ends so as not to disturb the soil during transport to the laboratory.

If an auger core sampler is used to draw an undisturbed sample, then following step-wise procedure is used to prepare the test specimen.

1. Remove the protective cover (paraffin wax) from the sampling tube
2. Place the sampling tube in the sample extraction frame, and push the plunger to get a cylindrical sample not longer than 35 mm in diameter and having a height equal to that of mould
3. The specimen shall be placed centrally over the porous disc to the drainage base
4. The annular space shall be filled with an impervious material such as cement slurry or wax, to provide sealing between the soil specimen and the mould against leakage from the sides
5. The drainage cap shall then be fixed over the top of the mould
6. Now the specimen is ready for the test.

Rest of the procedure is same as in the case of disturbed soils.

Observations and Calculations

A sample Performa for taking observations is given as under.

Test observation	Time (min)	Volume of water (cm^3)	Temperature (°C)	K_t (Eq. 9.10)	K_{27} (Eq. 9.11)
1	t_1	V_1	T_1	K_{t1}	$K_{27\text{-}1}$
2	t_2	V_2	T_2	K_{t2}	$K_{27\text{-}2}$
3	t_3	V_3	T_3	K_{t3}	$K_{27\text{-}3}$

Knowing the volume, V of the water flowing through the system during time t, the saturated hydraulic conductivity, K of the soil is determined as follows:

$$K_t = VL/(\Delta h \times t \times A) \qquad ...(9.10)$$

Here, A is cross-sectional area of the specimen. All other parameters are described earlier. Use similar units for the parameters in eq. (9.10). As per BIS, hydraulic conductivity values are reported at 27°C. Use the following equation to calculate the hydraulic conductivity at 27°C (BIS, 1984).

$$K_{27} = K_t \times \mu_t/\mu_{27} \qquad ...(9.11)$$

μ_t is the coefficient of viscosity at test temperature, T°C, and μ_{27} the coefficient of viscosity at 27°C.

Falling Head Method

The falling head test with the permeameter is primarily used to determine K value of fine-grained soils. Like the constant-head method, the falling-head test also operates in accordance with the direct application of Darcy's law to a one-dimensional, saturated soil column of a uniform cross-sectional area. The falling-head method slightly differs from the constant-head method. The liquid flowing through the saturated column is kept at an unsteady-state flow regimen in which both the head (Fig. 9.2 right) and the discharge vary during the test. Prepare the soil specimen as before and use the variable head apparatus to measure the hydraulic conductivity.

- Most of the procedural steps are similar to the permeameter test
- Open the outlet and note down the time, Δt $(t_2 - t_1)$ required for the water level in the burette to fall from initial head, h_1 to final head, h_2
- Fill the reservoir again to take at least 3 readings. A sample Performa of taking observations is given as under.

Test observation	Time (min)	Initial head (cm)	Final head (cm)	Temperature (°C)	K_t (Eq. 9.12)	K_{27} (Eq. 9.11)
1	Δt_1	h_1	h_2	T_1	K_{t1}	$K_{27\text{-}1}$
2	Δt_2	h_1	h_2	T_2	K_{t2}	$K_{27\text{-}2}$
3	Δt_3	h_1	h_2	T_3	K_{t3}	$K_{27\text{-}3}$

By measuring the change in head in the standpipe from h_1 to h_2 during a specified interval of time t, the saturated hydraulic conductivity is determined as follows (Klute and Dirksen, 1986):

$$K = [aL/(A \times \Delta t)]\ [Ln\ (h_1/h_2)] \qquad ...(9.12)$$

Here a is cross sectional area of the stand pipe, L is the length of the soil column, A is cross section area of the soil column, Δt is the time interval for the head to drop from h_1 to h_2. As per BIS, report the values of hydraulic conductivity at 27°C as in the constant head permeameter. Interpret the results as per the information provided in Table 9.2.

Table 9.2. *Interpretation guidelines for hydraulic conductivity*

Hydraulic conductivity (cm/hr)	Classes
<0.125	Very slow
0.125-0.5	Slow
0.5-2.0	Moderate slow
2.0-6.25	Moderate
6.25-12.5	Moderately rapid
12.5-25.0	Rapid
>25.0	Very rapid

Example 9.1. Flow from a locally arranged laboratory set-up shown in the following figure (constant head on left) was measured as 0.5 mL per minute. If *L*, *h* and d are 20, 5 and 10 cm respectively, calculate the hydraulic conductivity of the soil.

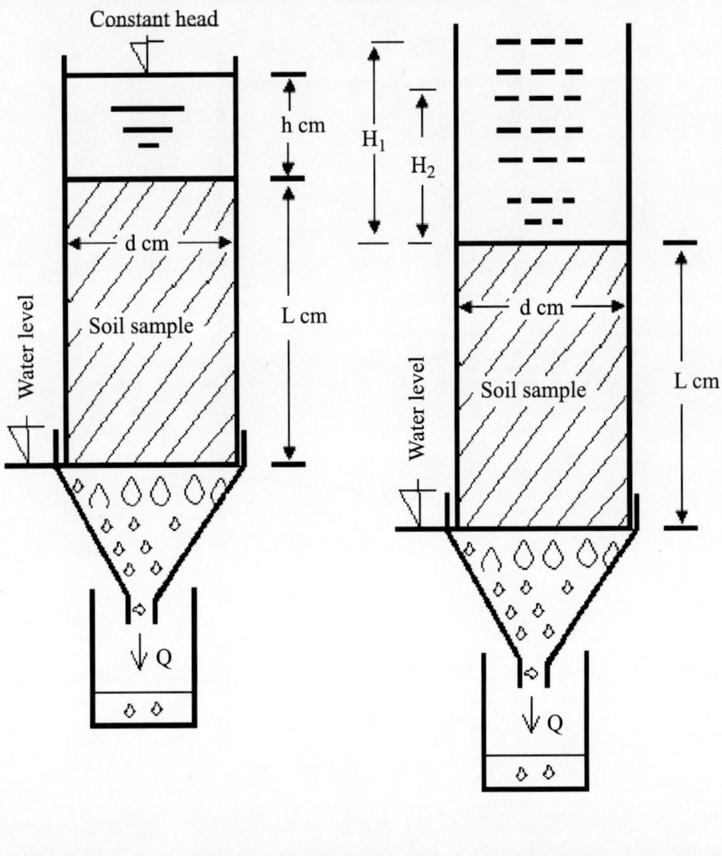

For the set-up shown on the left side, we take heads from the bottom of the column

$\Delta h = L + h$

Incorporating in eq. (9.10), we have

$K = VL/[(L + h) \times t \times A)]$

$K = 0.5 \times 20/[(20 + 5) \times 60 \times \pi(10/2)^2] = 8.5 \times 10^{-5}$ cm/s

Example 9.2. A falling head test was conducted in the laboratory, as shown in figure of Example 9.1 (right). The initial hydraulic head at 20 cm fell down to 15 cm after 10 hr. Calculate the hydraulic conductivity of the soil, if the length of soil sample is 20 cm.

For the set-up shown on the left side, the equation to determine hydraulic conductivity is written as (Stibinger, 2014):

$$K = [L/(t_2 - t_1)][\text{Ln}\ (H_1/H_2)] \qquad ...(9.13)$$

Here L is the length of the soil column and H_1 and H_2 are the water heads over the soil surface at times t_1 and t_2. Note that it is the same equation as eq. (9.12) except that $a = A$. Incorporating the values

$K = [20/(10 - 0)]/[\text{Ln}\ (20/15)] = 0.58$ cm/hr $= 1.6 \times 10^{-4}$ cm/s

Application in Rice Fields

An equation to measure the hydraulic conductivity of the rice soils (Fig. 9.3) has been independently derived in Taiwan and Japan. It takes the following form (Stibinger, 2014).

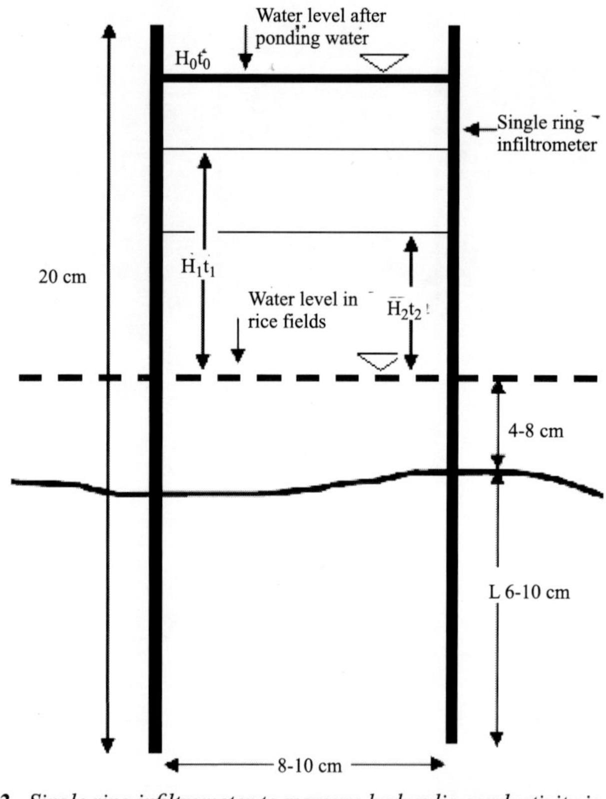

Fig. 9.3. *Single ring infiltrometer to measure hydraulic conductivity in rice fields*

$$K = [2L/(t_2 - t_1)][\text{Ln}\,(H_1/H_2)] \qquad ...(9.14)$$

Here L represents the depth of the infiltrometer under the soil surface. If we compare eqs. (9.13) and (9.14), the only difference is that L has been replaced with $2L$. Such a change characterizes the approximation of the supposed trajectory of the water flow in rice soils, which is different than the flow characteristics below the permeameter shown in Example 9.2 for derivation of eq. (9.13). Other parameters are same as for eq. (9.12) or eq. (9.13).

FIELD METHODS (SMALL SCALE)

Auger Hole Method

An auger-hole method is a simple, rapid, and reliable method of measuring hydraulic conductivity in the field. The method is commonly used to measure the hydraulic conductivity of the soil below the water table. The method, originally proposed by Diserens (1934), has been improved over time by various workers, Hooghoudt (1936) making the most important modifications. The general principle is to bore an auger hole to a certain depth below the water table. When equilibrium is reached with the surrounding groundwater, a part of the water in the hole is removed so that water seeps into the hole again. The rate at which water rises in the hole is measured to calculate the hydraulic conductivity with the help of an appropriate formula. This method is quite commonly used in the design of drainage systems in waterlogged land and in canal seepage investigations. For more details of this and inverse auger hole methods, readers may refer to Gupta (2019).

Equipment: Soil auger of suitable diameter normally more than 6 cm but less than 14 cm, Extension rods, Bailer or pump, Float gauge set consisting of measuring tape and a float

Procedure

1. Make an auger hole, and note down the depth at which groundwater level is encountered for the first time. Drill the hole going about 60-90 cm below the groundwater table (Fig. 9.4). It may be useful to observe the changes in soil characteristics (colour, water saturation, etc.)

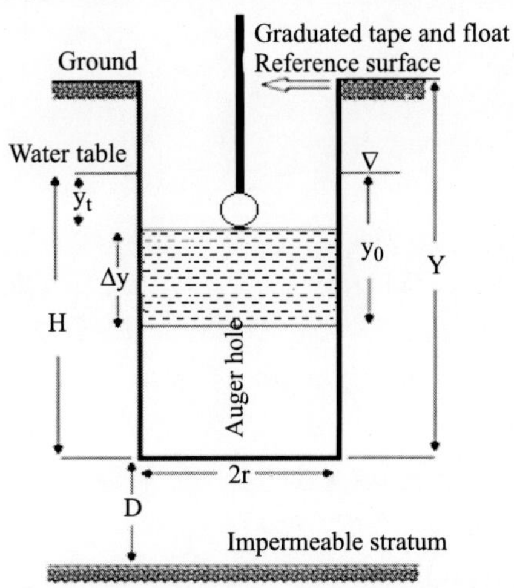

Fig. 9.4. *A schematic view of the auger hole to measure hydraulic conductivity*

2. Wait for sometime so that the water level in the auger hole is in equilibrium with the surrounding groundwater table. Measure the stable groundwater level
3. Remove the water from the hole and allow it to recoup. Repeat the process 2 to 3 times to remove soil smearing formed during auguring the hole and to open up the soil pores. Finally, remove the water as quickly as possible so that the water level is 20-40 cm below the static water level
4. Measure the rate of the rise of water level at frequent preferably regular intervals. One can use the float gauge with a measuring tape or any electrical device to measure the water levels. Take about 8-10 readings. In this process see that not more than 25-30% of the water removed is recouped at the time of the last measurement.

Calculations

Equations given by Ernst (1950) are now commonly used to determine the hydraulic conductivity. For $D > H/2$, hydraulic conductivity is calculated by the relation (van Schilfgaarde, 1974):

$$K = \frac{4000r^2}{(H+20r)[2-(y/H]y} \times \frac{\Delta y}{\Delta t} \quad ...(9.15)$$

Here K is in m/d, r, H, y and Δy in cm and Δt in seconds (Fig. 9.4). The parameter y is the average depth of the water table in the auger hole given by $y_0 - ½\,\Delta y$.

For $D = 0$, i.e. when the auger hole rests on the impermeable layer, hydraulic conductivity is estimated by the equation:

$$K = \frac{3600r^2}{(H+10r)[2-(y/H]y} \times \frac{\Delta y}{\Delta t} \quad ...(9.16)$$

The parameters are the same as described for eq. (9.15). Even today, there is no suitable formula to calculate hydraulic conductivity between the two limits i.e. $0 < D \leq H/2$. For such cases, either eq. (9.16) is used or values calculated with two equations are averaged to arrive at the value of hydraulic conductivity. For acceptable accuracies, following inequalities should be met with.

$2r > 6.25$ cm and < 14 cm

H > 20 cm and < 200 cm

$y > 0.2$ H

$\Delta y < 0.25y_0$

Example 9.3. Field observations for a bore hole test are reported in Table 9.3.Calculate the hydraulic conductivity, if the bore hole characteristics are as follows.

Radius of the auger hole (cm), $r = 4$ cm ($2r$ is more than 6.25 cm)

Depth to water table from reference of ground surface = 49 cm

The depth of the auger hole below the water table, H = 101 cm

Depth of impermeable layer below auger hole, $D = 70$

$y_t = 42.8$, $y_0 = 56.5$, $y = 56.5 - 13.7/2 = 49.65$ cm

$r/y = 0.08$, $H/r = 25.5$, $y/H = 0.49$

Table 9.3. *Water level observations from reference (y'_t) and below the water table (y_t)*

t (s)	y'_t	Reference point (cm)	y_t	$y_t - y_{t-1}$
0	105.50	49	56.5	–
10	102.50	49	53.5	3.00
20	99.10	49	50.1	3.40
30	96.20	49	47.2	2.90
40	93.30	49	44.3	2.90
50	91.80	49	42.8	1.50
60	89.70	49	40.7	2.10
70	87.50	49	38.5	2.20
80	85.60	49	36.6	1.90
90	83.90	49	34.9	1.70
100	82.30	49	33.3	1.60

Taking time t as 50 s from the beginning

$\Delta y = 56.5 - 42.8 = 13.7$

For the example case, use eq. (9.15)

$$= \frac{4000r^2}{(H+20r)[2-(y/H]y} \times \frac{\Delta y}{\Delta t}$$

$K = [(4000 \times 4^2)/(101 + 20 \times 4)(2 - 0.49) \times 49.65] \times (13.7/50)$

$K = 1.29\ m/d$

Inverse Auger Hole Method

Another variation of the auger hole test is the inverse auger hole method. It is used in areas where water table at the time of test is deep. There are two variations of this test.

- Constant head method
- Variable head method

The latter test is more commonly applied in field investigations because it dispenses with the need to keep constant water level in the hole. Therefore, only this variation is included in the current publication. The basic principle and few parameters that are to be observed / measured during the test are shown in Fig. 9.5.

Equipment: Soil auger of suitable diameter normally more than 6 cm but less than 14 cm, Extension rods, Bailer or pump, Float gauge set consisting of measuring tape and a float

Procedure

1. Make the auger hole to the desired depth of measurement. It is useful to observe the changes in soil characteristics (colour, water saturation, etc.)
2. Fill the auger hole with water and let it be absorbed. Repeat the process few times so that sufficient area around the auger hole gets saturated and infiltration attains more or less a constant value (Fig. 9.5 left)

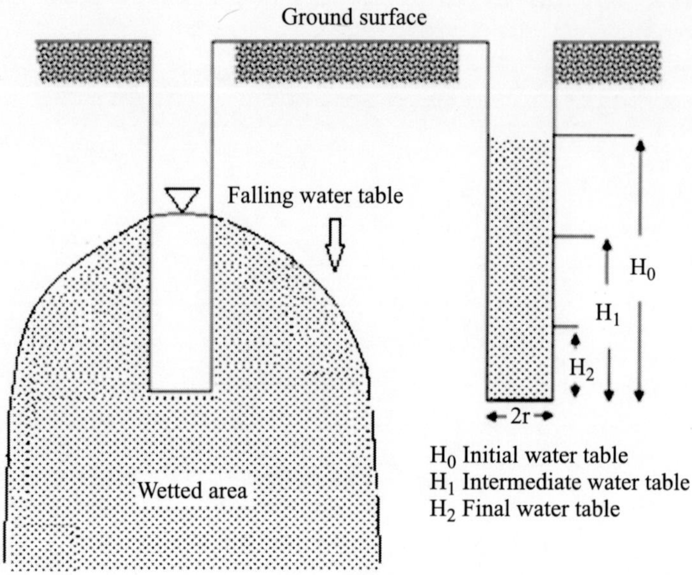

Fig. 9.5. *Inverse auger hole method to determine hydraulic conductivity, wetted area (left) and measurement parameters (right)*

3. Fill the water in the auger and take readings of the falling water table in the auger at regular intervals as in the case of auger hole method described in the previous section. A float gauge with a measuring tape or any electrical device can be used to measure the water level in the auger. Take about 8-10 readings. The Performa to collect data is depicted in Table 9.4. The hydraulic conductivity is then calculated with the help of an equation given in Ritzema (1994).

$$K = 1.15\text{r}\ \frac{\log(H_0 + r/2) - \log(H_t + r/H)}{(t - t_0)} \quad ...(9.17)$$

$$K = 1.15\ r \tan \alpha \quad ...(9.18)$$

Table 9.4. *The Performa to arrive at the values of $(H_t + r/2)$*

Time (s)	H_{tt} (cm)	$H_t = D - H_{tt}$	$H_t + r/2$
	The water level in the bore from the top of the auger hole or land surface	**D is the depth of the auger hole from top or ground surface, subtract values in column 2**	**Add r/2 in values of column 3**
0		H_0	$H_0 + r/2$
t_1			
t_2			
t_n			

Here H_o is the initial height of the water column in the hole (cm) at the initial time, t_o, H_t is the water level (cm) in the auger at any time *t*, *r* is the radius of auger hole (cm), *t* is the time (s) since the beginning of the test and *K* is the hydraulic conductivity (cm/s). The most appropriate way to use eq. (9.17) is to plot $(H_t + 0.5r)$ against *t* on a semi-logarithmic paper, the former being on the log scale. It will result in a straight line as shown in Fig. 9.6. The slope of this line is equal to tan α, which is used in eq. (9.18) to obtain K in cm/s. Multiplication of the *K* value by 864 gives *K* in m/d.

Piezometer Method

The piezometer method, like the auger-hole method, is applicable for determining the K of soils in an unconfined aquifer with a shallow water table (Fig. 9.7, Kirkham, 1946). Piezometer method has the following advantages.

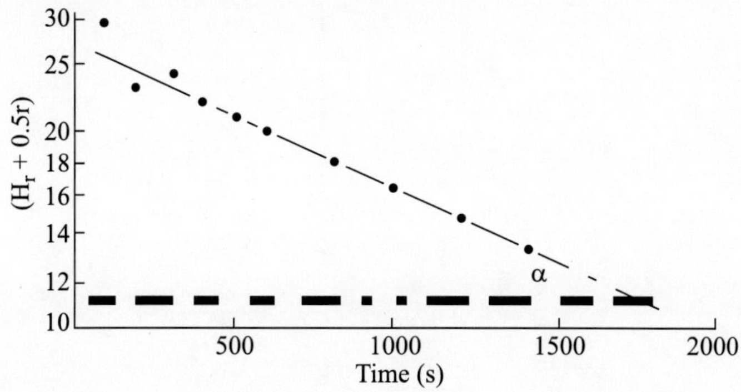

Fig. 9.6. *Plot of data against time for inverse auger hole method*

- The piezometer method is especially suitable to determine K of individual layers as the cavity can be dug in any layer of the stratified subsurface systems
- It can be used to determine either horizontal or vertical components of *K*. if the cavity length, *L* is large compared to the radius, r, the results reflect the horizontal component of *K*. If *L* is small compared to r, then the results reflect the vertical component of *K*. If *L* is zero, then it is pure vertical component of the hydraulic conductivity

Procedure

1. Install a piezometer tube (Fig. 9.7) into an auger hole drilled through the profile without disturbing the soil. The piezometer tube should at least partially penetrate the unconfined aquifer. The tube is fully closed except at its lower end
2. Using an appropriate auger, make a cylindrical cavity of radius, *r* and length, *L* within the aquifer. Similar to the auger-hole method, allow the water table to stabilize, remove the water from the pipe and then measure the rate of rise of the water within the pipe
3. K is then evaluated as a function of the geometrical dimension of the cavity in the piezometer tube, the dimensions of the aquifer, and the measured rate of rise of the water table in the tube.

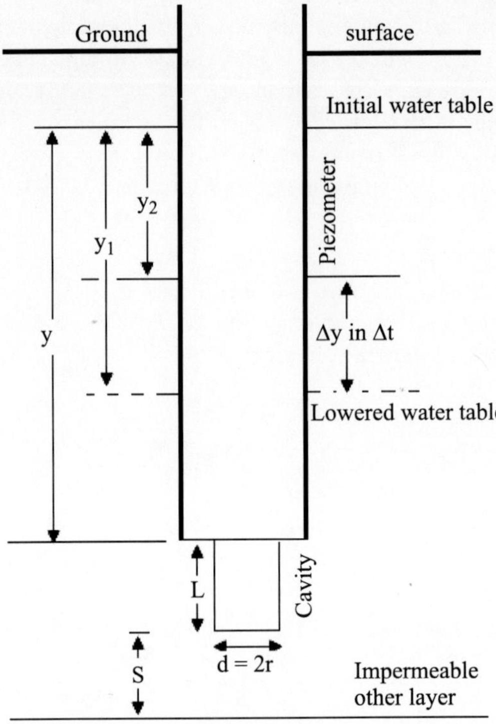

Fig. 9.7. *Schematic diagram of piezometer method to measure K*

Following relation is used

$$K = [\pi r^2 / C(t_2 - t_1)]\ [\ln(y_1/y_2)] \qquad ...(9.19)$$

Here y_1 is the distance from the initial water table to lowered water level in the tube or piezometer at time, t_1; y_2 is the distance from the water table to the water level in a tube at time, t_2; r is radius of the tube; $t_2 - t_1$ is the time for the water level to change from y_2 to y_1; C is a constant and is taken from appropriate tables one of which is given as Table 9.5. Christiansen (1964) suggested a modification to this method wherein a plexi-glass transparent pipe is attached to the top of the piezometer. Water is filled in the tube and the piezometer. The drop in water level in the plexi-tube is noted with respect to time. Channarasappa (1965) used this approach to measure relative hydraulic conductivity of various soils in a project area. Shape factor in these calculations was taken as 1.0

$$K = kr \ln (y_0/y_1)/(t_1 - t_o) \qquad ...(9.20)$$

Here k is the shape factor dependent upon the depth and size of the cavity at the bottom of the piezometer, r is the inside radius of the transparent tube, y_0 is the distance from the top of water in the transparent tube to water table at initial time (t_0) and y_1 is the distance from water level in the tube to the water table at time, t_1.

Example 9.4. A test was conducted to measure the hydraulic conductivity of the soil using piezometer method. Calculate the hydraulic conductivity of the soil layer, if the depth of the hole is 140 cm, depth of water table level 60 cm, depth of cavity 16 cm, the distance between the bottom of the cavity and the next layer (permeable layer) 80 cm, and the radius of the cavity 4 cm. Time drawdown data are given as follows.

t (s)	0	10	20	40
y (cm)	125	112	103	98

From the data given, we have

$L = 16$ cm, $r = 4$ cm, $S = 80$ cm, $t_1 = 0$, $t_2 = 40$

$y = 140 - 60 - 16 = 64$ cm, $y_1 = 120$ cm, $y_2 = 95$ cm

To calculate C, the shape factor of the cavity:

$S/r = 80/4 = 20$, $y/r = 64/4 = 16$, $L/r = 16/4 = 4$

Using Table 9.5, the value of C is 19.8.

Hydraulic conductivity is given as

$K = [\,\pi r^2 / C\,(t_2 - t_1)]\,[\text{Ln}\,(y_1/y_2)]$

$K = [3.142 \times 4^2/(19.8 \times 40)][\text{Ln}\ 125/98]$

$K = 0.015$ cm/s.

Table 9.5. *Shape factor, C of eq. (9.19) when cavity does not rest on an impermeable layer*

W/r	**H/r**	**S/r**					
		∞	**8.0**	**4.0**	**2.0**	**1.0**	**0.5**
0	20	5.6	5.6	5.8	6.3	7.4	10.2
	16	5.6	5.6	5.8	6.4	7.5	10.3
	12	5.6	5.7	5.9	6.5	7.6	10.4
	8	5.7	5.7	5.9	6.6	7.7	10.5
	4	5.8	5.8	6.0	6.7	7.9	10.7
0.5	20	8.7	8.9	9.4	10.3	12.2	15.2
	16	8.8	9.0	9.4	10.3	12.2	15.2
	12	8.9	9.1	9.5	10.4	12.2	15.3
	8	9.0	9.3	9.6	10.5	12.3	15.3
	4	9.5	9.6	9.8	10.6	12.4	15.4
1.0	20	10.6	11.0	11.6	12.8	14.9	19.0
	16	10.7	11.0	11.6	12.8	14.9	19.0
	12	10.8	11.1	11.7	12.8	14.9	19.0
	8	11.0	11.2	11.8	12.9	14.9	19.0
	4	11.5	11.6	12.1	13.1	15.0	19.0
2.0	20	13.8	14.1	15.0	16.5	19.0	23.0
	16	13.9	14.3	15.1	16.6	19.1	23.1
	12	14.0	14.4	15.2	16.7	19.2	23.2
	8	14.3	14.8	15.5	17.0	19.4	23.3
	4	15.0	15.4	16.0	17.6	20.1	23.8

4.0	20	18.6	19.8	20.8	22.7	25.5	29.9
	16	19.0	20.0	20.9	22.8	25.6	29.9
	12	19.4	20.3	21.2	23.0	25.8	30.0
	8	19.8	20.6	21.4	23.3	26.0	30.2
	4	21.0	21.5	22.2	24.1	26.8	31.5
8.0	20	26.9	29.6	30.6	32.9	36.1	40.6
	16	27.4	29.8	30.8	33.1	36.2	40.7
	12	28.3	30.0	31.0	33.3	36.4	40.8
	8	29.1	30.3	31.2	33.8	36.9	41.0
	4	30.8	31.5	32.8	35.0	38.4	43.0

MEASUREMENT IN TWO-LAYERED PROFILE

The method to determine K in a homogeneous profile can be easily extended to a two-layered profile. In a layered profile, initially a shallow àuger hole is made which terminates at least 20 cm above the interface of two layers but contains at least a minimum of 40 cm thick water body. A deep auger hole is also made going up to at least 50 cm below the interface into the second layer (Fig. 9.8). The minimum horizontal distance between the two bores should be about 3 m. The value of hydraulic conductivity from the deeper hole is measured first followed by the shallower hole. If the hydraulic conductivity of the shallower hole is K_1 and the hydraulic conductivity of the deeper hole is K_{12} then K_2, the hydraulic conductivity of the second layer is:

$$K_2 = [K_{12}(H_1 + H_2) - K_1H_1]/H_2 \qquad ...(9.21)$$

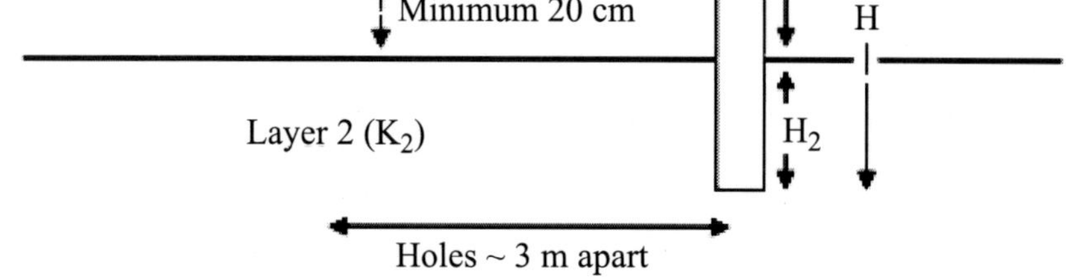

Fig. 9.8. *Set-up to measure hydraulic conductivity in a two-layered profile*

Ditch Method

Since auger holes cover a very small area, it may not represent the real average hydraulic conductivity of the area. A bigger ditch is therefore used to measure the hydraulic conductivity. Let us say that it has a bottom width, B and length L (Fig. 9.9). The ditch is filled with water. The water is allowed to infiltrate so that area around the ditch is saturated. The ditch is refilled with water. Measurements of the water levels in the ditch with time are observed. Following equation is used to calculate the hydraulic conductivity (Stibinger, 2014).

$$K(t_2 - t_1) = M\,\mathrm{Ln}[(y_1 + M)/(y_2 + M)] \qquad ...(9.22)$$

Here $M = BL/2\,(B + L)$ and y_1 and y_2 are the water levels in the ditch at times t_1 and t_2 respectively. The water requirement of this method to measure the hydraulic conductivity is quite high. At many places, it may prove to be a limitation.

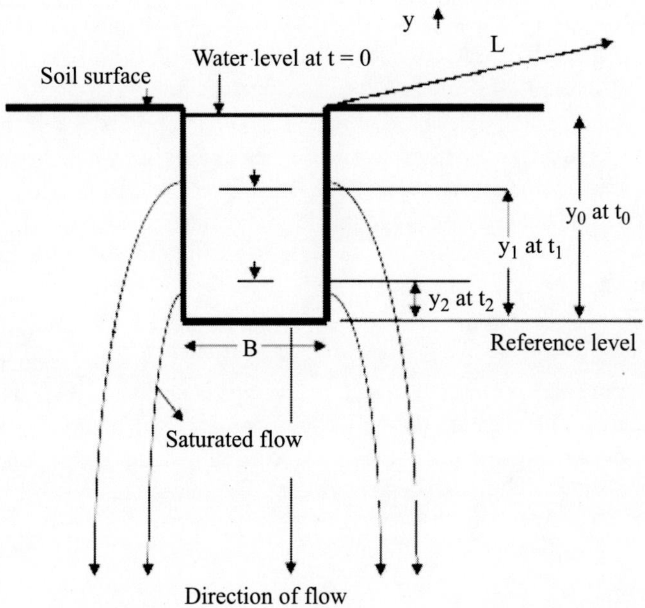

Fig. 9.9. *Measurement of hydraulic conductivity using ditch method*

AIR-ENTRY PERMEAMETER

The air-entry permeameter was devised to investigate the significance of flows in the capillary zone (Bouwer, 1966). It measures the soil-water pressure at which air entered the saturated voids. Assuming a relationship between this value and the pressure just above the advancing front of a wetted zone, the conductivity of a mass of soil absorbing water to the point of saturation can be calculated. Knowing that this conductivity value is closely equal to one-half the saturated hydraulic conductivity (K_{sat}/K_{unsat} =2) and that air entry pressure is about 2 times the water entry pressure (P_a/P_w =2), a new method of determining vertical hydraulic conductivity at saturation has been proposed. Although the method appears to make several assumptions, the results of this method compare favorably with other accepted methods. Moreover, this method has some distinct advantages. One of the advantages is that it uses much less water than other methods. Since its first use, several workers have improved the original version of this method (Topp and Binns, 1976, Watkins, 1987).

Equipment

It consists of a cylinder of about 20 cm diameter 15-20 cm long, with a sealable transparent lid (Fig. 9.10). It is connected with a supply reservoir, a vacuum gauge and an air escape valve.

Procedure

1. Drive the cylinder into the soil by about 10 cm. Place a layer of sand over the soil surface inside the cylinder. Also place a metal disk over it to break the energy of falling water. The water head of about 1 m is maintained.
2. Water is added through the supply valve with the air valve open until the embedded cylinder becomes full of water. Once the cylinder is filled, the air valve is closed and water is allowed to infiltrate, the reservoir being kept full
3. When the wetting front, W_f has reached the desired depth, which is dependent on soil texture and structure, no more water is added to the reservoir. Modified air-entry permeameter has a fine tensiometer probe that passes through the lid of the device. By setting the probe corresponding to the desired depth of the wetted zone, W_f (about 15 cm in sand and 5 cm in massive clay), it is possible to detect the arrival of the wetting front in real time
4. The drop in water level in the reservoir with time is measured to calculate the intake rate
5. Now the supply valve is closed. A vacuum develops inside the cylinder until air entry point is reached. The pressure on the vacuum gauge is noted periodically. At some point it will show the maximum value corresponding with minimum pressure. At this point, the pressure will begin to decrease again. The minimum pressure corrected for gauge height, G and depth of the wetted zone, W_f corresponds closely to the air-entry pressure, P_a, of the wetted zone.

Calculations

Referring to Fig. 9.10, the vertical hydraulic conductivity of the rewet zone, i.e., the zone being saturated is calculated from the following equation.

$$K = (Q/A)\,[W_f/(H_r + W_f + H_1)] \qquad \text{...(9.23)}$$

Here Q is the volumetric intake rate through area, A, of the permeameter. Q/A can be given as

$$Q/A = dH/dt\,[R_r^2/R_c^2) \qquad \text{...(9.24)}$$

Such that dH/dt is the rate of fall of water level in the reservoir just before closing the supply valve, R_r is the radius of the reservoir and R_c the radius of the cylinder. H_r is the height of the water level in the reservoir above the soil surface at the time the water supply is closed, W_f is the depth of the wetting front, H_1 is the matric potential of the soil just below the wetting zone, assumed to be $0.5P_a$. It is less than atmospheric pressure and therefore a negative quantity in eq. (9.15). P_a is the air-entry value, calculated as $P_{min} + W_f + G$. Here P_{min} is the minimum pressure (maximum vacuum) read from the vacuum gauge after stopping the water supply, and G is the height of the vacuum gauge above the soil surface. As stated previously, the vertical hydraulic conductivity at saturation, K_{sat} is assumed to be two times the value of K calculated from eq. (9.23), such that

$$K_{sat} = 2K \quad \text{...(9.25)}$$

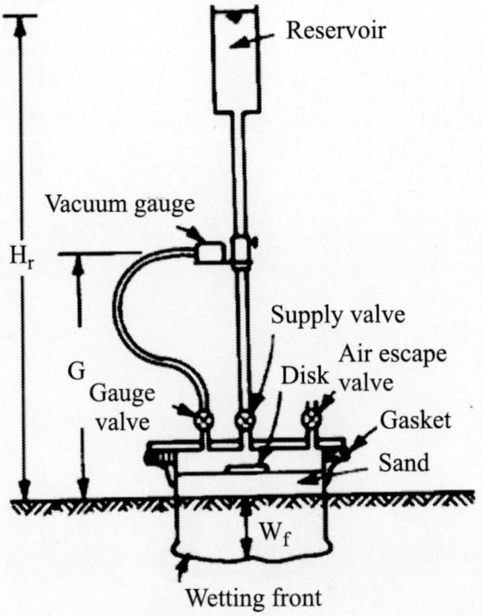

Fig. 9.10. *Diagrammatic representation of an air-entry permeameter*

AIR/WATER PERMEABILITY

Puddling or sealing of the soil surface caused by the impact of rain drops, trampling by livestock, or deposition of eroded clay is quite detrimental to soil structure. The degradation process in this manner accentuates splash erosion. One of the methods of detecting the puddle condition of the soil is to measure its permeability to air since this property is closely correlated with the infiltration capacity of soil.

Theory (Richard, 1954)

This method of measuring air permeability can be used for both the disturbed and undisturbed samples (Kirkham, 1947). The apparatus shown in Fig. 9.11 uses the compressed dry air (admitted through a calcium chloride drying tube) to an airtight tank of constant volume. An outflow tube leads to a water manometer and to the soil sample container. The soil sample container consists of tinned iron can with an extension made from a 4-cm section of brass tubing counter bored to give a snug fit on the top of the can. An outlet hole is punched at the bottom of the can. A disk of brass screen, 20- or 40-mesh, with 2 layers of fiber glass sheet is provided at the bottom of the can, which acts as a filter.

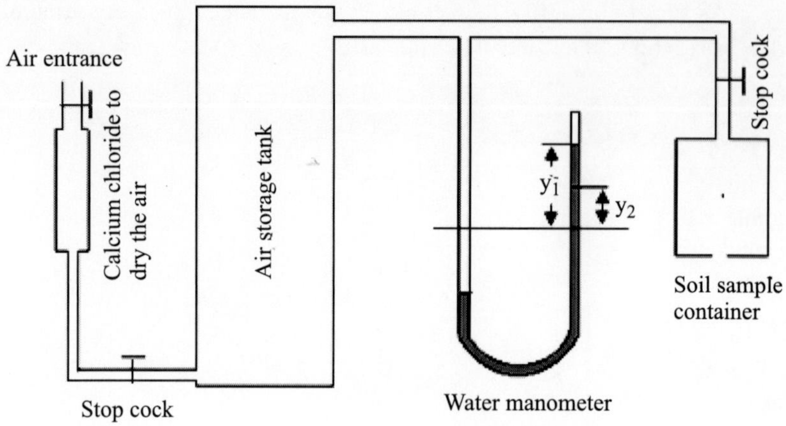

Fig. 9.11. *Apparatus to measure intrinsic permeability using air flow*

Procedure

1. Pass air-dried soil through a wire-mesh sieve having 1 mm openings. Attach the brass cylinder to the top of the can and fill the container about three-fourths full of soil. Dump the soil in the container rather than by pouring or scooping. With a spring-load of 3 kg weight on top of the soil, drop the container 200 times on a solid block of wood from a height of 2.5 cm
2. Remove the brass cylinder from the can and use a spatula to strike off the soil and level it with the top of the can. Cover the soil with a disk of filter paper, place a lid on the can, and complete the seal with a tight rubber band or beeswax
3. Connect to the air source by means of a short piece of copper tubing soldered to the lid of the can.
4. Pass compressed air through the drying tube to the tank until a manometer displacement of 40 to 50 cm of water is attained. Record the air temperature in the tank
5. Measure the rate of drop of the manometer level as air is allowed to flow from the tank through the soil sample. Allow an initial drop of about 10 cm in the manometer level before beginning to record the height and time.

To evaluate air permeability in the fields, a portable air-permeameter discussed in Iversen et al. (2003) can be used. It has similar features but uses nitrogen gas instead of air as the fluid.

Calculations

The intrinsic permeability of the soil using air is given by the equation:

$$K_i = 2.3 \text{ L V S } \eta/(\text{A P}_a) \qquad ...(9.26)$$

Here, K_i is intrinsic permeability with air, cm^2, L is the length of the soil column, cm, V is the volume of the tank, cm^3, η is the viscosity of air at the temperature at which determination is made, dyne-s/cm^2 (Poise), A is the cross-sectional area of the sample, cm^2, P_a is the atmospheric pressure, dynes/cm^2, S is the slope of log y v/s time curve, y is the displacement of the water surface in one arm of the manometer, cm, Δt is the time interval in seconds for the water surface in the manometer to drop from y_1 to y_2. S is calculated as:

$$S = (\text{Log}_{10} y_1 - \text{Log}_{10} y_2)/\Delta t$$

The time for a convenient and measurable drop in the manometer level can be controlled by the volume of the tank. For most soils, a 200 L drum gives a convenient time interval for a measurable manometer change. For soils of low permeability, smaller tank can be used. The soil conductivity for the test fluid can now be obtained from the intrinsic permeability. For the conductivity of soil to air, we have:

$$K_a = K_i \rho_a g/\mu_a \quad ...(9.27)$$

Here K_a is the conductivity of soil to air (m/s), K_i is the intrinsic permeability of the media (m^2), ρ_a is the density of the air (kg/m^3), g is the gravitational constant (m/s^2), and μ_a is the dynamic viscosity of the air (kg/m-s) (Freeze and Cherry, 1979).

Hydraulic Conductivity of Water

The intrinsic permeability of a stable soil is the same regardless of whether it is evaluated using water or air. The conductivity on the other hand is different for each fluid and requires an adjective to indicate the applied fluid. Thus, the soil saturated hydraulic conductivity of water, K_s can be calculated as follows (Huang et al. 2015):

$$K_s = K_i (g\rho_w/\mu_1) \quad ...(9.28)$$

It can also be derived from K_a calculated in eq. (9.27) as follows:

$$K_s = (\mu_a/\rho_a)(\rho_w/\mu_1)K_a \quad ...(9.29)$$

Here ρ_w is the water density (kg/m^3), and μ_1 is the dynamic viscosity of the water (kg/m-s).

Notes:

- In order to get consistent results or for comparing one soil with another or to determine the effect of various treatments, adopt a standard procedure for preparing and packing the soil
- Make determinations in triplicate.

Example 9.5. The intrinsic permeability of the soil is 3×10^{-10} cm^2. Calculate the hydraulic conductivity of the soil at 20°C.

Using eq. (9.28): $K_{20} = K_i (g\rho_w/\mu_1)$

$K_i = 3\times10^{-10}$ cm^2, $g = 981$ cm/s^2

$\rho_w = 0.99821$ g/cm^3, $\mu_1 = 1.0050$ centipoise or 1.00050×10^{-2} poise

$K_{20} = 3 \times 10^{-10} \times 981 \times 0.99821/1.00050 \times 10^{-2} = 0.2936 \times 10^{-4}$ cm/s

$K_{20} = 0.2936 \times 10^{-4} \times 3600 = 0.106$ cm/hr

Portable Sphygmomanometer Equipment

Air permeability of the soil can also be determined rapidly by portable sphygmomanometer equipment, a modification of the apparatus used for taking blood pressure. This equipment consists of a brass cylinder, 15 cm long and 3-5 cm diameter. It is provided with an air intake corrugated connection at the top to attach rubber tubing (Fig. 9.12). The cylinder is sharpened at the open end. The rubber tubing connects the brass cylinder with the constant pressure rubber chamber and either an aneroid or mercury sphygmomanometer by means of a "T" tube. The rubber chamber is provided with a rubber bulb, which is operated by hand and has a provision of air release value. The manometer is placed at a safe and convenient position for taking readings and attached by means of a three way connector to the rubber chamber.

Clear the test area of coarse fractions of litter and crop residues, if any. The tube is inserted into the soil to a depth of 5 cm marked on the cylinder. The rubber tubing leading to the brass cylinder is tightly clamped. The clamp is placed approximately halfway between the cylinder and the T tube. The air release valve on the rubber bulb is closed and the air is pumped steadily until the manometer shows a reading of about 210 mm. The clamp is released and the air is allowed to flow for exactly 30 s. At the end of this period, the tubing is clamped off and the manometer readings are noted (Wilde, 1950). Air resistance readings below 15 mm are obtained for highly porous material such as coarse sandy soils or well aggregated loam and clay soils. Readings exceeding 100 mm for surface soil layer indicate puddling or sealing of soil caused by the action of raindrops, grazing livestock, heavy farming machinery or logging equipment, or deposition of fine colloidal material. The procedure of calculations is same as discussed in the previous method. Being locally made equipment, it needs to be calibrated with standard equipment before being used on a large scale.

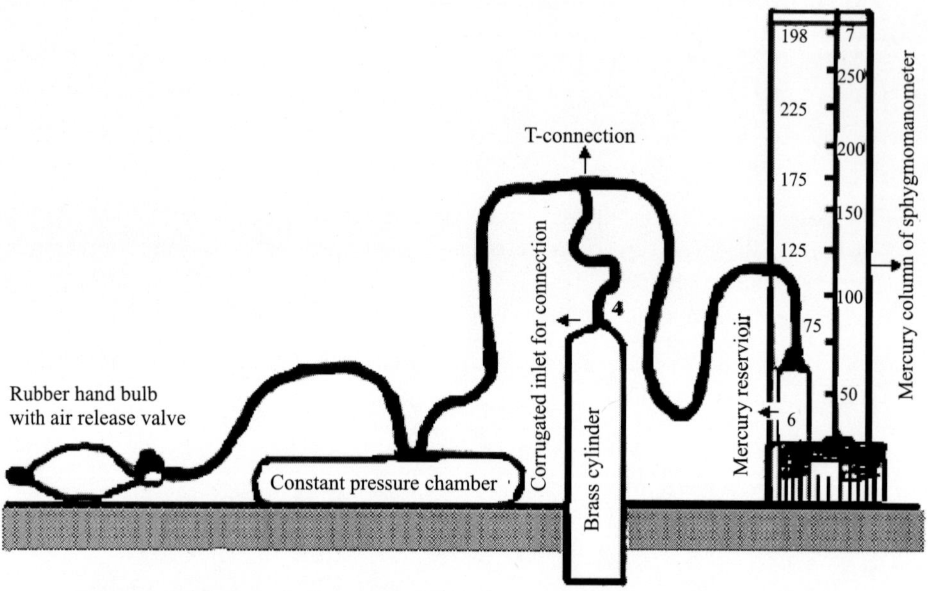

Fig. 9.12. *Portable sphygmomanometer unit to determine air permeability of soils*

FIELD METHODS (LARGE SCALE)

The hydraulic conductivity of soils can be determined by analyzing the records of drain discharge and water table head at mid-spacing between the two adjacent laterals of subsurface drainage projects (Fig. 9.13). A minimum of 3 drains are required. While the discharge is measured in the manhole of the middle lateral, observation wells are used to measure the head at mid-spacing of the drains. Ideally, the water table levels are recorded under controlled steady state conditions. This state is best achieved during periods of low evaporation after the end of few good rainy days. The recharge to the water table must cease while these measurements are taken (no standing water on the land surface). It ensures that readings are being taken on the recession side of the drain flow hydrograph. Since such situations might occur rarely, the convenient way is to compile such a curve from a series of continuous or sometimes occasional single measurements provided all data are obtained during the recession phase. While analyzing the data, it is assumed that the relation between q and h obeys the Hooghoudt's equation:

$$q = (8K_b d_e h + 4K_a h^2)/L^2$$
...(9.30)

Here q is the discharge from the lateral, K_b is the hydraulic conductivity of the layer below the drain and K_a of the layer above the drain, d_e is the equivalent depth to impermeable layer, L is the drain spacing and h the hydraulic head. The first term ($8K_b d_e h/L^2$) represents the flow contribution from below the drain depth, while the second term ($4K_a h^2/L^2$) represents the flow contribution from above the drain depth. Equation (9.30) can also be written as:

$$q = A\,h + B\,h^2 \qquad ...(9.31)$$

Such that $A = 8K_b d_e/L^2$ and $B = 4K_a/L^2$

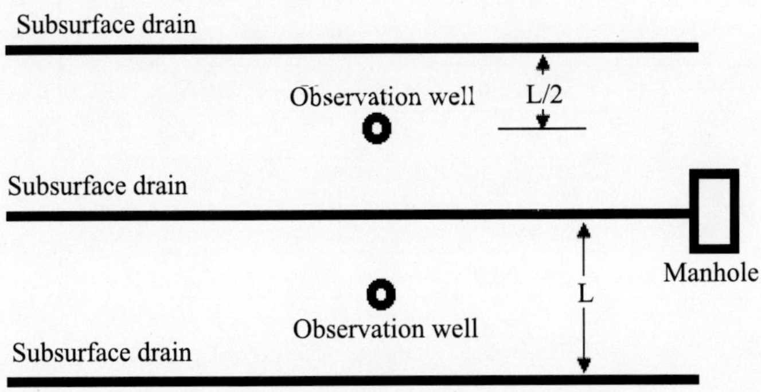

Fig. 9.13. *A field layout of subsurface drains*

Procedure

1. Collect data on drain discharge and hydraulic head, *h*' (Initially h is denoted by *h*')
2. Plot data on q and h' with h' on the *x*-axis. The shape of this plot (eq. 9.22) on a normal arithmetic graph paper will depend upon the relative contribution of the two terms. The plot is linear for $B = 0$ but becomes increasingly non-linear with increasing B or *h*'. It is because of the increasing contribution of the second term of the equation (Fig. 9.14 left).
3. Since field data are quite heterogeneous, the shape of the curve might vary and one may end up in some peculiar situations. For example, a *q-h*' line might intersect the

x-axis at some point. The errors in the measurement of h', depth variation in the drains along the length due to the slope of the drains, the error in measurements of the height of the observation wells above the soil surface and the hydraulic head required to compensate the entrance resistance of the drains together cause such a situation (Ritzema, 1994). Practically, it means that the drains are not discharging even when there is some positive hydraulic head. Although, we know that some minimum head is needed for flow to take place, but the Hooghoudt's equation assumes ideal drains. Therefore, it is advisable to subtract this value of the head from the measured h' values. The new values are real h values, which are used for further analysis. Data in Table 9.6 show that this correction in the head is 17 cm for an operating drainage system at Sampla, Haryana (Rawat *et al.*, 2001).

4. If $B = 0$, i.e. there is no flow from above the drains, the plot of q-h as per eq. (9.31) will be a straight line. If extended, it should pass through the origin (Fig. 9.15). The slope of this line will be equal to A and therefore, $K_b d_e$ can be calculated. If d_e is known, then K_b can be calculated straightway.
5. The other approach is to divide both sides of eq. (9.31) by h, so that we have

$$q/h = A + Bh \qquad ...(9.32)$$

Table 9.6. *Hydraulic head and drain discharge data from Sampla, Haryana*

h'(m)	q (m/d)	h (m)	q/h (1/d)
	Deep system		
0.98	0.0132	0.81	0.0163
0.75	0.0102	0.58	0.0176
0.66	0.0081	0.49	0.0165
0.55	0.0060	0.38	0.0158
0.46	0.0048	0.29	0.0166
0.40	0.0039	0.23	0.0170
0.35	0.0030	0.18	0.0167

h' is actual and h the corrected hydraulic head

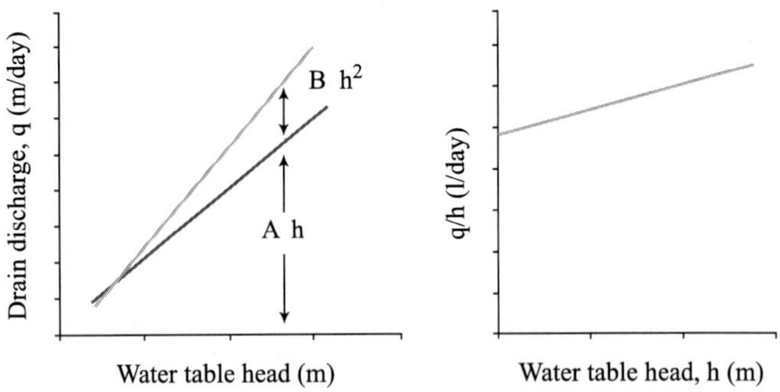

Fig. 9.14. *The q-h plot showing non-linearity in the plot with increasing h (left) and q/h - h plot with contribution from the layer above the drains (right)*

6. If we use this equation, *q/h* versus *h* plot should be a straight line (Fig. 9.14 right). The slope of this line yields $4K_a/L^2$ while the intercept gives $8K_bd_eh/L^2$. Thus, it is possible to calculate K_b and K_a. If K_b and K_a are equal, then equivalent depth can also be calculated explicitly. If $B = 0$, then also the plot would be a straight line (eq. 9.32) shown schematically in Fig. 9.15. The straight line plot is parallel to *x*-axis (Fig. 9.15). It cuts the *y*-axis at a value equal to A. Therefore, K_bd_e and K_b can be calculated as discussed in step 4.

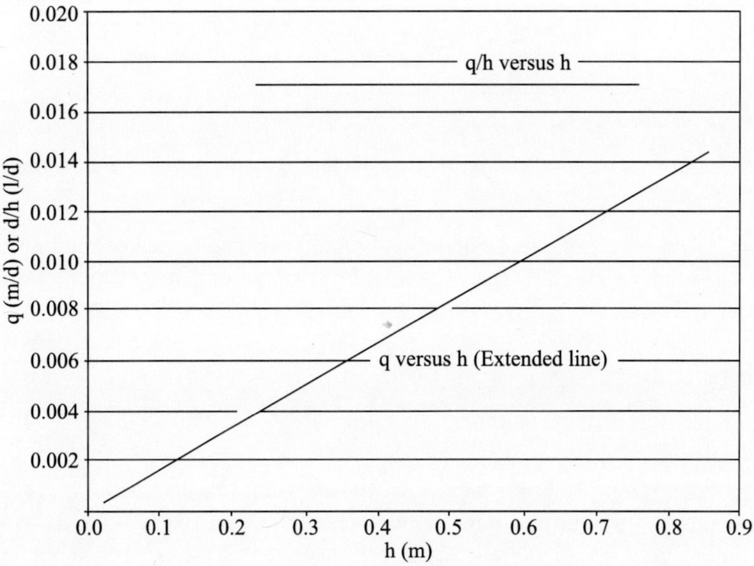

Fig. 9.15. *Drain discharge and hydraulic head plots for a subsurface drain*

SOIL WATER DIFFUSIVITY

Soil-water diffusivity, sometimes referred as hydraulic diffusivity, is a function of the soil wetness and has the dimensions of L^2/T. Soil-water diffusivity $D(\theta)$ has been defined as the ratio of the hydraulic conductivity $K(\theta)$ to the specific water capacity $C(\theta)$ as follows.

$$D(\theta) = K(\theta)/C(\theta)$$

Where $C(\theta) = d\theta/d\Psi$

Such that $D(\theta) = K(\theta)/(d\theta/d\Psi) = K(\theta)\, d\Psi/d\theta$ ...(9.33)

Here Ψ and θ represent soil water suction and soil water content, respectively.

Measurement of hydraulic diffusivity under transient flow conditions was first made by Bruce and Klute (1956) using horizontal infiltration technique. The schematic view of the apparatus used by them and others is shown in Fig. 9.16 (Donghao et al., 2016).

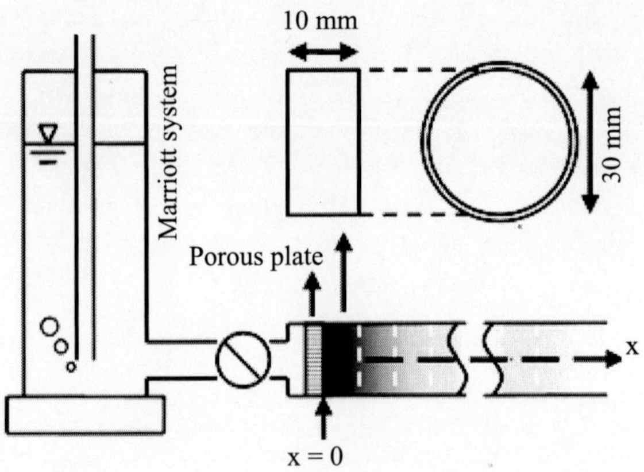

Fig. 9.16. *Horizontal absorption of water for evaluating soil water diffusivity*

Apparatus: Sectional acrylic cylindrical column 30 cm long (each section of 1 cm height and 3-5 cm internal diameter), Porous plate with filter paper, Glass funnel with Marriott device, Hot air oven, Balance, Aluminum moisture cans and Boiled and cooled water

Procedure

1. Assemble the sectional cylindrical column (Fig. 9.16) with the help of bolts and nuts with adequate support
2. A known quantity of soil (($\pi d^2 L/4$) × BD) passing through 2-mm sieve is filled in the column to the desired bulk density. Here d, L and BD are the diameter, length of the column and soil bulk density respectively. Soil is packed in the column in installments. It is compacted with the help of a 5 cm diameter rubber stopper attached to a long wooden stick. The surface soil is scratched after compacting each installment to avoid layering of soil along the length of the soil column
3. Mark scale on the column by pasting a strip of graph paper. It helps to measure the wetting front advance
4. The glass funnel is filled with air-free water (boiled and cooled water) and attached with the Marriott device. The air bubbles are removed from the funnel and the water supply unit is adjusted to apply a 2 cm pressure at the centre of the sintered disc
5. The water is allowed to infiltrate into the soil. Distance of the advancing wetting front (x) is noted in cm and time in minutes. The easier way is to note the corresponding time (t) in min as soon as water reaches the middle of each segment (Column 3, Table 9.7)
6. Discontinue the water supply as soon as water has moved to more than half of the column length
7. Immediately thereafter, sectional column is broken to determine the soil gravimetric moisture contents of each section, θ_m. The moisture content is then converted to volumetric moisture content (θ) by multiplying each value by the bulk density of the soil (Column 2 of Table 9.7)

8. Plot θ versus x curve (Fig. 9.17). The curve is used to evaluate the integral of the equation required to determine soil water diffusivity as per the following equation.

   $D(\theta) = -(1/2t)\,(dx/d\theta) \int x d\theta$ ...(9.34)

   Here, x is the wetting front length, cm and t is the total time of infiltration, s. It is accomplished by dividing the area under the curve into a finite number of segments. The integral is obtained by adding up the areas of these segments up to θ value for which $D(\theta)$ is to be evaluated (Bandyopadhyay et al., 2012).
9. Calculate $K(\theta)$ values as product of $D(\theta)$ times $d\Psi/d\theta$, if Ψ versus θ curve is available.

Since it may be difficult to use eq. (9.34), following step-wise procedure is used to solve the same for assessing soil water diffusivity.

1. Prepare a graph between experimentally measured moisture content (θ) and wetting front advance (Column 2 & 3 of Table 9.7 and Fig. 9.17)
2. Divide the whole soil moisture content range from air-dry to saturation into equal intervals (say 0.03 cm^3/cm^3 as in column 4 Table 9.7). Designate it as θ_r
3. Find out wetting front depth, x (cm) from the graph for each moisture content of column 4. it is determined from the experimentally measured x and θ curve. These values are noted in column 5
4. From the plot of θ versus x, evaluate the integral by dividing the area under the curve into finite number of segments. Add up the areas of the segments to θ values up to which $D(\theta)$ is required. It is achieved by multiplying the wetting front depth in column 5 with difference in θ (from column 4) i.e. θ of the next row minus θ of the current row to get $x\Delta\theta$ in column 6. Add all values of column 6 in column 7 to get the cumulative values of $\int x d\theta$

Find out $dx/d\theta$, from column 5 and 4 using the equation given below (Also see footnote 5 below Table 9.7). Note that dx/dθ is negative because θ decreases with increasing x.

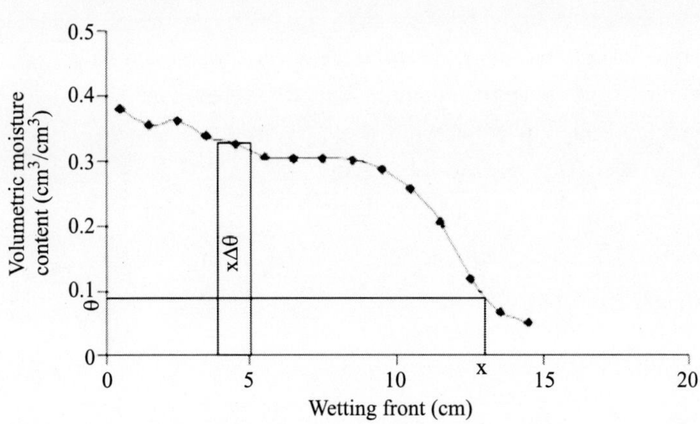

Fig. 9.17. *Schematic presentation of volumetric soil moisture content and wetting front for a horizontal column study*

Table 9.7. *Description of step-wise procedure in tabular form*

Segment	Moisture content (θ)	Wetting front (cm)	θ_r	x (cm)	xΔθ	∫xdθ	dx/dθ	D(θ)
1	2	3	4[1]	5[2]	6[3]	7[4]	8[5]	9 (Eq. 9.25)
1	Highest	0.5	0.06	x_1	X_1 $(\theta_{r2}-\theta_{r1})$	$(x\Delta\theta)_1$	$(x_2-x_1)/(\theta_{r2}-\theta_{r1})$	
2		1.5	0.09	x_2		$(x\Delta\theta)_1 + (x\Delta\theta)_2$		
3		2.5	0.12					
4		3.5	0.15					
5		4.5	0.18					
6		5.5	0.21					
7		6.5	0.24					
8		7.5	0.27					
9		8.5	0.30					
10		9.5	0.33					
11		10.5	0.36					
12		11.5	0.39	X_1 $(\theta_{r2}-\theta_{r1})$				
13		12.5						
14		13.5						
15		14.5						
16		15.5						
17	Lowest	16.5						

1. Arbitrary distribution (Approximately): (Highest – Lowest)/number of segments
2. The value of x for moisture content in column 4 (From Fig. 9.17)
3. Column 5 multiplied by column 4 (Δθ =(θ (i+1 row) – θ (i row))
4. Cumulative successive values of column 6
5. $dx/d\theta = (x_{i+1} - x_i)/(\theta_{i+1} - \theta_i)$ [From Column 5/ From Column 4]
   $dx/d\theta = (x_{i+1} - x_i)/(\theta_{i+1} - \theta_i)$

Calculate $D(\theta)$ using eq. (9.34).

Let us assume the following values from Table 9.7

Total time = 5400 s (Experimental value), dx/dθ = 33 (Column 8), ∫xdθ = 0.8 (Column 7)

Since $dx/d\theta$ is negative, θ decreases with increasing depth, we have

$D(\theta) = (1/2t)\,(dx/d\theta)\int x d\theta$

$D(\theta) = (1/(2 \times 5400)) \times 33 \times 0.8 = 2.44 \times 10^{-3}\ \text{cm}^2/\text{s}$

Additional Determinations

Following additional data sets are obtained to determine few additional soil characteristics.

Weighted mean diffusivity (Gardner, 1959, Gardner, 1970)	Cumulative infiltration at the end of experiment
Soil sorptivity (Philip, 1957)	Water infiltrated versus time
Soil penetrability (Jackson, 1973)	No additional data are required

Weighted mean diffusivity

$$\bar{D} = I^2\pi/[(\theta_s - \theta_i)^2 t] \qquad ...(9.35)$$

Here $\bar{D}$ is the weighted mean diffusivity, cm^2/min, I is the cumulative infiltration, cm, θ_s is the volumetric water content at the beginning near the source, θ_i is the initial volumetric water content/air-dry moisture content.

Soil sorptivity

According to Philip (1955), the cumulative infiltration of water, I for initially air-dry soil is proportional to the square root of time such that

$$I \propto t^{0.5}$$

$$I = St^{0.5} \qquad ...(9.36)$$

Here S is the proportionality constant and is termed as soil sorptivity. Plot square root of time versus cumulative infiltration data on an arithmetic paper. The slope of the line is the soil sorptivity, cm/min$^{0.5}$.

Soil penetrability

According to Jackson (1963), the advance of wetting front, x from the source in an air-dry soil is proportional to square root of time.

$$x \propto t^{0.5}$$

$$x = \lambda t^{0.5} \qquad ...(9.37)$$

Here proportionality constant, λ is the soil penetrability, cm/min$^{0.5}$. Plot square root of time versus water front advance (cm) data on arithmetic paper. The slope of the line is the soil penetrability, cm/min$^{0.5}$.

UNSATURATED HYDRAULIC CONDUCTIVITY (WIND'S METHOD)

Since its development, Wind's evaporation method (Wind, 1968) has been widely used by researchers to measure unsaturated hydraulic conductivity. The main merit of this method lies in the fact that it permits the simultaneous determination of both the water retention characteristic curve and the unsaturated hydraulic conductivity on one soil sample. The water retention characteristic curve is obtained in the whole tensiometer range, i.e. from 0 to -900 cm, but the range of the conductivity depends on the soil type. For sandy soils the range is approximately from -90 to -500 cm while for loamy and clay soils from -5 to -800 cm. Schindler (1980) and later Schindler and Müller (2006) simplified the method that employs two tensiometers. They also proposed few modifications that extend the range of measurements. These modifications also obviate the need to assume that the matric suction from top to bottom of the soil column is linearly distributed.

Procedure

1. Undisturbed soil sample is taken in a stainless steel cylinder equipped with iron cutting head (inner diameter 110 mm, height 50-80 mm). The cylinder is gently pushed into the soil using a hydraulic jack. A straight knife is used to remove projecting soil on both sides of the cores. A circle of mesh or similar hydrophilic close woven material is placed at the bottom of the core. The mesh is secured with an elastic band and the core is placed on a supporting sheet to retain the soil sample
2. Holes are excavated for fixing the tensiometers using a special drill and a template. The modified evaporation method (Schindler, 1980, Schindler et al., 2010.) discussed in this section uses only two tensiometers at two depths i.e. $z_1 = 0.25$ L and $z_2 = 0.75$ L from top, where L is the height of the column (Fig. 9.18). For 5 cm high soil core (for two tensiometers smaller the length better it is), tensiometers are installed at 1.25 cm and 3.75 cm below the top of the core. For clay cores, thin plastic pipes, reaching only up to the beginning of the ceramic cup of the tensiometers, are inserted in the holes. The pipes help to safely remove the tensiometers at the end of the experiment. Note that the tensiometers are not installed one above the other but at some angle to each other (Fig. 9.18)
3. Cores are slowly saturated in the laboratory by placing them in a pan of water (tensiometer holes are taped at this time). To minimize air entrapment, the water level is increased from an initial height of about 0.5 cm above the sample bottom to a final height of about 0.2 mm below the sample surface
4. After saturation, the tensiometers are installed while the core is still in the pan. The assembly (core with inserted tensiometers) is removed from the pan. All free water is removed from the assembly, and the sample bottom is fully sealed with a lid so that there is no leakage. Tensiometers can now be connected to data loggers for continuous monitoring, if such facilities are available
5. The assembly is now placed on an electronic weighing scale. The water from the soil core evaporates only from the top in a controlled environment
6. The tensions and the total weight of the soil sample are measured at regular intervals. Time between the readings should neither be too large nor too small. The pressure difference between atmospheric pressure and the water pressure inside a tensiometer will be referred to as tension (Ψ, kPa) and expressed as a positive quantity
7. The experiment is terminated as soon as air enters the top tensiometer
8. After removing the tensiometers, weight of the tensiometers is noted. The soil sleeve is dried at 105°C. The weight of the tensiometers plus the weight of the dried soil sleeve, m_d is used to determine the average water content at the end of the experiment.

   Weight loss upon oven drying = m_d – weight at the time of termination of the experiment

   $$\theta_d = \text{Weight loss upon oven drying} / \rho_w V \qquad ...(9.38)$$

   Here V is the sample volume minus the volume of the tensiometer holes, and ρ_w is the density of water taken as 1 g/cc.

The total soil water content at the beginning of the evaporation experiment is calculated as the sum of the amount of water at the end of the evaporation experiment and the decrease in water mass during the evaporation experiment. Dry bulk density is derived from the dry soil mass divided by the core volume, V.

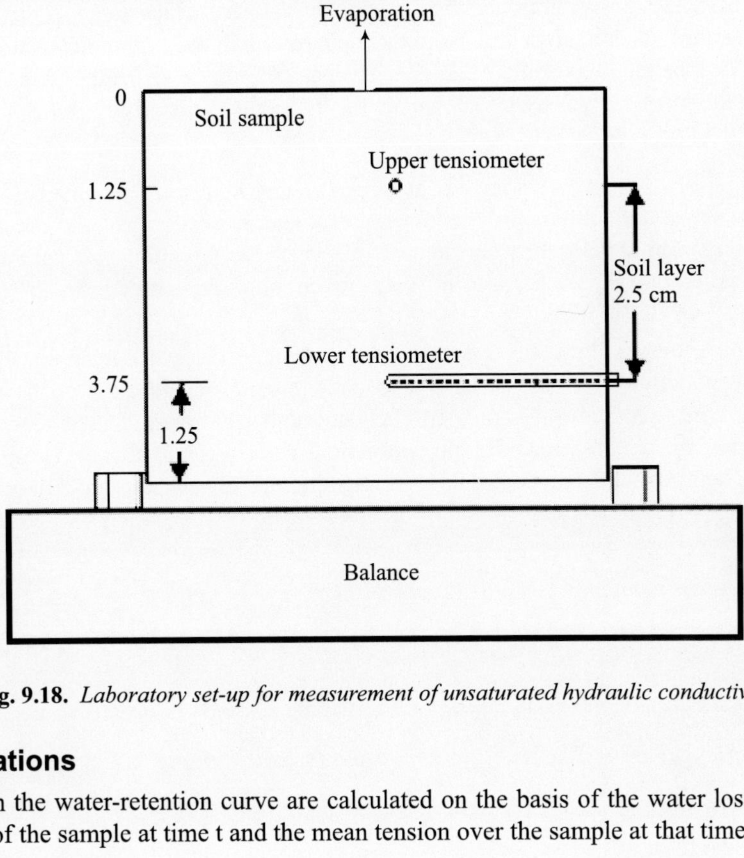

Fig. 9.18. *Laboratory set-up for measurement of unsaturated hydraulic conductivity*

Calculations

Points on the water-retention curve are calculated on the basis of the water loss per unit volume of the sample at time t and the mean tension over the sample at that time.

$$\theta_i = (m_i - m_d)/\rho_w V \qquad ...(9.39)$$

Here m_i is the mass of the soil core at i^{th} step. The following equation can also be used for this purpose

$$\theta_i = (m_i - m_{i-1})/(\rho_w V) + \theta_{i-1} \qquad ...(9.40)$$

Corresponding mean tensions Ψ_{mi} at the ith step is calculated as

$$\Psi_{mi} = 0.5(\Psi_{tui} + \Psi_{tli}) \qquad ...(9.41)$$

Here Ψ_{tui} and Ψ_{tli} are the upper and lower tensiometer values at i^{th} time step. Plot of θ_i versus Ψ_{mi} gives the soil water characteristic curve.

The mean hydraulic gradient (I_m for the ith time interval) is calculated on the basis of the tension values in that interval.

$$I_m = 0.5[(\Psi_{tui} - \Psi_{tli})/\Delta z + (\Psi_{tui-1} - \Psi_{tli-1})/\Delta z] \qquad ...(9.42)$$

Here Ψ_{tui}, and Ψ_{tli} are the upper and lower tensiometer values at time t_i, Ψ_{tui-1}, and Ψ_{tli-1} at t_{i-1} are the upper and lower tensiometer values at time t_{i-1} and $\Delta z = z_1 - z_2$, the vertical distance between the tensiometers.

The flux (J) is derived from the soil-water volume difference ΔV (1 cm^3 of water = 1 g), equal to the sample mass difference, Δm, in the interval per surface area (A) per time unit (Δt). The factor ½ is used since flux from the zone Δz is half the evaporated water volume per time. This factor is the ratio of $\Delta z/H$ such that H is the total height of the cylinder. Darcy-Buckingham's law is used to calculate the hydraulic conductivity (K).

$$K(\Psi_{mi}) = \Delta V/(2A\ \Delta t\ I_m) \qquad ...(9.43)$$

Here Ψ_{mi} (for K) is the mean tension averaged across the upper tensiometer at position z_1 (3.75 cm above the bottom) and the lower tensiometer at position z_2 (1.25 cm above the bottom), and average across the time interval Δt. A is the cross-sectional area of the sample. It may be noted that Ψ_{mi} to be used for water retention curve and Ψ_{mi} for the hydraulic conductivity will be different.

Computer programs have been developed to calculate the water retention curve and hydraulic conductivity for cases where more than 2 tensiometers are used. Normally 4 tensiometers are used and soil is divided in equal number of layers. The tensiometers are fixed in such a manner that either the compartments are of equal height or the compartments decrease in height towards the top of the core. Readers are advised to refer Halbertsma and Veerman (1994) for details description of the multi-layer method, being beyond the scope of this book.

Chapter 10

Drainable Pore Space

The drainable pore space is defined as the volume of water released or taken into storage in an unconfined aquifer per unit cross sectional area per unit decline or rise of the water table as a fraction of the total volume of soil in the absence of evaporation. It is also known as effective porosity or specific yield. Drainable pore space (f) is one of the main parameters used in designing subsurface drainage systems under unsteady-state flow conditions. In a tile drained field, water in the drainable pore space is released to the tile drain until the water table is lowered to the depth of the drain. Clearly, drainable pore volume is a dynamic quantity although this dynamics is difficult to evaluate. Since drainable porosity is a function of the soil moisture tension represented as $f(h)$, the quantity of water draining out of unit soil column during lowering of the water table from h_1 to h_2 is given by

$$Q = \int_{h_2}^{h_1} f(h)dh \qquad ...(10.1)$$

The function f(h) depends upon many soil properties. As a result, it is quite complex in nature and poses insurmountable problems while solving real-world problems. Nikam (1989) and Pandey et al. (1992) used a variable drainable pore space to design subsurface drainage systems. Although a step forward in scientific investigations, practical field application are difficult. Therefore, drainable pore space in drainage design is assumed to be a constant parameter. Even then, determination of this parameter, f has been both a difficult and time-consuming process. Researchers have developed and used several methods to determine this parameter. Three most commonly used methods are as follows.

- Correlation between drainable pore space and other soil properties
- Laboratory methods
- Field methods

CORRELATION METHODS

The basic principle of these methods is to correlate drainable pore space with one or more than one soil properties that are easier to evaluate. A definite correlation has been established between drainable pore space (specific yield) and hydraulic conductivity. It is believed that such a relationship is also practically justified because drain spacing is directly related to hydraulic conductivity while inversely to drainable pore space. A curve prepared by the USDA is commonly used to determine the drainable pore space with known hydraulic conductivity (K) (Fig. 10.1). Since the fluctuation of the water table in a drained area takes place in the zone between the drains and the maximum allowable water table height, the weighted average hydraulic conductivity of this zone should be used to evaluate drainable pore space.

Procedure

1. Evaluate the hydraulic conductivity by any method described in Chapter 9
2. Take this value on the x-axis of Fig. 10.1
3. Read the value of drainable pore space on the y-axis.

If this figure is unavailable, few mathematical relations are also available to calculate drainable pore space using the known values of hydraulic conductivity. Note that most of these relations are site specific. Some relations are given as under.

$f = K^2$ (f is in percent and K in cm/d, Smedma and Rycroft, 1983)

$f = 0.025 + 0.006$ K (f is in fraction and K in m/d, Chossat, 1987)

$f = 3.86\ K^{0.37}$ (f is in fraction and K in m/d, RAJAD, 1995)

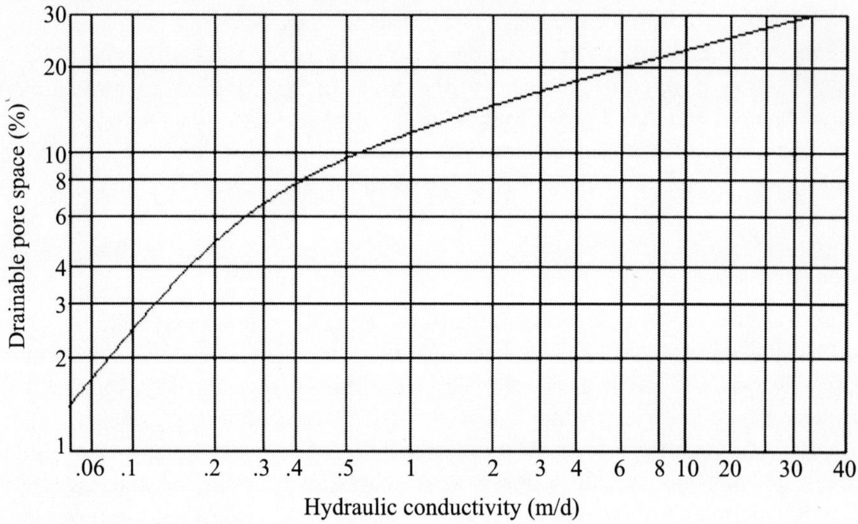

Fig. 10.1. *Relationship between hydraulic conductivity and drainable pore space*

Another approach utilizes the relation between drainable pore space and soil texture and soil structure, both of which have considerable influence on the drainable pore space. From practical applications, the representative values of the volumetric drainable porosity for different soil textures can be taken as 3-5 % for heavy clays, 6-9 % for medium textured soils and 10-15 % for sandy soils. Exact estimates can be made from the data reported in Table 10.1.

LABORATORY METHODS: SOIL CHARACTERISTIC CURVES

Amongst the independent measurement procedures, soil-water characteristic curves (SWCC) are used to assess drainable pore space. SWCC describes a relationship between the amount of water held by the soil (soil water content expressed on a weight or volume basis) and the tension by which it is held in the soil i.e. capillary pressure or matric tension expressed in bars or kPa or MPa. Since SWCC depends upon the hydraulic properties of the soil, its shape is impacted by size and connectivity of pore spaces. Besides SWCC is strongly affected by soil texture, soil structure and some soil constituents notably soil organic matter.

Table 10.1. *Influence of soil texture and structure on drainable pore space*

Texture	Structure	Drainable pore space (%)*
Clay Heavy clay loam	Massive, very fine or fine columnar	1-2
Clay Clay loam Silty clay Sandy clay loam	Very fine or fine prismatic, angular, blocky or platy	1-3
Clay Silty clay Sandy clay Silty clay loam Clay loam Silty loam Silt Sandy clay loam	Fine and medium prismatic, angular, blocky and platy	3-8
Light clay loam Silt Silt loam Very fine sandy loam Loam	Medium prismatic and sub-angular blocky	6-12
Fine sandy loam Sandy loam	Coarse sub-angular blocky and granular, fine crumb	12-18
Loamy sand Fine sand	Medium crumb and single grain	15-22
Medium sand	Single grain	22-26
Coarse sand Gravel	Single grain	26-35

* Based on the data from the Water and Power Resources Survey (Formerly United States Bureau of Reclamation).

Procedure

1. Use the available SWCC or prepare the same in the laboratory for the site specific soil (Chapter 6)
2. The volume of water in percent of the total volume between saturation and 100 cm suction or -9.806 kPa is taken as the representative drainable pore space
3. Some changes have been suggested in the upper value of the suction. It can be taken into account while assessing the parameter depending upon the type of the test soil.

Example 10.1. A soil (bulk density 1.40 g/cc) contains 30% moisture by dry weight at saturation. What is its drainable pore space, if the moisture content at 100 cm tension is 20%?

At saturation 1 cc of soil contains = $1.4 \times 0.30 = 0.42$ cc of water

At 100 cm tension soil contains = $1.4 \times 0.20 = 0.28$ cc of water.

Assuming the density of water to be 1.00, the drainable porosity at 100 cm tension is 0.42-0.28 = 0.14 cc of water per cc of soil. It means drainable pore space is 14%.

As the water table drops from the soil surface to 100 cm below the soil surface, 14% or 14cc of water will be drained per cm² of the area of soil surface.

HANGING WATER COLUMN METHOD

The hanging water column or Haines apparatus (Haines, 1930) is commonly used to determine water retention curves, being easiest requiring no external vacuum source and pressure regulators. The hanging water column is very similar to the tension table. The only difference is that the porous bed of sand or kaolin of the tension table is replaced with a Buchner funnel and a porous plate. Funnels with sintered plates of porosities such as porosity 3 for suctions up to 80 cm, porosity 4 for suctions up to 120 cm, and porosity 5 for suctions up to 150 cm are used. The sintered plates having fine pores that work at negative pressures up to 2 m are now available. In such cases, flexible transparent tube should be quite rigid so that it does not collapse under the negative pressure. An advantage of the hanging water column method is that the air in the soil sample is always at atmospheric pressure. Thus, the shape of the air–water interface is not likely to be impacted because of the air entrapment as might be the case with the pressure plate method.

Apparatus: Glass funnel with a porous plate, Burette, A flexible transparent tube connecting the glass funnel and the burette (Fig. 10.2), Moisture boxes and Oven

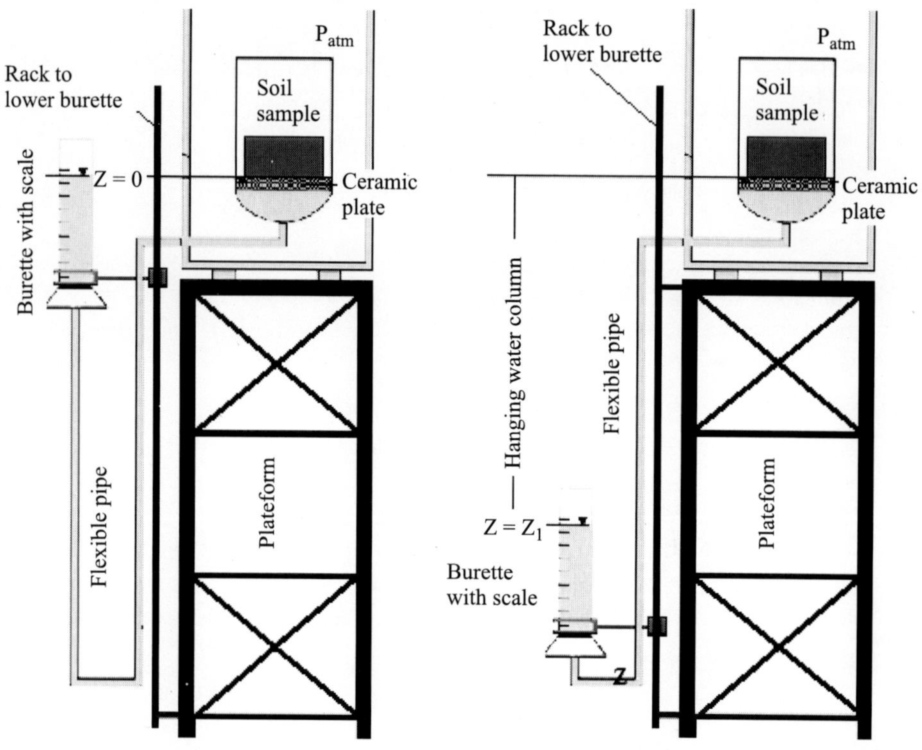

Fig. 10.2. *Hanging-water column apparatus: Initial saturation of the soil column (left) and the same set-up with lowered burette (right)*

Procedure

1. Submerge the assembly in de-aerated distilled water to remove the air from the porous plate, the space below the plate in the funnel, and the tube leading to the burette. Allow it to remain overnight and then pump the air out with an aspirator or vacuum pump attached to the plastic tube, so that the water enters through the plate
2. Place the undisturbed soil sample drawn from the test fields in metal rings of known volume over the sintered plate in the apparatus. The soil sample is saturated by maintaining the water level in the burette at the same level as the top of the porous plate (Fig. 10.2 left). The saturation may take 24 hr or more depending upon the soil type
3. Cover the Buchner funnel to prevent evaporation from the soil sample and the porous plate. Make sure that the soil air remains at atmospheric pressure
4. Upon saturation, the burette level is set at Z = 0. For thin samples, it is the top of the porous plate. For thick samples, it is the middle of the sample
5. Adjust the burette until outflow stops. It is the starting point of measurements
6. To impart suction to the saturated soil sample, the burette is lowered by a certain distance (usually in steps of 10 cm) resulting in water draining slowly (Fig. 10.2 right). The drained water enters the burette and raises the water level. At a given suction, if the rise in water level is more, the burette is readjusted such that the originally applied suction is closely maintained. As soon as the water level in the burette stops rising further, note down the elevation difference between the top of the porous plate and the water level in the burette. It gives the average value of suction applied to the soil sample
7. The difference between the initial and the final burette readings is the volume of water drained from the soil sample due to the applied suction
8. This process is repeated by lowering the burette in steps of 10 cm initially and more afterwards until the desired suction (corresponding to the maximum possible depth of the subsurface drain or any other criteria) is obtained. Since the drain depths are within 200 cm, the maximum suction may be around this value although the method is most suitable in the tension range of 0-150 cm
9. Based on the cumulative volume of water drained, drainable porosity is calculated by dividing the volume of water drained with the volume of the soil sample. For the data in Table 10.2, drainable pore space for 100 cm suction is given as: 100 (V_{10}/V)
10. Upon reaching the desired matric tension, remove the sample from the plate. Weigh and oven-dry the soil at 105°C for approximately 48 h to determine its gravimetric and volumetric water content.

Bulk density of the soil sample = Oven-dry weight of the soil divided by the volume of soil, O/V

Moisture content = Weight of water divided by oven day mass of the soil, W/O

Since the weight of water and its volume are equal, assuming its density as 1 g/cc, we prepare column 5 of Table 10.2. The moisture content is calculated by dividing the respective weight with the oven-dry mass of the soil (Column 6).

Table 10.2. *Analysis procedure of a hanging water column data (Sample volume, V cm³)*

Capillary pressure (cm)	Matric potential	Cumulative water drained (cc)	Water drained in each step (cc)	Volume of water in the sample	Gravimetric moisture content
1	2	3	4	5	6
H_0	0	V_0	0.0	$W_0 = W + \Delta V_1$	W_0/O
H_1	$\Psi_1 = H_0 - H_1$	V_1	$\Delta V_1 = V_1 - V_0$	$W_1 = W + \Delta V_2$	W_1/O
H_2	$\Psi_2 = H_0 - H_2$	V_2	$\Delta V_2 = V_2 - V_1$	$W_2 = W + \Delta V_3$	W_2/O
H_3	$\Psi_3 = H_0 - H_3$	V_3	$\Delta V_3 = V_3 - V_2$	$W_3 = W + \Delta V_4$	W_3/O
H_4	$\Psi_4 = H_0 - H_4$	V_4	$\Delta V_4 = V_4 - V_3$	$W_4 = W + \Delta V_5$	W_4/O
H_5	$\Psi_5 = H_0 - H_5$	V_5	$\Delta V_5 = V_5 - V_4$	$W_5 = W + \Delta V_6$	W_5/O
H_6	$\Psi_6 = H_0 - H_6$	V_6	$\Delta V_6 = V_6 - V_5$	$W_6 = W + \Delta V_7$	W_6/O
H_7	$\Psi_7 = H_0 - H_7$	V_7	$\Delta V_7 = V_7 - V_6$	$W_7 = W + \Delta V_8$	W_7/O
H_8	$\Psi_8 = H_0 - H_8$	V_8	$\Delta V_8 = V_8 - V_7$	$W_8 = W + \Delta V_9$	W_8/O
H_9	$\Psi_9 = H_0 - H_9$	V_9	$\Delta V_9 = V_9 - V_8$	$W_9 = W + \Delta V_{10}$	W_9/O
H_{10}	$\Psi_{10} = H_0 - H_{10}$	V_{10}	$\Delta V_{10} = V_{10} - V_9$	$W_{10} = W$	W_{10}/O

Soil Water Retention Curve

Gravimetric water content of soil (column 6) as a function of matric potential (ψ, column 2) is plotted to get the soil water retention curve. It is also possible to obtain the data for a wetting curve with step-wise increase in the level of Z following Step 8 of the procedure

FIELD METHODS

Water Balance Approach

Water balance parameters in the field are observed to measure drainable pore space using the following step-wise procedure.

1. Select a short period of time say for few days when there is continuous heavy rainfall
2. Measure the soil water content before the rainfall, if possible or make a good guess from experience
3. Measure the water table depth at the beginning of the monitoring period, say d_1
4. Measure total rainfall, runoff, evaporation and field soil moisture storage (At the end of the selected period). Use normal measurement procedures for each parameter
5. Measure the water table depth at the end of the monitoring period, say d_2
6. Conduct water balance and estimate the total rainfall percolated to the groundwater
7. Calculate the rise in the water table during this period
8. The drainable pore space, f is then given by the following relation

f = Change in groundwater storage/Change in groundwater level $(d_2 - d_1)$ ...(10.4)

Example 10.2. The drainable pore space of the soil is 5%. What will be the fall in groundwater table if 10 mm water is extracted from the groundwater?

Using eq. (10.4), f = (Change in groundwater storage)/(Change in groundwater level)

Change in groundwater level = Change in groundwater storage/(f/100)

The change in groundwater level = (10/5)100 = 200 mm or 20 cm.

Drain Flow Methods

Drain discharge and water table drawdown data of an operating drainage system or pilot drains (Fig. 9.11) are used to determine drainable pore space. Depending upon the data in hand, simple or theoretically elegant methods can be used (Gupta, 2019). In the simple method, the drain flow during a given duration is used to calculate drainable pore space as follows:

The length of the drain = L, m

Number of drains contributing to flow = n

Drain spacing = S, m

Initial depth to water table = h_1 (Observed), m (Flow measurement begins at this time)

Final depth to water table = h_2 (Observed), m (Flow measurement ends at this time)

The total area drained = $L \times n \times S$

The total volume of soil drained = $L \times n \times S \times (h_1 - h_2)$, m^3

The volume of water drained during the period (Observed) = Q, m^3

The drainable pore space = $Q/[L \times n \times S \times (h_1 - h_2)]$, in fraction ...(10.5)

The calculated value of drainable porosity using eq. (10.5) is the average of the soil layer given by $(h_1 - h_2)$ and over the time period for which the volume of drain flow is measured. The computed value of drainable porosity is approximate, as it has been assumed that the entire depth of the soil has experienced a uniform water table decline of $(h_1 - h_2)$ cm. It is not true in subsurface drained lands, since the water table is curved between the adjacent parallel drains. Provisional values of the drainable pore space can be improved later as and when more data are available.

Use of Non-steady State Equations

Glover and Dumm equation for the falling water table as reported by Dumm (1954) is reproduced as follows.

$$L^2 = [\pi^2 K d_e\, t/f] / [\text{Ln}\,(1.27 h_0/h_t]$$

Here L is the drain spacing, m, K is the hydraulic conductivity, m/day, d_e is the equivalent depth to impermeable layer, f is the drainable pore space, h_0 is the initial water table and h_t the water table at time t, days. The above equation can be written as:

$$f = [\pi^2 K d_e/L^2]\,[(t_0 - t_h)/\,[\text{Ln}\,1.27 + \text{Ln}(h_0 - h_t)] \quad ...(10.6)$$

If we neglect Ln 1.27, we have the following equation given by Dieleman (1974) (Buckland et al., 1986).

$$f = [\pi^2 K d_e/L^2]\,[(t_0 - t_h)/\,[\text{Ln}(h_0 - h_t)] \quad ...(10.7)$$

Thus, we can either use eq. (10.6) or simplified version eq. (10.7) to calculate the values of f. Plug in various values of the parameters and any two values of h_0 and h_t measured at the mid-point of the drains with corresponding values of time t_0 and t_h. Numbers of values are calculated using several pairs of h and t. is the average value f. The geometric mean of all the values.

Another approach to determine drainable pore space using the same data sets consists of writing the equation in a slightly different form. Without going into this detail, eq. (10.6) can be written as

$$h_{t+1} = 1.16\, h_t\, e^{(-\alpha\, \Delta t)} \tag{10.8}$$

$$\text{Ln}\,(h_{t+1} / 1.16\, h_t) = -\alpha\, \Delta t \tag{10.9}$$

Here $\alpha = \pi^2\, Kd/f\, L^2$, and $\Delta t = (t + 1) - t$, h_{t+1} is the head at t+1 days and h_t the head on a previous day. Note the slight difference in constant 1.16 instead of 1.27. It is because in deriving eq. (10.6), initial water table was assumed as horizontal. For eq. (10.8), it is assumed as fourth degree parabola. The idea to use different equations is to emphasize that any of the equation can be used to determine the values of f.

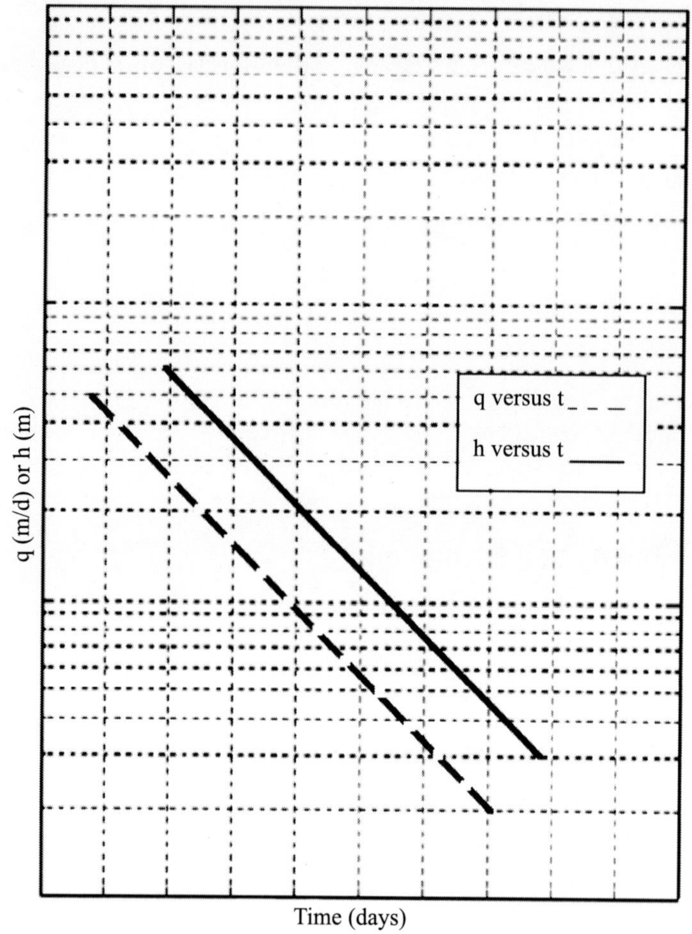

Fig. 10.3. *Plots of hydraulic head and discharge with time on semi-log paper*

Since *q* and h are linearly related so without making much error by retaining the constant 1.16, we can also write

$$q_{t+1} = 1.16\, q_t e^{(-\alpha\,\Delta t)} \quad ...(10.10)$$

$$\text{Ln}\,(q_{t+1}/1.16 q_t) = -\alpha \Delta t \quad ...(10.11)$$

Here q_{t+1} is the discharge on $t + 1$ days and q_t the discharge on a previous day.

With this background, following step-wise procedure is used (Kumar, 1994).

1. Observe q_t and h_t at mid-point of the drains on an almost daily basis following a heavy rainfall period (Fig. 9.11)
2. Calculate (h_{t+1} / 1.16 h_t), for each time step. Use these values in eq. (10.9) to calculate α to asses drainable pore space for each step. Neglecting few outliers, drainable pore space is the geometric mean of the values so calculated. A similar procedure is used with (q_{t+1} / 1.16 q_t) values with eq. (10.11)
3. A better way is to plot (h_{t+1} / 1.16 h_t) or (q_{t+1} / 1.16 q_t) on a log scale against time on a normal scale. Time is given by the days 1, 2 and 3 taken for each step. For example, for 3 days we calculate (h_4 /1.16 h_3) or (q_4/1.16 q_3) and plot against time = 3 days. If the system obeys the assumptions underlying the non-steady state situations, the observed values will more or less plot as a straight line
4. Calculate the slope of the resulting straight line (Fig. 10.3). It is the value of α. The negative sign is taken care of as the h or q decreases with increasing time
6. Calculate f from the equation: $\alpha = \pi^2\, Kd/f\, L^2$ since all other parameters are known from drainage design

Notes:

1. It is recommended that analysis should be done using both the steady and non-steady state equations (See Chapter 9 for steady state case). It will help to determine all the three parameters i.e. hydraulic conductivity, drainable pore space and depth to impermeable layer.
2. The drainable pore space, like hydraulic conductivity, varies widely in space and is known to be log-normally distributed. It is recommended that the geometric mean of the values should be used as the mean value of the drainable pore space.

Chapter 11

Aeration or Excess Water Stress

The chain reactions that follow flooding of the soil results in profound changes in physical, chemical and biological processes, which in turn alter many physico-chemical properties of the soil. Foremost amongst the physical processes is restricted atmospheric gas diffusion in the soil. It results in the accumulation of carbon dioxide with corresponding depletion in soil oxygen in the root zone. It follows a chain of soil chemical changes that include reduced soil pH, denitrification, reduction of iron, manganese and sulphate, and low soil redox potential (Eh). Since these changes have profound implications on crop production, oxidation and reduction measurements are made to understand the behaviour of flooded soils. The information obtained through these measurements provides important inputs to manage flooded or wet soils. For example, it should be possible to assess whether aerobic or anaerobic conditions prevail. This will reveal whether chemical compounds such as Fe oxides or nitrate are present in their oxidized forms or have been chemically reduced. This chapter first takes a look at air-filled porosity and then discusses the following three measurements relevant to aeration stress.

- Oxygen diffusion rate
- Redox potential
- Actual depth to water table

AIR-FILLED POROSITY

Air-filled porosity (F_a) is the fraction or percentage of the total bulk volume that contains air at the given moisture content, i.e. cm^3 of air per cm^3 of soil. For non-swelling soils it is found by the difference between the total porosity and the volumetric moisture content as:

$$F_a = F_t - F_{\theta.}$$

Here F_a is the air-filled porosity at a volumetric moisture content θ such that F_θ is the volumetric moisture content and F_t is the total porosity. The macro-porosity is sometimes used to denote the air-filled porosity at field capacity although terms such as air capacity and air-space porosity have also been used for the latter; air capacity is favoured. Most plants cease to grow when air porosity is <10%; air-filled porosity of 10% being suitable for most plants (Hazelton and Murphy, 2016). Air-filled porosity is rated as per guidelines given by Wesseling (1974) as follows.

Air filled porosity (%)	Remark
1	Very low
1-5	Low
5-20	Moderate
20-30	High
>30	Very high

OXYGEN DIFFUSION RATE

An adequate exchange of gases between the root and aerial atmosphere, often called soil aeration, is essential to achieve optimal plant growth. Plant roots, for optimum plant metabolism, require sufficient afflux of oxygen that balances the oxygen supply rate to the oxygen consumption rate. Letey and Stolzy (1967) reported that plant growth was not impaired, if oxygen diffusion rate (ODR) values were above 0.4 μg/cm^2-min. Roots of most plant species (corn, barley, grasses, sunflower, cotton, sugar beet, tomato) were inhibited or failed to grow when ODR values were below about 0.2 μg/cm^2-min. These values are not sacrosanct, as critical values of ODR for root growth vary with plant species, stage of growth, fertilizer, irrigation, tillage management, and type of soil. For example, Letey and Stolzy (1967) reported that plant's requirement of soil aeration is much greater at the emergence stage, being at least 0.5–0.7 μg/cm^2-min ODR for good emergence.

The transport of oxygen is much faster in air than in water. Thus, in a given soil type, the diffusion rate will vary strongly on the air to water ratio in the soil pores. This mobility has great significance to the availability of oxygen to the plant roots. An ODR meter is used to measure the mobility of oxygen in the soil. In principal, ODR meter measures the electric current needed to reduce whole oxygen available at the surface of a cylindrical Pt-electrode in the soil (Lemon and Erickson, 1952). Thus, the oxygen flux through the air-filled pores and the water film on the electrode to reach a zero concentration at the electrode surface can be measured.

Equipment

Commercially available instruments consist of a carrying case containing an indicating instrument (read-out unit), a plastic bottle with an Ag-AgCl reference electrode with a KCl salt bridge and a brass or calomel or copper counter electrode (Anode). The Pt electrode is 10 mm long, 1.2 mm diameter Pt wire, hardened with 10% addition of Ir. The wire is mounted at the tapered end of a 750 mm long rod that helps to place the electrode at desired depths in the soil. The reference electrode is used to measure and check the potential between the Pt-electrode (Cathode) and the soil. A brass electrode is used to close the electrical circuit. Cable and appropriate connectors and an auger are supplied along with the instrument. The instrument can also be used as a redox meter. The measuring range for oxygen diffusion is 0 - 999 μA and for redox potential 0 - 999 mV (resolution 1 μA and 1 mV). It can operate in the temperature range of 0 and 50°C and air humidity between 30-80%. Schematic diagram of an ODR meter is Fig. 11.1.

Procedure

1. Take out the electrode assembly and connect the signal cable to the connector
2. Auger a hole shorter by few mm than the desired depth to accommodate the wire. Insert the electrode-rod into the hole. Exert slight pressure on top of the assembly to bring the electrode to its correct position. It is quite important to position the electrode in undisturbed soil
3. Check that the switch on the read-out unit is in "OFF" position. Connect the signal cables with Pt-electrodes to the input connectors on the read-out unit

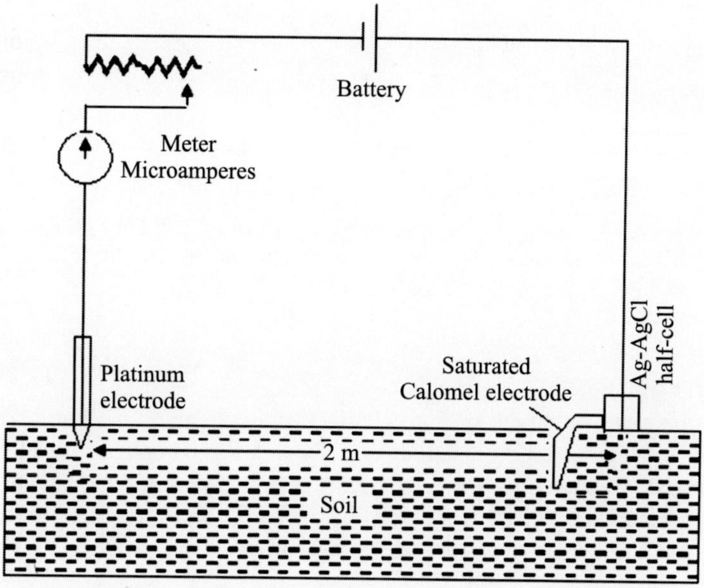

Fig. 11.1. *Schematic of an ODR meter with calomel anode*

4. Select an appropriate place for the positioning of reference and the brass electrodes. A distance of about 2 m between these electrodes and the Pt-electrode is usually satisfactory in the field, although it is not critical. It is now known that the distance between them is immaterial as long as the three electrodes i.e. cathode, reference and anode are in the same soil water system. Besides, the salt bridge and the brass electrode should be buried in the top soil layer. The electrodes in some horticulture crops have been positioned within few cm without experiencing any problem. Connect the signal cables from the reference electrode and the brass electrode to the corresponding inputs on the read-out unit
5. Allow 2 - 4 min for the system to stabilize
6. Take the observation. Repeat the process at 10-12 places around the measurement site.

Calculations

Soil ODR is given by the following relation (Feng et al., 2002):

$$ODR = 60\ M\ I/\ n\ F\ A \qquad ...(11.1)$$

Here I is the current readout, μA, A the surface area of the electrode, m^2, M is the molecular weight of the oxygen, 32 g/mole, n is the number of electrons used per molecule of oxygen, usually taken as 4, F is the Faraday, 96500 Coulombs. Constant 60 is used to convert second to minutes. ODR is given in $\mu g/cm^2$-min.

$$ODR = 0.0049\ I/A \qquad ...(11.2)$$

For a Pt wire 4 mm long and 0.65 mm diameter, the above equation reduces to

$$ODR = 0.059\ I$$

A readout of 8 μA means

ODR = 8 × 0.059 = 0.472 μg/cm^2-min

Precautions

- The electrode should not be twisted hard lest it breaks or gets damaged. Besides, the ODR-sensor must not compress the soil in any manner
- The platinum electrode should be completely covered by a water film. In practice, it means that readings should be taken only if the moisture content of the soil is at or more than the field capacity. Redox measurement is a better alternative to characterize the dry soils
- Average values of several measurements up to ten or more will only give a reasonable estimate of the soil environment
- The soil is a very complex system. The ODR measurements are highly sensitive to the voltage between the Pt-electrode and the soil. For comparable results, the measurement conditions should be well defined and closely guarded. The ODR meter has been designed to provide a stabilized voltage of 0.65 V between the Pt-electrode and the Ag-AgCI reference electrode. It is equivalent to 0.428 V between the Pt-electrode and the soil (at 25 °C)
- In spite of being easy to use and inexpensive, the method has few limitations. Currently used meters need some improvements to overcome these limitations (McIntyre, 1970).

SOIL REDOX POTENTIAL

Soil redox potential (Eh) or oxidation-reduction potential (*ORP*) is a semi-quantitative indirect measurement of *in-situ* soil oxygen status. It is regarded as a simpler more reliable indicator of the *in-situ* status of soil oxygen (Light, 1972, Fiedler et al., 2007). The measurements in reality are recordings of voltage (Eh) over time between a reference electrode and a sensor electrode inserted into the soil. It is usually measured by the difference between the potential of the hydrogen electrode (or Calomel or Ag/AgCl electrode) and the potential of a platinum electrode immersed in the medium (Fig. 11.2). Aerobic (oxidized) soils, for good root development and to efficiently perform other functions, have Eh values exceeding +300 mV. Anaerobic (reduced) soils on the other hand, have values between +200 mV and –400 mV. Severity of the problem (water logging) increases with decreasing values. Phytotoxic concentrations of several elements such as Fe, Mn and Al begin to appear in the root zone with negative Eh values. This creates additional stresses in the soil besides the poor aeration adversely impacting crop productivity.

Construction

A redox potential meter comprises a non-reacting contact with the medium, normally the stable metal platinum (Pt) or any other noble metal electrode. A stable potential reference source, most reference probes now-a-days have an Ag/AgCl combination liquid, the toxic calomel type becoming obsolete. Since the potential of the standard reference electrode is temperature dependent, temperature is also measured along with Eh. A stable measurement device, a milli-volt meter with an input resistance not less than 10 GΩ, is used to measure the voltage. Laboratory mV/pH meters can easily be used for this purpose. A data logger is also used to take and store simultaneously measured temperature, pH and redox potential (Fig. 11.2). Although, this simple set-up is sufficient in normal circumstances, a salt bridge is used for long-term monitoring. It helps to avoid several problems that may be encountered with electrodes remaining in place for long durations (Fig. 11.2, Vepraskas, 2002).

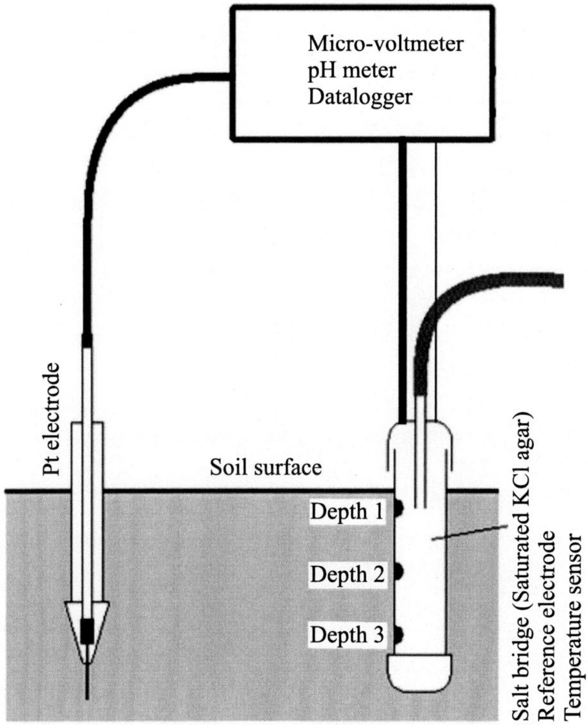

Fig. 11.2. *A schematic diagram of a redox potential meter with a salt bridge and temperature sensor*

Installation and Measurements

To measure the redox potential, the inert electrode is introduced directly into the system to the desired depth. Alternatively a bore hole is augured to the desired depth and filled with mud slurry to install the electrode. The mud slurry provides a close grip to the electrode. The reference electrode is inserted at a distance of 10-100 cm away from the inert electrode. Voltage measured through the instrument indicates the comparative chemical potential at the redox electrode and the chemical potential of the reference electrode. The reading may be initially high but it does decrease over time. Therefore, it is desirable to wait for the reading to stabilize before it is recorded. Also note down whether the voltage is plus or minus. Convert the voltage to Eh or redox potential. It is accomplished by adding to the voltage reading a correction factor appropriate to the reference electrode. The potential of the common Ag/AgCl 3M KCl reference electrode at 20°C equals +213 mV versus the standard hydrogen electrode, which has zero- chemical potential. For most situations, a correction factor of 200 mV can be used for Ag/AgCl electrodes, and 250 mV for calomel electrode. As an example, if the voltage measured from a redox electrode against the Ag/AgCl 3M KCl reference is +60 mV, the redox potential equals +273 mV or approximately +260 mV versus the standard hydrogen electrode.

Precautions

- Electrodes installed by direct insertion can be used to measure Eh as soon as the electrode tip reaches the desired depth. If electrodes are installed by using a rigid rod or in mud slurry, then the observations should be made only after about 24 hr. This much time is required for the solution around the Pt wire to equilibrate with the soil solution
- The temperature and the pH of the medium should be measured to calculate the standardized redox potential
- The electric flow rate between the reference electrode and the soil is greatly reduced at low soil moisture contents. This causes continuous drifting of the potentiometer or requires quite large stabilization times (Veneman and Pickering, 1983). A salt bridge can help to overcome this problem
- Although redox electrodes are strong, but still require careful handling. Too much force on the probes may bend and possibly break the probes

Interpretation Guidelines

Tentative guidelines for transformation of redox couples, organisms and metabolism are given as under. The values may differ from soil to soil and may be affected by several other environmental factors.

Redox potential (mV)	Reaction	Type of organism and metabolism
+300 - +700	$O_2 \rightarrow H_2O$	Aerobes, aerobic respiration
+300 - +200	$NO_3^- \rightarrow N_2$, N_2O $Mn^{4+} \rightarrow Mn^{2+}$	Facultative anaerobes, nitrate reduction
-100 - -200	$SO_4^{2-} \rightarrow S^{2-}$	Anaerobes, sulphate reduction
-200 - -300	$CO_2 \rightarrow CH_4$	Anaerobes, methanogenesis

MEASUREMENT OF WATER TABLE AND HYDRAULIC HEAD

Groundwater investigations are performed to obtain information on status of water logging in an area, which indirectly indicates the aeration problem, if any. For this purpose observation wells and piezometers are used. Normally, a network of observation wells on a grid is established to measure the water table. To investigate artesian pressure or hydraulic head, a cluster of piezometers at appropriate locations is installed (Gupta, 2019).

Observation Wells

Uncased and cased observation wells are used to measure the depth to water table. These wells are installed at strategic locations usually on a grid. An uncased well, used to take short-term measurements, is an auger hole left uncased in the field. A cased well, usually made of plastic or galvanized iron pipe, is a shallow miniature well. The pipes are perforated throughout the length except at the top 0.5 m (Fig. 11.3). The perforations drilled in five to six rows are covered with jute or coir or nylon rope. Now synthetic filters are available in the market for this purpose. This arrangement is necessary to prevent clogging of the tiny holes and minimize the entry of fine particles into the pipe. The pipe is placed in the center of a bore hole made with a hand or power auger. The annular space between the pipe and wall of the hole over the entire length of the perforated portion is filled with gravel or coarse sand. The gravel pack allows unrestricted flow of water from the soil into the pipe and vice versa. The rest of the annular space is filled with fine materials (bentonite, clay, clayey fine sand) whose permeability is low enough to prevent leakage along the pipe. To keep the pipe in position and to prevent any direct contact between the surface water and the backfill material, a concrete seal is provided around the pipe, which extends from above the ground surface to a depth of about 10 cm, a total of about 15 cm. The lower end is sealed while the upper end is provided with a cap, which can be opened while recording the observations. A rough guide to decide the number of wells/observation wells is given in Table 11.1. The number of observation wells decreases with increasing size of the project area.

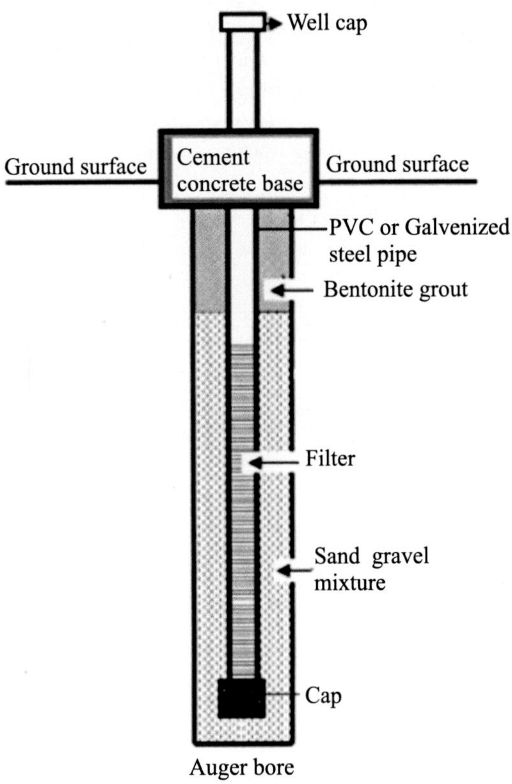

Fig. 11.3. *A schematic view of an observation well (not to scale)*

Table 11.1. *Number of observation wells as influenced by the size of the project area*

Project area (ha)	Number of observation wells	Number per 100 ha
100	20	20
1000	40	4
10000	100	1
100000	300	0.3

Piezometer

Groundwater always moves from a point of high hydrostatic pressure to one of low pressure. Therefore, to chart the flow of groundwater, hydrostatic pressure investigations are carried out simultaneously with water levels. A piezometer is used to determine the hydrostatic pressure at a point in the soil profile mainly to detect artesian pressure and to assess differences in pressure between various strata. Normally, a battery of piezometers is installed at one location with the lower end of each piezometer terminating at different depths to detect the vertical movement of groundwater. Piezometers spaced at horizontal intervals are used to detect horizontal seepage or movement. A piezometer is an open-ended non-perforated galvanized iron (or sometimes plastic) pipe driven or jetted into the ground to the depth at which the piezometric head is proposed to be measured. Water enters only through the open bottom or through the perforations made only at the bottom end for about 5-10 cm length (Fig. 11.4). The water level in the pipe corresponds with the hydraulic head at the lower end where the pipe terminates or at the center of the perforated portion of the pipe.

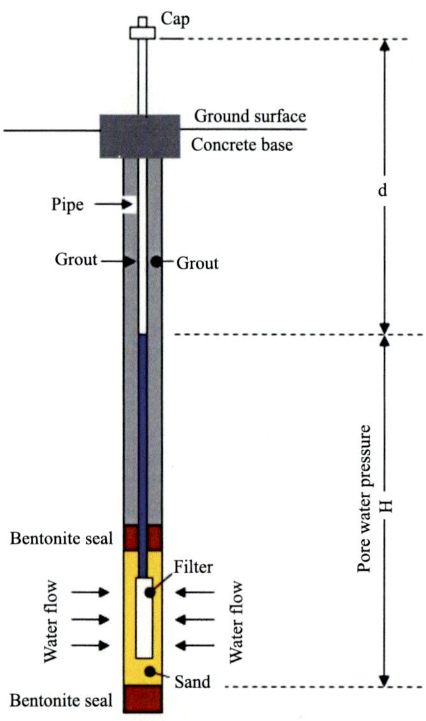

Fig. 11.4. *Schematic cross-section through a piezometer (not to scale)*

Measurement of Water Level

Water levels in observation wells or piezometers are measured using steel tapes, electric sounders, bubblers, pressure transducers, float-operated water level recorders, and acoustic equipment. Manual measurements are taken using a steel tape marked with a water-soluble marker. For this purpose, simple equipment such as a tape, a marker, and a piece of rough cloth to wipe the tape dry after each reading will suffice. Height of pipe above the ground surface is noted every time the instrument is read because pipes may move up (human interference) or down in a season. Another simple instrument is the chain and bell- sounding device. The device is a 1-1.5 cm galvanized iron pipe of about 15-20 cm length. The bottom end is sealed with a rubber cork leaving about 1 mm gap (Fig. 11.5). Some arrangement at the top is made to attach it with a rope or a thin steel or metallic tape. When lowered into the well, it makes a clearly audible "plop" sound as it contacts the water surface. This device can be easily used to draw water samples from these wells. Besides this advantage, the device is inexpensive and can be prepared locally.

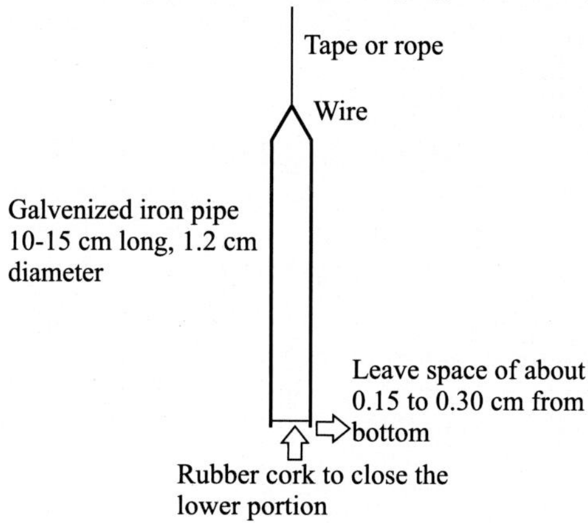

Fig. 11.5. *Chain and bell sounding device to measure groundwater levels*

Amongst the electric instruments, one commonly used device (pair of wires, battery, open electric junction, and light or meter) is an open electric circuit that is completed when the junction makes contact with water. A bell sounder or any other electro-optic device is used to identify the completion of the circuit. Water levels are read with a steel measuring tape, since the flexible wire will not give accurate measurements. Automatic recording devices record water levels with down-well transducers or capacitance-based sensors. The cost of these instruments is much higher than manually read instruments. These may be used only if high degree of accuracy is required or studies require continuous records.

Analysis of Data: Observation Wells

The data of an observation well is used to plot a well hydrograph. It is prepared by plotting depth to water table against time on a graph paper. A hydrograph provides insight into the causes of the rise or fall in the water level at a given site. In irrigated areas, the water table tends to build-up through the irrigation season, reaching a peak at the end of the season. In monsoon climatic regions, the water level is at its maximum at the end of the season (generally October, in India) and minimum before the onset of monsoon (May or June).

The hydrograph can also be used to assess the rise in the water table during the monsoon season.

Water table contour map shows the phreatic surface (water table). It is prepared by plotting absolute water levels or elevations of the water table for all observation wells on a topographical map. The map shows the relationship between the ground surface and the water table. A water table contour map provides direct visual observation of the slope of the water table, horizontal groundwater movement being in the direction of the slope of the water table. Radical changes in direction or slope may indicate unusual barrier formation, perched water table, artesian conditions, or areas with high and low permeability. Such information provides clues for further investigations in the project area.

Depth to water table at any point is the difference in groundwater contour and the surface contour. An isobath map is prepared to know the extent and degree of water logging. It shows lines of equal depth to water table. This map is generally used to justify a project. It can also be used to assess the yield losses due to the shallow water table, if the crop yield-depth to water table relation is known.

Analysis of Data: Piezometers

Observations from piezometers are used to decipher artesian conditions, if any. Artesian pressure is indicated, if water level elevation in the piezometer is above the true water table. The hydraulic gradient is the change in hydraulic head per unit distance in the direction of the maximum rate of decrease in head. The vertical component of the hydraulic gradient is obtained from the battery of piezometers by subtracting the difference in equilibrium elevations of the water surface in two pipes divided by the difference in elevations of the depths at which respective piezometers terminate. A battery of three piezometers installed in two aquifer configurations shown in Fig. 11.6 reveal some natural drainage (Fig. 11.6a), presence of hydrostatic pressure (Fig. 11.6b), hydrostatic pressure in one of the stratum with water being forced up and down from the stratum (Fig. 11.6c) and natural drainage with water moving out of the stratum (Fig. 11.6d). Similar situations have to be analyzed and interpreted in reference to the stated objectives. From the drainage point of view, artesian conditions influence the layout of a subsurface drainage system as well as the kind of drainage system to be adopted. A normal subsurface drainage may be unable to control water logging or salinization in areas with upward seepage. Thus, extra measures are necessary to strengthen drainage in these areas. On the other hand, natural drainage reduces the required intensity and even makes subsurface drainage unnecessary.

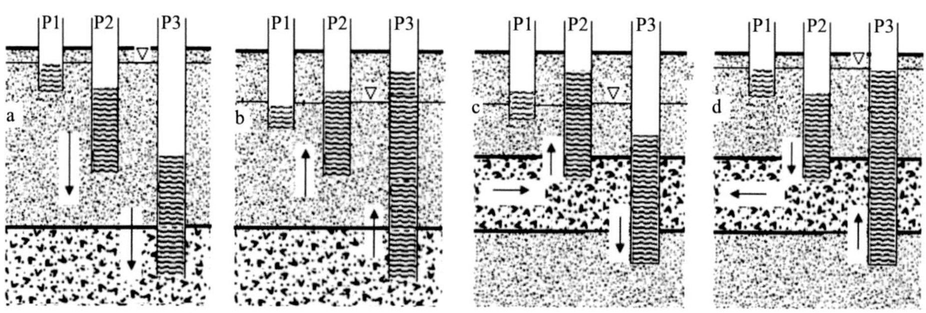

Fig. 11.6. *A battery of 3 piezometers (P1, P2 and P3) showing the presence of artesian pressure and natural drainage in two aquifer configurations*

Chapter **12**

Engineering Properties

Soil is an important engineering material used by the mankind since time immemorial. The characteristics of the soil that describes its usefulness as a construction material besides many physical properties discussed in the previous section are: compressibility, consistency, modulus of rupture, cohesion, angle of repose, capillarity, permeability, elasticity etc. This chapter mainly deals with 3 of these properties namely compressibility, consistency and modulus of rupture. Besides, compaction and land slope, which are commonly required in many engineering applications, are also discussed.

COMPRESSIBILITY: PROCTOR COMPACTION TEST

Most engineering properties of the soil, such as strength, stiffness, resistance to shrinkage, and imperviousness, all improve with increasing soil density. A common and cost effective method of increasing soil density for its stabilization is mechanical compaction. In most design specifications, the required density is usually mentioned as a percentage of the maximum density measured in a standard laboratory test. Besides, moisture content at which such densities can be obtained is also mentioned. This section deals with this test. The objective is to obtain a graphical relationship of the dry density of the soil to its moisture content. The graph is used to determine the value of the optimum moisture content (OMC) to achieve maximum dry density (MDD) of a given soil type. A light compaction test as per BIS standard is described as follows (BIS, 1980b).

Apparatus: Proctor mould (Fig. 12.1 left), mould volume 944 cm^3 with 10.2 cm diameter and height 11.6 cm and/or mould volume 2250 cm^3 with 150 mm diameter, Rammer weight 2.5 kg (Fig. 12.1 right), Balances of capacity 10 kg with a least count of 1 g and other of 200 g capacity with a least count of 0.01 g, IS Sieve 19 mm, Oven thermostatically controlled to maintain a temperature between 105^0 to 110^0C, Steel straight edge for trimming, Mixing tools such as a tray or pan, Spoon, Trowel or spatula or suitable mechanical device, Airtight containers to determine moisture content

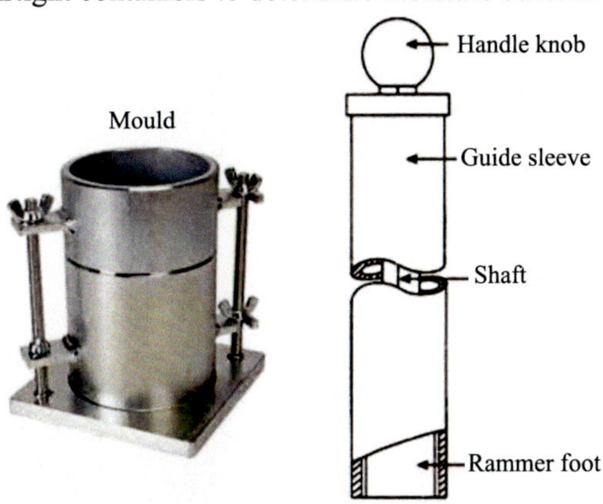

Fig. 12.1. *Mould and rammer used in proctor test*

Procedure

1. Take a representative portion of air-dry soil to yield about 5 kg of material passing through 19 mm IS sieve (for soils not susceptible to crushing during compaction). For soils susceptible to crushing during compaction, it should be about 15 kg passing the 19 mm IS sieve. Reject the coarse fraction after its proportion to the total sample is recorded
2. Add water to the soil and mix it thoroughly. For sandy and gravelly soil add 3% to 5% water. For cohesive soil, the amount of water may be 12% to 16% below the approximate plastic limit
3. Weigh the mould with base plate attached, to the nearest 1 g and record the weight, W_1. Attach the extension collar with the mould
4. Compact the moist soil into the mould in 3 layers. For this purpose divide the required soil in approximately 3 equal parts. Each layer should be given 25 blows with a 2.5 kg rammer dropped from a height of 30 cm above the soil. The blows should uniformly cover the whole surface of each layer
5. After compaction, remove the extension collar. Carefully level the top of the mould by means of straight steel edge. Weigh the mould with the compacted soil to the nearest 1 g and record this weight, W_2
6. Remove the compacted soil from the mould and place it in a mixing tray. Determine the water content of a representative sample and record the moisture content in percent as 'M'
7. The remainder of the soil shall be broken up and used to repeat steps (4) to (6) by adding suitable increments of water to the soil. For sandy and gravelly soils the increment is generally 1% to 2% while for cohesive soils the increment is 2% to 4% (Prakash, 2015)
8. At least five determinations shall be made. The range of moisture content should be selected in such a manner that the optimum moisture content at which the maximum dry density occurs is within that range.

Calculation

Bulk density, ρ in g/cm³ of each compacted specimen is calculated from the following equation.

$$\rho = (W_2 - W_1)/V_m$$

Here, W_1 is the weight in g of mould + base plate, W_2 is the weight in g of the mould + base plate + soil, V_m is the volume of mould. Dry density, ρ_d in g/cm³ is

$$\rho_d = \rho/(1 + M/100)$$

Graphical Presentation

The dry densities, ρ_d, obtained in a series of tests are plotted against the corresponding moisture content (Fig. 12.2). A smooth curve is then drawn through the plotted points. The position at which maximum dry density (MDD) occurs is marked. The corresponding moisture content is the optimum moisture content (OMC). Report the dry density corresponding to the maximum point to the nearest 0.01 g/cm³. Report the optimum moisture content corresponding to the maximum dry density to the nearest 0.2% for values below 5%, to the nearest 0.5% for values in the range of 5 to 10% and to the nearest whole number for values exceeding 10%. In engineering projects, 95-100% of MDD of standard Proctor test needs to be achieved.

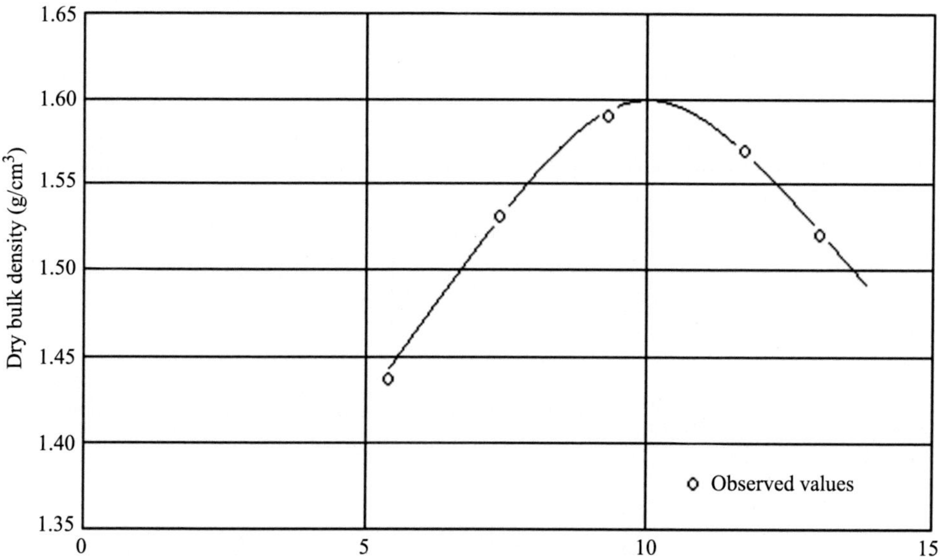

Fig. 12.2. *A plot of dry bulk density and moisture content*

Notes:

- Soil and water should be thoroughly mixed, since inadequate mixing gives rise to variations in test results especially for cohesive soils. Store the mixed sample in a sealed container for a minimum period of about 16 hr before the test is performed
- It is necessary to control the total volume of compacted soil. The test results are vitiated, if the amount of soil struck off is more following the removal of extension from the mould
- For higher accuracies, the increments of water additions in the region of the optimum moisture content shall be small
- The tube of the rammer should always be clear of any obstruction including soil so that the rammer falls freely on the soil
- For compacting soil containing coarse material up to 37.5 mm size, 2250 cm³ mould should be used. In this case, sample weighing about 30 kg passing the 37.5 mm IS sieve is used. Soil is compacted in five layers, each layer is given 55 blows using the 4.9 kg rammer
- For compacting soils containing coarse material up to 40 mm size, 2250 cm³ mould of 15.0 cm diameter should be used. A sample weighing about 6 kg passing the 40 mm IS sieve is used for the test. Soil is compacted in three layers, each layer being given 55 blows of the 2·5 kg rammer (BIS, 1980b)

- For soils susceptible to crushing during compaction, fresh samples are used for each determination. Soils not susceptible to crushing during compaction, the same sample can be used for all the determinations.

Modified Proctor's Test: Heavy Compaction

Apparatus: Cylindrical metal mould shall be either 100 mm diameter of 1000 cm^3 volume or 150 mm diameter and 2250 cm^3 volume, Rammer weight 4.9 kg, other equipment same as in the light compaction test

Procedure

The procedure essentially consists of steps similar to the light compaction test. Some of the steps are briefly described as under (BIS, 1983).

1. Step 1 to 3 same as before in light compaction test
2. Mould with its extension attached is used to compact the moist soil in 5 layers of approximately equal thickness. Each layer is compacted giving 25 blows with 4.9 kg rammer dropped from a height of 45 cm. Distribute the blows uniformly on each layer
3. Remove the extension and carefully level the compacted soil. Use a straight edge to strike off any soil above the top of the mould. Weigh the soil-filled mould to the nearest gram
4. Take other steps as in light compaction test
5. For compacting soil containing coarse material up to 37.50 mm size, take a representative sample of material passing through 37.50 mm IS sieve. Use 2250 cm^3 volume mould to compact the soil in five layers. Each layer is given 55 blows with a 4.90 kg rammer dropped from a height of 45 cm above the soil

Calculations and Graph

The same procedure is adopted as in the light compaction test. In engineering projects, 90-95% of MDD of heavy compaction proctors test needs to be achieved.

Standard versus Modified (heavy) Proctor Test

The test procedure in both the tests is almost similar except for some minor variations. The main difference is that a heavier rammer weight and longer drop distance are used in the modified Proctor test (Fig. 12.3). More compaction energy, as per the following equation, is used in the heavy compaction test.

$$E = n\, h\, P/V \text{ kg-m/m}^3$$

Here n is the number of blows, h the height of drop, P is the weight of the rammer and V is the volume of the mould. Note that 1 J = 0.1 kg-m. Apparently, more compaction energy creates higher bulk density or unit weight. Higher unit weight means less void space. Moreover, it also means that the maximum unit weight can be achieved with less water. For example, a soil for the same density in a standard Proctor test having an OMC of 18.75% may have an OMC of 14.75% in the modified Proctor test. The question commonly asked is whether one should go for a standard Proctor or a modified Proctor test? It has emerged that soils susceptible to high dynamic or very high static loads should be tested with the modified Proctor test. On the other hand soils not susceptible to high loading, a standard Proctor test should suffice. Alternative tests as per ASTM standard are included in Annexure III.

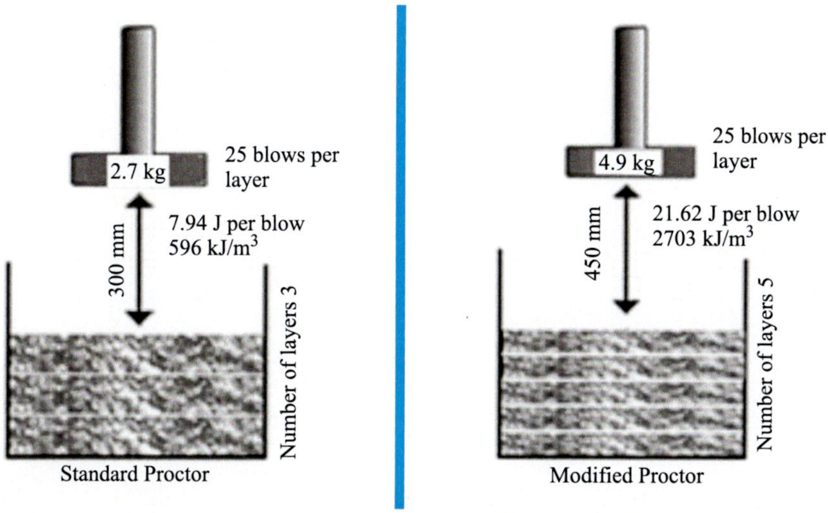

Fig. 12.3. *A comparison of standard and modified Proctor tests*

SOIL CONSISTENCY

Soil consistency of the soil is the strength with which soil materials are held together. It is also the resistance of the soil to deformation or rupture. Since the soil consistency varies with water content, it is measured for wet, moist and dry soils. The consistency of a soil ranges from solid (dry) to semi-solid to plastic to liquid depending upon its water content. While some limits of soil consistency are briefly described in this section, other three are included in the next section on Atterberg limits.

Flocculation Limit (FL): It is the moisture content at which soil suspension is transformed from liquid state to a semi-liquid state. Viscosity increase is quite appreciable.

Sticky Limit: Minimum moisture content at which soil paste adheres to a steel spatula drawn over its surface. Maximum adhesion occurs at this moisture content and a normal soil will scour during tillage.

Friable (soft) Consistency: The soil crumbles easily at this consistency. It is why; this consistency represents the optimum conditions for tillage and preparation of seed-bed. Moisture content at this consistency is slightly less than lower plastic limit discussed in the next section.

Harsh Consistency: Consistency is said to be harsh, if soil becomes hard due to clay cementation upon dehydration. More power is required to plough the soil at this water content. Moreover, the soil becomes cloddy upon ploughing at this moisture content.

ATTERBERG LIMITS

The water contents at which the consistency changes from one state to the next are called consistency limits or Atterberg limits (Fig. 12.4). Such changes in soil consistency can be accurately measured in the laboratory using standard procedures. These tests are mainly performed on clayey or silty soils, which have the tendency to swell and shrink with change in moisture content. The soil consistency limits help to distinguish between silts and clays.

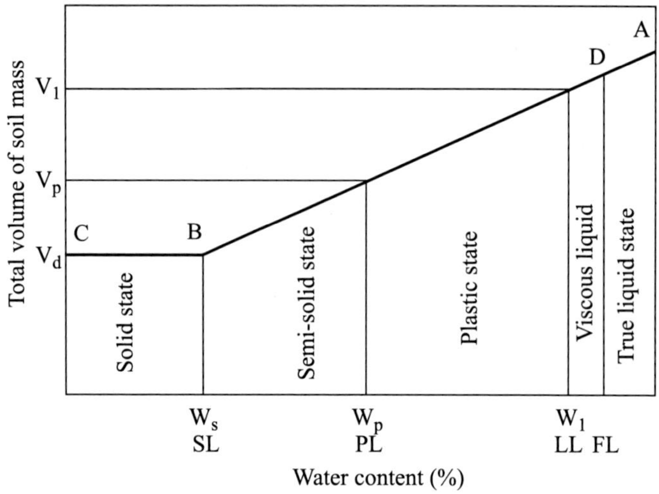

Fig. 12.4. *Consistency limits of soils*

Liquid Limit (LL): Also known as the upper plastic limit, it is defined as the moisture content at which soil begins to behave as a liquid material and begins to flow. Soil changes from the liquid to the plastic consistency with slight decrease in moisture content and from the plastic to the liquid consistency with slight increase in moisture content. Soil is near saturation, and behaves like softened butter. The tension of the water at the upper plastic limit is equivalent to pF of about 0.5. For a fine grained soil, it is the moisture content at which shear strength of the soil is about 2.5 kN/m^2.

Plastic Limit (PL): Also known as lower plastic limit, it is defined as the moisture content at which soil begins to behave as a plastic material. At the water content of the plastic limit, soil begins to break apart and crumbles when rolled into threads 3 mm in diameter. Between upper and lower plastic limits, it is possible to mould the soil into various shapes without breaking. The tension of the water at the lower plastic limit is equivalent to pF values of 2.8 to 3.3. The plasticity index is the range of moisture in which soil passes from a semi-solid state to liquid state still remaining in a plastic state.

Shrinkage Limit (SL): Moisture content at which soil changes from semi-rigid to a rigid solid. It is the maximum water content below which the volume of the soil remains constant irrespective of the change in the water content. In practice, it is the ratio of a given volume change expressed as a percentage of the dry volume to the corresponding change in water content above the shrinkage limit. Volumetric shrinkage is the decrease in volume of a soil mass when the water content is reduced from given percentage to the shrinkage limit and is expressed as percentage of dry volume of the soil mass.

If these three limits are known, it is possible to determine plasticity index, consistency index, liquidity index, toughness index, and flow index.

LIQUID LIMIT TEST

Liquid limit of a soil is directly proportional to the compressibility of a soil and hence its ability to support a load. It also governs its trafficability potential in the wet state. It also indicates shrink-swell potential of the soil (Hicks 2007). The objective of this test is to determine the liquid limit of the soil using Casagrande liquid limit apparatus (Fig. 12.5 top).

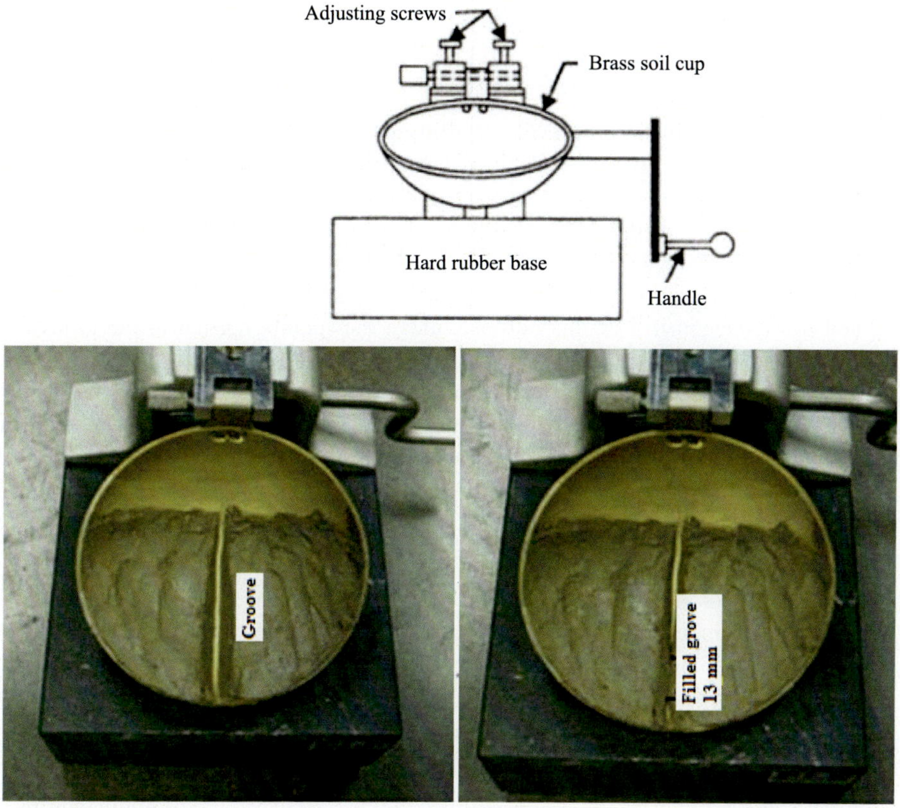

Fig. 12.5. *Casagrande apparatus (top), formation of groove (bottom left) and filling of groove (bottom right)*

Apparatus: Casagrande liquid limit apparatus, Casagrande grooving tool, Glass plate 10 mm thick and about 45 cm^2, Spatula, Balance- sensitive to 0.01 g, Hot air oven, Containers for moisture content determination, Wash bottle, distilled water and IS 425 μm sieve

Adjustment of Casagrande Apparatus

Use the gauge on the handle of the grooving tool to adjust the height through which cup of the Casagrande apparatus is lifted and dropped. The point on the cup that comes in contact with the base should fall exactly through one centimeter for every revolution of the handle. Tighten the adjusting screws after adjustment.

Procedure

Take about 120 g of soil sample passing through 425 μm IS sieve and mix it thoroughly on the glass plate with distilled water so as to form a uniform paste. Allow sufficient time for the moisture to distribute throughout the soil mass. Remix the soil thoroughly.

1. Take a portion of the soil paste with the help of a spatula and place it in the central portion of the cup. Spread it into position with the spatula. While doing so, ensure that the soil surface is parallel to the rubber base. The depth of the soil at the centre of the cup should be 1.0 cm maximum
2. With the help of a grooving tool, divide the soil paste in the cup along the diameter of the cup (Fig. 12.5 bottom left). Have a clean, sharp groove of proper dimensions
3. Turn the handle of the apparatus at a rate of 2 revolutions per second. Continue observing the soil groove formed in step 2, while counting the revolutions. Stop rotating the handle, once the two parts of the soil paste come in contact at the bottom of the groove for a distance of about 12 mm (Fig. 12.5 bottom right). Record the number of revolutions (drops) to achieve this.
4. Collect a representative sample of the soil from the portion of the groove where the soil flowed together. For this purpose, move the spatula normal to the groove i.e. width wise. Put the sample in a container and determine its water content by oven-drying method as discussed in previous chapters of this book
5. Transfer the remaining soil in the cup back on to the glass plate. Dry the soil by kneading the wet soil using spatula
6. Repeat steps 1 to 4 for a minimum of 5 trials. The trials are conducted in such a manner that the number of blows required to fill in the groove are in the range 25 ± 10
7. Plot a flow curve on a semi-log sheet with moisture content on y-axis (arithmetic scale) and number of blows on x-axis (log scale). Draw a well defined straight line through the points (Fig. 12.6). Record the moisture content corresponding to 25 blows. Round off to the nearest whole number and report it as the liquid limit of the soil
8. The slope of the flow curve prepared in step 7 gives the flow index (I_f). It is calculated from Fig. 12.6 as follows:

$$I_f = (W_1 - W_2)/\log (N_2/N_1) \qquad ...(12.1)$$

Such that W_1 is the moisture content at N_1 blows and W_2 is the moisture content at N_2 blows. BIS (1985) suggest that a better way is to extend the straight line and use the moisture contents at 10 and 100 blows to calculate I_f. The flow index is a measure of soil plasticity, latter being well correlated with the flow index. Soils having higher flow index (steeper flow curve) will have lower shear strength for the same plasticity index. These soils are unable to sustain heavy loads as they rapidly lose their shear strength with increase in water content.

Note that plotting of the data is necessary if one proposes to find out both the LL and I_f. If only LL is required, it is possible to do so even without plotting the data. Use any of the following formulae to calculate the LL. For blows in the range of 15 to 35, the liquid limit water content is given as (BIS, 1985)

$$\text{LL} = W_n/(0.3215 - 0.23 \log N) \qquad ...(12.2)$$

Here W_n is the moisture content for given blows, N.

Norman (1959) has proposed another formula to calculate LL, which is reproduced as follows.

$$LL = W_n\,(N/25)^{0.121} \quad ...(12.3)$$

Eq. (12.3) works well for blows in the range of 20-30. The only limitation is that the exponent in eq. (12.3) is not a constant and may vary with soil types.

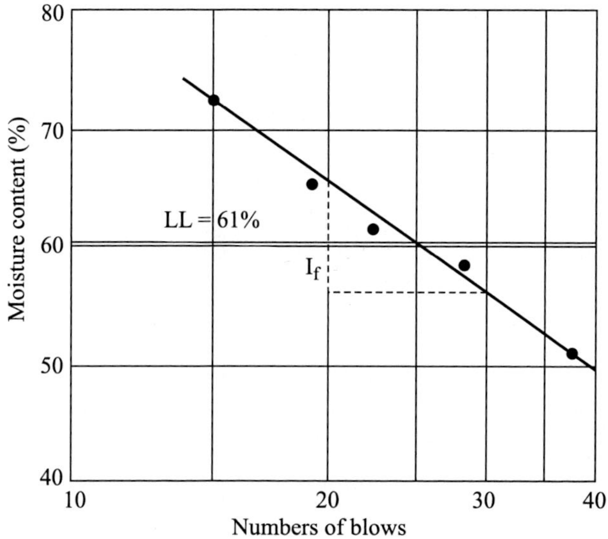

Fig. 12.6. *Plot of the Casagrande test data on a semi-log paper to calculate LL and* I_f

Liquid Limit (Cone Penetrometer)

Apparatus: Cone penetrometer (Fig. 12.7), Marble or glass plate, Spatula, Balance, Hot air oven, Containers for moisture content determination, Wash bottle containing distilled water, 425 μm IS sieve.

Procedure

1. Take about 250 g of soil passing through 425 μm IS sieve. Mix it thoroughly with distilled water on the glass plate to form a uniform paste. Allow sufficient time for soaking. It ensures uniform distribution of moisture throughout the soil mass. Remix the soil thoroughly
2. Transfer the wet soil into the cylindrical cup of the cone penetrometer apparatus. Ensure that no air is entrapped within the soil mass during this process. The top surface of the wet soil mass is levelled off corresponding to the top of the cup
3. Place the cup filled with soil on the base of the cone penetrometer apparatus. Adjust the cone tip so that it just touches the surface of the soil paste in the cup. The initial reading of the dial is adjusted to zero
4. The vertical clamp is released at this time allowing the cone to penetrate into soil paste under its own weight for 5 s. Note down the penetrometer reading to the nearest first decimal of a millimeter

Fig. 12.7. *A view of the cone penetrometer: note the brass cup on the base*

5. Collect a representative sample of the soil from the cup and determine its gravimetric moisture content by drying in an oven
6. Repeat steps 2-5 to have at least 4 to 5 sets of penetrometer readings in the range 14- 28 mm
7. A graph representing moisture content on y- axis and the cone penetration on x-axis (log scale) is prepared, which should be a near straight line (Fig. 12.8). Draw best fit straight line through the points. Read the moisture content corresponding to a cone penetration of 20 mm. It is the liquid limit moisture content of the soil
8. Report the value to nearest first decimal place of a millimeter

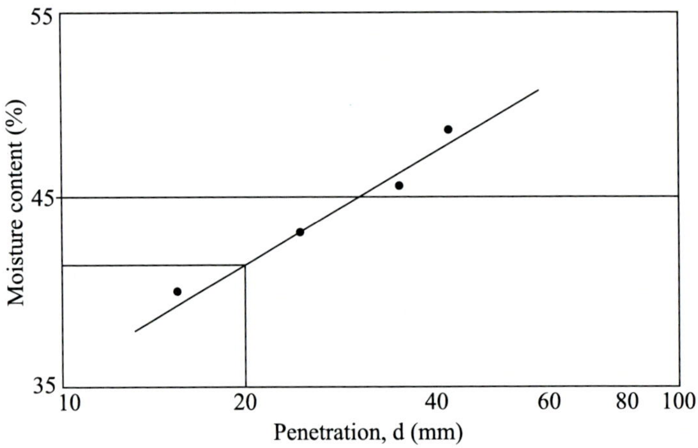

Fig. 12.8. *A semi-log plot of penetration depth and moisture content*

The liquid limit water content can also be computed from any one of the following relationships (BIS, 1985):

$$LL = W_n / (0.77 \log d) \quad ...(12.4)$$

$$LL = W_n/(0.65 + 0.0175\, d) \quad ...(12.5)$$

Here W_n is the water content at d mm depth of penetration (Fig. 12.8).

Observations

The liquid limit is a measure of the compressibility of the soil, i.e. the decrease in volume of saturated soil under loads exerted by the structure. The compressibility of the soil having higher liquid limit is high, and its shear strength is less compared to a soil with lower liquid limit, although their plasticity index may be the same. More observations on this issue will follow after discussions of other Atterberg limits.

PLASTIC LIMIT

The objective of this test is to determine the plastic limit of the soils and to calculate the plasticity index and toughness index of fine-grained soils.

Apparatus: Flat glass plate 10 mm thick and about 45 cm^2, Spatula, Balance- sensitive to 0.01 g, Oven, Containers for moisture content determination, Wash bottle containing distilled water, IS 425 μm sieve, 3 mm diameter rod of about 10 cm length.

Procedure

1. Take about 25 g of soil sample passing 425 μm IS sieve. Mix with sufficient distilled water on the glass plate, so that the soil becomes plastic enough to be shaped into a ball. Allow the soil to remain as such for a sufficient time so that moisture gets uniformly distributed throughout the soil mass
2. Make three balls of about 8 g each from the soil so prepared. Roll one of these on the glass plate with hand (Fig. 12.9). While rolling, the pressure on the soil mass should be just sufficient to roll it into a thread of uniform diameter throughout its length
3. Continue rolling until a threaded roll of 3 mm diameter is made. Collect the threaded roll; knead it and make a uniform ball again. Roll this ball again to form the threaded roll

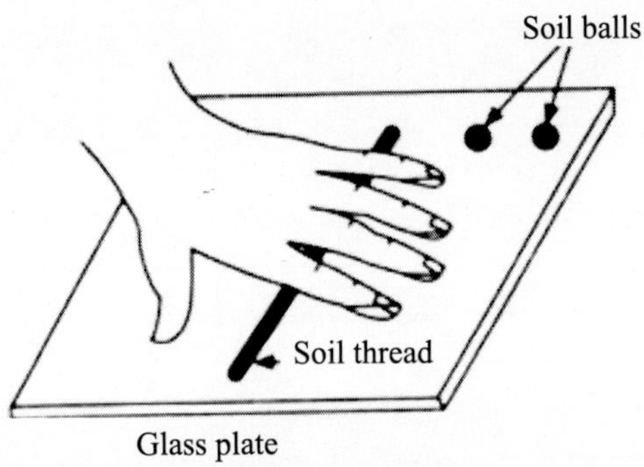

Fig. 12.9. *A view of rolling the soil ball on a glass plate*

4. Continue step 3 until the soil thread just crumbles at 3 mm diameter
5. Collect the crumbled soil thread in a container and determine the corresponding water content by oven drying
6. Repeat the test to have three trials
7. Report the average water content rounded off to the nearest whole number as the plastic limit of the soil.

Guidelines reported in Table 12.1 are used to describe the soil plasticity.

Table 12.1. *Criteria for describing soil plasticity*

Description	Criteria
Non-plastic	Not possible to roll a 3 mm thread at any water content
Low	The thread is difficult to roll, a cohesive mass cannot be formed when moisture content is lower than the plastic limit
Medium	It is easy to roll the thread and the plastic limit is reached in a little time. The thread cannot be re-rolled as the mass crumbles after the plastic limit is reached
High	Considerable time is needed for rolling and kneading the sample to reach the plastic limit. The thread can be rerolled and reworked several times before reaching the plastic limit. It is easy to form a mass when the sample is drier than the plastic limit

Observations

LL and PL of the soil depend upon the amount of clay as well as the type of clay present in the soil. Few observations are:

- A soil with high clay content has high LL and PL
- Colloidal clays have high LL and PL compared to non-colloidal clays
- For sands and gravel, PL = 0

Calculations

Plasticity index is the difference in moisture contents between the liquid limit and lower plastic limit.

Plasticity index = PI = LL – PL

It indicates clayeyness or potential plasticity of soil. Soils with a low plasticity index can change from a solid to a liquid with little change in moisture content. These soils may be prone to mass movement (Mills et al., 1980, Hicks, 2007). Ratings for compressibility and shrink-swell potential of soils are assessed as per following guidelines either with plasticity index or the liquid limit. The degree increases from low to very high.

Rating	Plasticity index (%)	Liquid limit (%)
Low	<25	<45
Medium	25-35	45-55
High	35-45	55-75
Very high	>45	>75

For soils with the same plasticity index, the toughness, a measure of the shear strength of the soil at its plastic limit, is inversely proportional to flow index as follows.

Toughness index, TI = PI/I_f ...(12.6)

The liquidity index is calculated as follows:

Liquidity index, LI = 100 (w_0 – PL)/(LL – PL) ...(12.7)

Here w_0 is the natural moisture content of the soil, PL is the plastic limit of the soil, and LL – PL is PI, plasticity index of the soil. When natural soil is at the plastic limit, the liquidity index is 0. If $0 < \text{LI} < 1$, soil is in plastic state, if $\text{LI} < 0$, soil is in semi-plastic or solid state and if $\text{LI} > 0$, it is in liquid state.

The consistency index is a measure of the shear strength of the soils. Higher the values of consistency index, higher will be the unconfined compressive strength of the soil. The consistency index is calculated as follows.

Consistency index, CI = (LL – w_0)/PI ...(12.8)

Activity, A

Soils with high activity are very reactive chemically. The values of activity reveal the dominant clay type present in the soil. High activity signifies large volume expansion and shrinkage upon wetting and drying of the soil. Normally, values of the activity of the clay are in the range of 0.75 and 1.25. Within these values clay is called normal clay. The activity, A is calculated as follows:

A = PI/Clay fraction (%) ...(12.9)

The following guidelines are used to interpret the data. Typical values of activities for various clay minerals are given in Table 12.2.

$A < 0.75$	Non-active clay
$0.75 \leq A < 1.25$	Normal clay
$A \geq 1.25$	Active clay

Table 12.2. *Clay types and their activity values*

Clay mineral	Activity
Muscovite	0.25
Kaolinite	0.33–0.46
Glacial clay and loess	0.5–0.75
Illite	0.9
Organic estuarine clay	>1.25
Montmorillonite (Ca)	1.5
Montmorillonite (Na)	7.2

Plasticity Chart

Indian Standard Soil Classification System (ISSCS) describes a plasticity chart that aids in soil classification. It is drawn using the values of LL and PI (Fig. 12.10). The ‘A’ line in this chart is expressed by the relation PI = 0.73 (LL - 20). The chart is also divided vertically into three divisions of plasticity by drawing vertical lines from LL =35% and LL =50%. Thus, soils in total are grouped in 6 divisions, fine soils being divided into clays (C), silts (M), or organic soils (O). Herein organic content is expressed as a percentage of the mass of organic matter in a given soil mass to the mass of the dry soil solids. Besides, H, I and L respectively indicate high, intermediate and low plasticity.

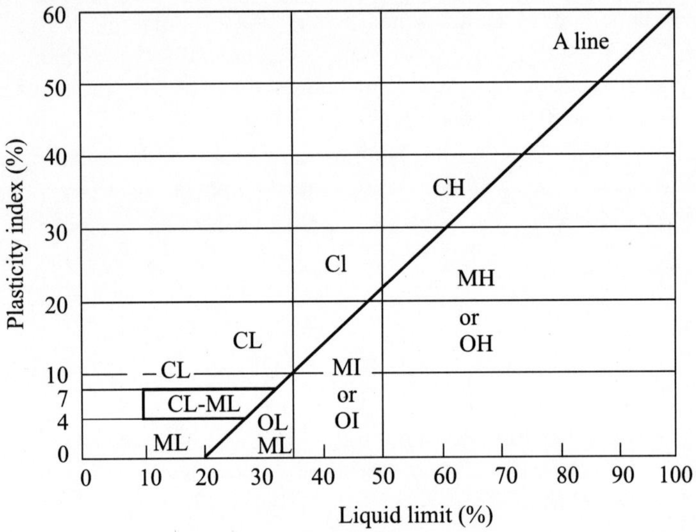

Fig. 12.10. *Plasticity chart for soil classification*

Observations

- The change from a liquid state to a semi-solid state in sandy soils occurs quite abruptly. These soils do not posses plasticity
- Earth works can be carried out easily with the least effort when the soil water content is at plastic limit
- The plasticity index is a good measure to assess the suitability of clay for potteries, construction of the clay core in earth dams, and construction of clay liners to minimize leaching of polluted materials from landfills. High plasticity index means better workability and low permeability, the two desirable properties required in all such uses
- Higher the values of LL and PI more will be the problems related to the soil's compressibility and swelling and shrinkage. These characteristics need to be kept in view while designing foundations and structures on such soils
- Soils have higher toughness and dry strength with increasing plasticity index
- For the same LL, the plasticity index of the soil increases with increase in organic matter content of the soil
- The liquid limit of a clay compared to its plastic limit decreases at a faster rate with increasing silt content. The plasticity index, therefore, decreases with the increasing silt content in a clayey soil

- The plastic limit of soils increases with increasing organic matter content, without any significant increase in the liquid limit. Therefore, soils with high organic matter content have a low plasticity index.

Example 12.1. Liquid limit, flow index and plastic limit of a soil are 24%, 12.5% and 15% respectively. Determine the plasticity index and toughness index. If the water content of the soils in its natural state in the field is 20%, find the liquidity index and relative consistency.

Since we know

LL = 24% , I_f = 12.5%, PL = 15%, Natural water content, w_0 = 20%

PI = 24 – 15 =9%

Toughness index = 9/12.5 = 0.72 or 72%

Liquidity index, LI = 100 (w_0 – PL)/PI = 100(20 – 15)/9 = 55.5%

Consistency index, CI = (LL – w_0)/PI = (24 – 20)/9 = 0.444 or 44.4%

SHRINKAGE LIMIT TEST

The objective of this test is to determine the shrinkage limit and calculate the shrinkage ratio of the soil (Prakash, 2015).

Apparatus and Reagents: Evaporating porcelain dish about 12 cm in diameter with a pour out and flat bottom, Shrinkage dish with a flat bottom 45 mm in diameter and 15 mm height internally, Glass cup of 50 mm diameter and 24 mm height, Plain glass plate of dimensions 75 mm × 75 mm × 3 mm, Pronged glass plate of dimensions 75 mm × 75 mm × 3 mm with three prongs fixed to the plate at 120° from each other and spacing of 30 mm center to center, Spatula, Straight edge, 425 µm IS sieve, Balance sensitive to 0.1 g to 0.01 g, Thermostatically controlled hot air oven, Wash bottle with distilled water, Desiccator

- **Mercury**

Procedure

1. Take about 100 g of soil sample passing through 425 µm IS sieve. Place about 30 g in an evaporating dish and make a creamy paste. The water content of the paste shall be nearly equal to the liquid limit of the soil
2. Apply a thin coat of Vaseline to the inside of the shrinkage dish. Fill the dish with the soil paste in three layers, each time putting approximately 1/3 rd of paste with the help of a spatula. Strike off the excess soil paste with a straight edge and wipe off all the soil paste adhering to the outside surface of the shrinkage dish
3. Immediately weigh the dish with wet soil and record its weight
4. Air- dry the wet soil cake for 6 to 8 hrs, until the colour of the paste turns from dark to light. Then oven-dry the cake at 105°C to 110°C for about 12 to 16 hr
5. Remove the shrinkage dish from the oven, cool it in a desiccator and weigh the dish containing the dry sample. Calculate the moisture content of the soil paste used to form soil pat.
6. To determine the volume of shrinkage dish, place it in an evaporating dish and fill the shrinkage dish with mercury until it overflows slightly. Place the plain glass plate on its top and firmly press so that excess mercury spills out. Pour the mercury in

the shrinkage dish into a measuring jar and determine its volume. This volume is equal to the volume of wet soil pat. Alternatively, determine the mass of mercury in the shrinkage dish. The mass divided by the unit mass of mercury will also give the volume of the shrinkage dish

7. Fill a glass cup to the brim with mercury. Remove any excess by pressing the glass plate with the help of three prongs. Place the cup with mercury in an evaporating dish such that no drop of mercury is spilled out of the cup. Place the oven-dry soil pat on the surface of the mercury in the cup. Carefully force the pat into the mercury by means of a glass plate with prongs. Collect the displaced mercury and record its mass. The volume so determined is the volume of the dry soil pat
8. Conduct three trials for each soil, calculate the shrinkage limit as per eq. (12.10) and report their average value as the shrinkage limit
9. If any individual value varies from the average by ± 2%, discard the test result and repeat the test.

Calculations

Shrinkage limit, $W_s = [W_1 - (V - V_d)\,\rho_w/w_d]100$ ...(12.10)

Here W_1 is the moisture content of wet soil pat, V is the volume of the wet soil pat, cm^3, V_d is the volume of dry soil pat, cm^3, ρ_w is the specific weight of water, w_d is the mass of the oven-dry soil pat, g.

Shrinkage Index (I_s): Numerical difference between the plastic limit (PL) and the shrinkage limit (W_s) as well as between the liquid limit (LL) and the shrinkage limit (W_s) have been used to define shrinkage Index depending upon the objective for which it is being defined. Thus, it is given as:

$I_s = \text{PL} - W_s$ ...(1211)

It describes the semi-solid state in the consistency. For describing and classifying the potential swelling of soil, it is defined as (Ranganatham and Satyanarayana, 1965)

$I_s = \text{LL} - W_s$ ...(12.12)

Potential swelling classification guidelines as per eq. (12.12) are given as under.

Shrinkage index	Potential swell classification
<20	Low
20-30	Medium
30-60	High
>60	Very high

Shrinkage Ratio (R): The shrinkage ratio is defined as the ratio of a given volume change, expressed in terms of percentage of the dry volume, to the corresponding change in water content of the soil sample. Shrinkage ratio is calculated using the following formula.

$R = 100[(V_1 - V_d)/V_d]/(W_1 - W_s)$ ...(12.13)

Here, V_1 is the volume of the soil mass at water content W_1, and V_d is the volume of the soil mass corresponding to the shrinkage limit, W_s.

Volumetric Shrinkage (Volumetric change): Volumetric shrinkage represents the change in volume expressed in percentage of the dry volume when the water content of the given soil mass is reduced from an initial value to the shrinkage limit. Hence,

$$V_s = 100[(V_1 - V_d)/V_d] \quad ...(12.14)$$

Or it can also be given as:

$$V_s = (W_1 - W_s)R \quad ...(12.15)$$

Here W_1 is the moisture content, W_s is the shrinkage limit, and R is the shrinkage ratio. It can be shown that the shrinkage limit, shrinkage ratio and the mass gravity of the soil at the dry state (G_m) are related as:

$$W_s = (1/R) - (1/G_m) \quad ...(12.16)$$

Linear Shrinkage (LS): It is defined as the ratio of the decrease in length to the original length when the water content is reduced from some initial value to the value at the shrinkage limit. It is usually expressed as a percentage.

$$\text{LS} = [(L_1 - L_d)/L_1] \times 100 \quad ...(12.17)$$

Here L_1 is the initial and L_d the length at the shrinkage limit. Following equation describes the relationship between the linear shrinkage and the volumetric shrinkage.

$$\text{LS} = 100\,[1 - [100/(V_s + 100)]^{1/3}] \quad ...(12.18)$$

Here V_s is the volumetric shrinkage.

Observations

The shrinkage limit is useful to estimate the expected settlement of structures due to drying of soils with the change in seasons. The possible settlement of structures will be high for soils with low shrinkage limit. Note that expansive soils have low values of shrinkage limit.

Note: Apparatus made of Perspex may be used instead of glass to avoid damage during the tests.

Result Summary

Summarized results may be reported as per the following format.

Liquid limit	Flow index	Plastic limit	Plasticity index	Toughness index	Liquidity index	Consistency index	Shrinkage limit

STICKY POINT

For wet soils, consistency is denoted by terms stickiness and plasticity. Stickiness is grouped into four categories namely non sticky, slightly sticky, sticky and very sticky.

Non-sticky: If no soil or practically no soil sticks to the fingers.

Slightly Sticky: If the soil begins to stick to fingers but comes off cleanly and does not stretch when the fingers are opened.

Sticky: If the soil sticks to both the thumb and forefinger and tends to stretch a little and pull apart rather than pulling free from the fingers.

Very Sticky: The soil is very sticky, if the soil firmly sticks to both thumb and forefinger. It also stretches when the fingers are opened.

Atterberg (1911) employed the sticky point as a soil moisture constant to separate the plastic consistence of soils into two parts constituting the sticky and non-sticky ranges. Atterberg defined the sticky point as the moisture content of a drying clay paste when it just fails to stick to a nickel spatula drawn across the surface. Hardy's criterion was the failure of the drying soil mass to adhere to the fingers (Hardy, 1923). From a practical standpoint, the sticky point gives an estimate of the maximum water content at which normal soils scour during tillage. Some of the ways by which a sticky point can be determined are as follows.

Correlation Methods

Olmstead (1937) observed that sticky point is a function of silt, clay and organic matter (OM) contents of the soils. Accordingly, he described the relation as per the following regression.

$$\text{Sticky point} = a + b\,(\text{silt}) + c\,(\text{clay}) + d\,(\text{OM}) \pm e \qquad \text{...(12.19)}$$

Here silt, clay and OM designate the % contents of these constituents in the soils. Also a, b, c, d and e are regression constants. He also observed that sticky point may also be affected by the speed at which tillage equipment is drawn. He gave the following correlations for 5 and 10 cm/s speeds of the roller specially designed for the test.

The sticky point (5) = 11.8 +0.115 (silt)+0.408 (clay)+0.08 (OM) ± 2.9

The sticky point (10) = 8.9+0.135 (silt)+0.409 (clay)+0.31 (OM) ± 2.7

Laboratory Method

Apparatus: Flexible steel spatula, Porcelain tile, Balance, Moisture boxes

Procedure

1. Take 15 to 20 g soil sample passing 2-mm round holes sieve. Moisten with distilled water and thoroughly puddle by kneading with the hands or by mixing with a spatula on a porcelain tile. Ensure that the mixture is wet and sticky
2. Scrape the soil from the tile and knead thoroughly between the fingers, until the soil reaches the stage at which it no longer sticks to the fingers or to the spatula. Besides finger test, Olmstead (1937) reviewed and commented upon various tests to decide the sticky point. Only two of these are discussed in this step
   - At sticky point, it should be possible to cut cleanly through the moulded soil. It should not stick to the knife when drawn quickly across the sample (speed of about 5 cm/s). By quickly means the speeds approaching ones that are ordinarily attained by tillage instruments. When this state is reached the soil is at the sticky point
   - In the glass-plate test, 5 or 6 g soil sample is moulded into a cube, and then one corner is pressed against a clean dry glass plate until the diagonal of the cube is

shortened about one-third its initial length. The end point is reached when only about 5 to 10 mg of soil adhered to the plate.

3. Whichever the method used, when sticky point has reached, place the soil block in a covered aluminium dish, the weight of which has been previously noted (W_1). Weigh without undue delay, to obtain the weight of the dish and moist soil at the sticky point (W_2). Dry in an oven at 105°C for 16 hours, cool in a desiccator and weigh again (W_3). Calculate the moisture content per 100 g of soil using the weight of oven-dry soil ($W_3 - W_1$) and the weight of water lost ($W_2 - W_3$) at the sticky point.

   Sticky point = $100\,(W_3 - W_1)/\,(W_2 - W_3)$

The sticky point is slightly higher than the plastic limit. On the basis of a sticky point, soils are categorized as: High > 35, Medium 20-35 and Low <20%.

SOIL COMPACTION (PENETROMETERS)

Different kinds of cone penetrometers are used in geotechnical and engineering applications. Some of these are mechanical, mechanical friction, electrical friction, and piezocone penetrometers. Whereas mechanical cone penetrometers are low cost and easy to operate, the electrical cone and piezocone penetrometers can potentially extend the measurement range of soil engineering properties. The application of penetrometers in agriculture began with the realization that soil compaction impacts root growth and root penetration depth is associated with penetration resistance (mechanical impedance) registered by penetrometers. Seedling emergence was also observed to be associated with the mechanical impedance resulting from the weight of soil on seedling and crust strength. These observations brought into focus the design and development of non-invasive instruments to measure soil penetrability or penetration resistance. Dynamic and static cone penetrometers are now commonly used in agriculture engineering, soil science and agronomy (Baver et al., 1972). While a cone penetrometer is used to measure the penetration resistance of soils at various depths, a pocket penetrometer is used to determine the penetration resistance of soil-surface crusts. This section discusses these three types of penetrometers. It may be noted that there could be several manufacturer specific features in various kinds of penetrometers. Therefore, readers must consult the manufacturer manual while using a specific penetrometer. The following sections only describe their general use procedures.

Pocket Penetrometer

Pocket penetrometer is popularly used to measure soil compaction by agronomists, botanists, engineers, and geologists. A small, handy and comparatively inexpensive device is designed to quantify surface penetration resistance (soil crust resistance). It measures the extent to which an area of soil can withstand the pressure exerted downward at a right angle to the soil surface (Laryea, et al., 1997). A pocket penetrometer comprises an outer tube, an inner tube, end cap, indicator ring, spring, and retaining ring (Fig. 12.11 left). Maximum reading of the penetrometer is normally 4.5 kg/cm^2. An adapter of 25 mm diameter is supplied along with the instrument to test weak soils and sediments. Since this adapter increases the effective area by 16 times, readings are multiplied by a factor of 0.0625 to arrive at the correct reading. Drop cone penetrometer, also a kind of dynamic penetrometer, is also used to estimate surface soil strength or soil compaction.

Procedure

1. Reset the indicating ring to zero
2. Grip the cylinder of the instrument. Push the tip of the penetrometer into the soil so that the groove marked on the tip is even with the level of the soil
3. The piston compresses a calibrated stainless steel spring, forcing the indicating ring to slide forward on the piston
4. The user stops exerting downward pressure as soon as the piston calibration mark sinks to the same level as the soil surface. Record the reading (kg/cm^2) indicated by the new location of the indicating ring.

Errors

Pocket penetrometers are not individually calibrated. An error of up to 1/2 divisions on the scale, equivalent to 0 to 0.124 kg/cm^2, is possible.

Cone Penetrometer

Two types of hand-held penetrometers namely the dynamic and the static penetrometers are available in the market. Principally, both of them measure soil resistance offered during vertical penetration of a probe or cone. The distinction between the two lies in the way in which force is applied to the cone. The basic principle involved in the operation of this equipment is to measure the resistance offered by the soil to the penetration of a standard cone of 20 mm diameter having an apex angle of 60^0 driven by 8 kg hammer (Varies from 7-9 kg) freely falling through a height of 575 mm. The amount of penetration (in mm) of the cone is generally reported in terms of the average penetration per blow, D_d (mm/blow). Dynamic cone penetrometers (DCP) as against the static penetrometers do not rely on constant penetration velocity. While using dynamic penetrometers, one measures either the number of blows to penetrate a pre-decided depth, or the penetration depth per blow. The results in any case are converted to a cone index. The weight of the hammer, slide distance, and cone angle determine the energy delivered and can be adjusted to local conditions (e.g. soft versus hard soils).

The device, comprising parts namely handle, top (driving) rod, hammer (8 kg), anvil, hand guard, bottom rod, meter rule and cone, is assembled as shown in Fig. 12.11 (middle). Spanners and Tommy bar are used to screw the joints and to keep them tight at all times. The DCP is driven into the soil by fall of the hammer, which provides nearly 26.88 J of kinetic energy (It will change with the mass of the hammer and the drop height). The DCP index or reading is defined as the penetration depth (D_d) in mm per drop of the hammer. Greater this value, softer the soil and vice versa. The number of blows to drive the DCP 50 mm into the sample is averaged and taken as the reading of DCP. Relative compaction of different soils or different soil layer is assessed by comparing these values.

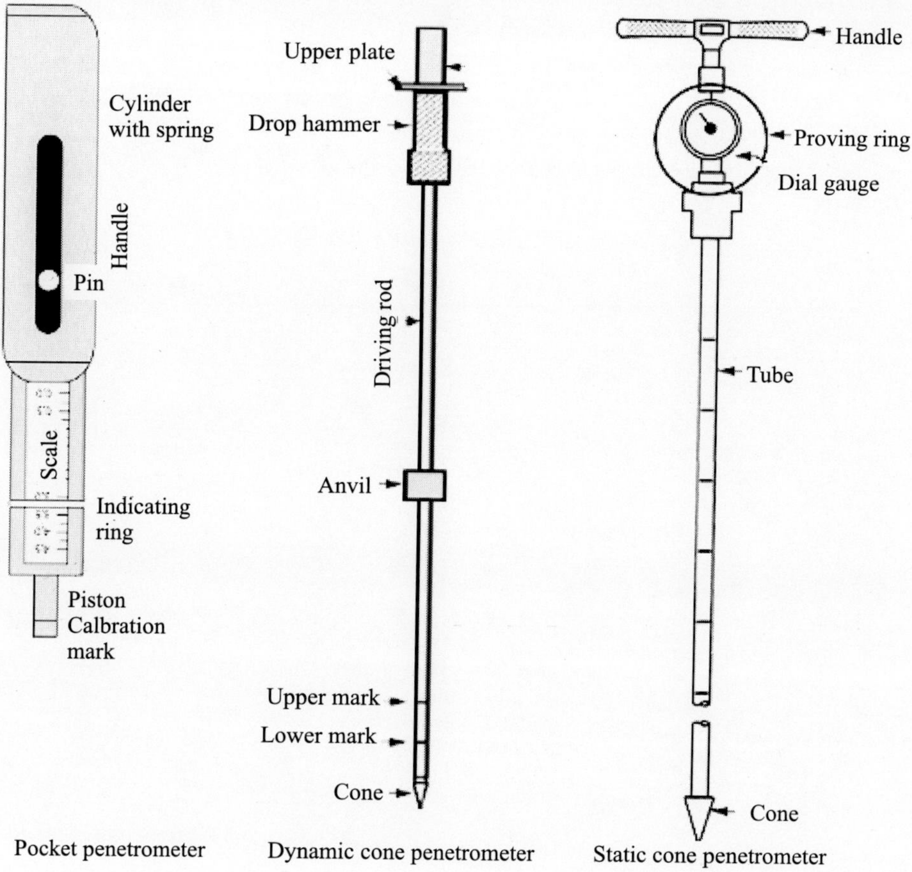

Fig. 12.11. *Schematic diagrams of various cone penetrometers: pocket (left), dynamic (middle) and static (right)*

Procedure

1. Place the cone on the soil surface with the shaft upright
2. Press the cone into the soil until the soil is level with the base of the cone. The reading at this stage is recorded as the initial reading
3. The slide hammer is raised until it touches the collar (upper plate) and released
4. The depth of penetration is recorded for each blow. The penetration depth (mm) for each blow is calculated by subtracting the previous recorded reading (i.e. reading on the scale) from the current reading. Continue until a maximum or desired depth is reached
5. Calculate the cumulative depth of penetration (in mm) for each blow by adding all the previous individual depths of penetration

6. Using the data of cumulative depth of penetration and the cumulative number of blows, draw the graph (Fig. 12.12)
7. Identify the slope changes in the penetration curve. Every change in slope represents a change in layer type (For Fig. 12.12, note layer 1 is above 27 cm, layer 2 is between 23-73 cm and layer 3 is below 73 cm)
8. After identifying each slope change, calculate the average penetration/blow for each layer.

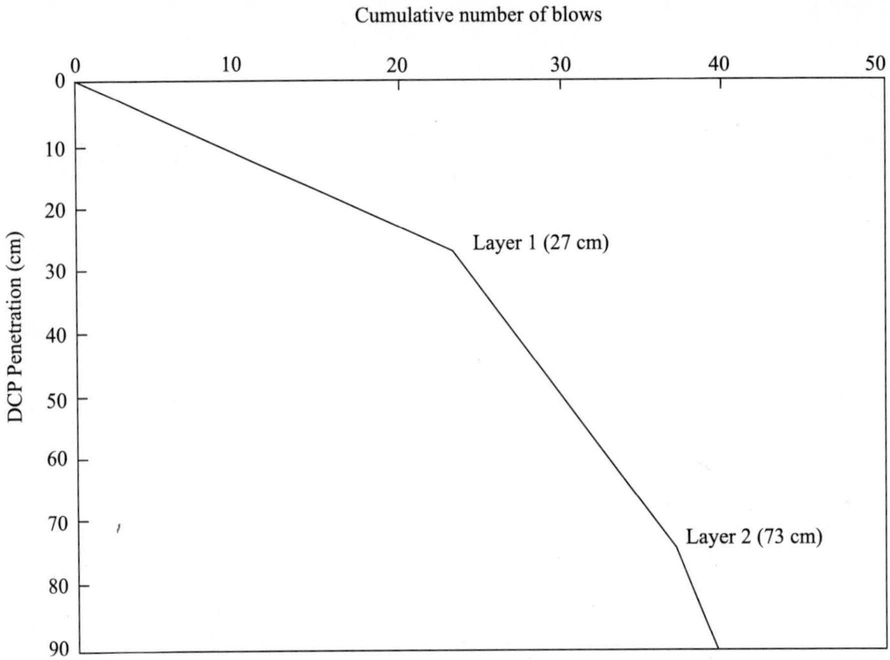

Fig. 12.12. *Plot of DCP penetration with number of blows*

Calculations of cone resistance

Resistance offered to the cone is calculated as the work done by the soil in stopping the movement of the penetrometer divided by the distance

$$R_s = W_s/P_d \qquad ...(12.20)$$

Here R_s is the soil resistance (N), W_s is the work done by the soil (J), and P_d is the distance the penetrometer travels per blow, m. The resistance calculated by eq. (12.20) represents the average resistance offered by the soil layer in the respective zone. Substitution of kinetic energy, KE of the hammer into eq. (12.20) for W_s allows one to calculate the soil resistance for each blow of the hammer.

$$\text{KE} = 0.5 \text{ m } V^2 \qquad ...(12.21)$$

Here m is the mass of the hammer and $V = (2gx)^{1/2}$ is the velocity of fall at the anvil. Here g is acceleration due to gravity, m/s^2 and x is the distance travelled by the hammer from the bottom of the hammer to the top of the anvil.

Calculations of California Bearing Ratio

In engineering studies, California Bearing Ratio (CBR) defined as the ratio expressed in percentage of the force per unit area required to penetrate a soil mass with a standard circular plunger of 50 mm diameter at the rate of 1.25 mm/min to that required for the corresponding penetration in a standard material. The ratio is usually determined for two penetration depths of 2.5 and 5 mm. If the ratio for 5 mm depth is consistently higher than at 2.5 mm, the ratio at 5 mm is used. CBR can be calculated with the help of DCP and is denoted as CBR-DCP.

$$\text{CBR, \%} = 3700/(P_5)^{1.3} \qquad \text{...(12.22)}$$

The P_5 is the penetration in mm, for every 5 blows interval. Relationship between the DCP reading and the CBR can also be found through empirical relations. One of the commonly used relations to determine CBR values corresponding to the penetration depth of each layer is given as:

$$\log_{10}\text{CBR} = 2.48 - 1.057 \log_{10}(\text{D}_\text{d}) \qquad \text{...(12.23)}$$

Here D_d is the ratio between the depth of penetration in mm and the number of blows required to achieve such penetration. Note that the values of the regression constants may vary for different soils, different cone angles (30^0 and 60^0) and other factors (Livneh, 1987).

Caution

Normally, it may be difficult to take out the penetrometers from depths greater than 30 cm below the soil surface. Therefore, investigation depths may be limited on this account.

Static Cone Penetrometer

Static cone penetrometer measures the force required to push a metal cone through the soil driven at a constant velocity. In one class of these devices, a proving ring or pressure gauge is provided. It requires a note keeper to log force and depth as the operator pushes the cone tip into the soil (Morrison and Bartek, 1987). The second class of manual push cone penetrometers consists of hand-pushed units with force and depth recorded by a data logger. In later versions, graphic features are also included. Advanced versions of this kind are the mechanical cone penetrometers for engineering applications. These are mounted on a tractor, truck, or trailer. They employ mechanical, hydraulic, or electric power to force the cone into soil, while cone index (CI) and depths are recorded using a microprocessor-based data acquisition system.

Equipment

The hand-pushed cone penetrometer is a portable handy tool for simple, quick and accurate assessment of soil condition. It consists of a handle, proving ring and dial gauge equipped with drag pointer (It records the maximum force encountered during the test irrespective of depth), 100-cm rod graduated at 10-cm intervals and stainless steel cones (NRCS, 2012, Fig. 12.11 right). Conical tips of 2 cm^2 (D = 16 mm), 4 cm^2 (D = 22 mm), and 6 cm^2 (D = 28 mm) areas having tip angles of 30^0 or 60^0 are used. A static cone penetrometer with a 30°cone is recommended by the American Society of Agricultural Engineers (ASAE, 1999), as the standard measuring device for characterizing the penetration resistance of soils. While such configurations suit a wide variety of soils, these are not critical and all instruments do not adhere to these standards. The standard equipment is supplied with a cone wear gauge to check the dimensions of the conical tips. Optional accessories include extra sounding rods for testing to a maximum depth of 3 m and a hand operated auger.

Penetration resistance

The cone index, an index of the shear resistance of the soil, is the force required per unit base area to push the cone penetrometer through a specified small increment of the soil (ASAE, 1999). The force is commonly expressed in kilopascals (kPa) or kg/cm^2 or psi. For high values, a small and for low values, a larger conical tip is used. The tool is designed to allow a maximum force of 2000 kPa. It is in line with ASAE recommendations for soft soils, considering possible *in-situ* soil strength values and being a hand tool. Cone index, which was originally defined to mean the average penetration resistance of the top 150 mm of soil, has since been broadened to include penetration resistance values of any depth. Values may be reported as: CI_Y, cone index at (Y_m) depth or CI_{Y-Z}, average cone index in the (Y to Z m) depth. Cone base size should be stated while reporting these values. Cone indices taken at different depths are used to plot cone index curve, which is a plot of penetration force versus depth of penetration. It can then be correlated with other soil physical properties and/or with crop yields. Soil properties such as bulk density, soil strength, pore size distribution, and cementation have been observed to be well correlated with CI.

Procedure

1. Select the test location and prepare a flat, clean soil surface for the test
2. Set the drag pointer as well as the measuring needle to zero. Once both the pointers are at zero, select a cone appropriate for the expected density of the soil. The measuring needle indicates the force required to push the cone, which varies as the cone encounters various soil types or hard pans
3. Hold the penetrometer in vertical position and push the cone point slowly downward. The surface reading is measured when the base of the cone is flush with the soil surface
4. The cone is pushed into the soil at a uniform rate of approximately 30 mm/s, but somewhat slower rates will not result in significant errors. Subsequent readings should be made continuously or as frequently as possible while maintaining a 30 mm/s penetration rate. If automatic recording of CI is not possible, manual readings should be made by a recorder (second person) at depth increments of 50 mm or less. Readings of the dial gauge are normally recorded at vertical increments of 5 or 10 cm
5. Repeat the test at several locations around each test site to obtain an average of at least three sets of consistent and reliable readings
6. Measure depth wise soil moisture contents because it strongly affects the cone index
7. For spring-type penetrometers, convert the dial-gauge reading, r (mm) for each depth into a corresponding resistance value (kgf) using the calibration curve established between the dial-gauge reading and the applied force F (kgf). Calculate the cone index q_c (kgf/cm^2) using the following equation, which takes into account the weight, W (kgf), of the penetrometer (Nakayama, 1983):

$q_c = (F + W)/A = (Kr + W)/A$

Here A (cm^2) is the cone base area. If K, the response coefficient of the dial gauge, is provided by the manufacturer, the same equation is used to calculate q_c with $F = Kr$.

8. In many versions, the dial gauge is calibrated to give the direct value of the force. The resistance of the soil is determined by dividing the reading by the surface area of the selected conical tip. In that case, a dial reading of 300 N and a cone base area of 5 cm^2 mean a cone resistance of: 300/5 = 60 N/cm^2 or 600 kN/m^2 or 0.6 MPa (mega-Pascal = 10^6 Pascal)
9. Plot the cone index curve for the test location.

Interpretation

Mobility/Trafficability: Mobility of the machinery at various moisture contents can be assessed. Most traction models typically use the average CI for the top 150 mm (6 in) of soil depth (Kennedy and Rush, 1968).

Depth of Root-impeding Layers: CI values are used to determine the depth of soil layers that inhibit root growth. These compacted soil layers may be due to natural causes or produced by external sources such as the movement of tractors and farm machinery. Depth to maximum cone index or critical cone index value (2 MPa) characterizes the hardpans. Researchers (Schuler et al., 2000, USDA-NRCS, 2003) have recommended that soil cone penetration measurements should be taken under wet soil conditions i.e. when the soil is at about field capacity.

Root Growth: The penetrometer is capable of projecting potential spots where root growth may be inhibited. Root growth decreases linearly with increasing penetration resistance, and practically stops at above 2 MPa or about 300 psi (Bradford, 1986). Remember that roots may still penetrate in soils with natural cracks and pores even if penetration resistance is greater than 2 MPa,. Guidelines reported by Murdock et al. (1995) are used to assess the need of sub-soiling (Table 12.3).

Table 12.3. *Interpretation of penetration resistance measurements*

Percentage of measuring points having cone index > 2 MPa in top 45 cm	Compaction rating	Recommendation for sub-soiling
< 30	Little to none	No
30–50	Slight	No
50–75	Moderate	Yes
>75	Severe	Yes

Precautions

- The rod and cone should be wiped clean after each test
- Space the individual penetrations so that they do not interfere with one another. At the same time, they should not be too far apart causing variations in test results due to spatial variations in the soil strength characteristics
- Keep the apparatus dry and clean. Apply few drops of oil to the screw threads to avoid rusting

- Regularly check the wear and tear of the cones using cone check. If measurements are made using a caliper gauge, the cones should be replaced as soon as their base area is 5% less than the prescribed base area
- The cone should also be replaced as soon as any damage to its conical surface is noticed.

Additional Analysis

Regression analysis is performed to establish relationships between CI, the independent variable and other soil parameters, the dependent variables. For example, is CI varyies linearly with depth, D then the following regression can be used to determine CI at any soil depth.

$$\mathrm{CI} = a + b\,D \quad \text{...(12.24)}$$

Similarly, variation of CI with bulk density, BD and moisture content, θ can be described in any one of the following forms (See Chapter 6).

$$\mathrm{CI} = a + b\,BD + c\theta \quad \text{...(12.25)}$$

$$\mathrm{Ln\,CI} = \mathrm{Ln}\,a + b\,\mathrm{Ln}\,BD + c\,\mathrm{Ln}\,\theta \quad \text{....(12.26)}$$

Here a, b and c are regression coefficients in the respective equations.

MODULUS OF RUPTURE

A measure of the ultimate load carrying capacity of a beam, the modulus of rupture in engineering is equal to the ratio of the bending moment at rupture to the section's modulus of the beam. It is widely used in agriculture to characterize the crust strength (Reeve, 1965). Carnes (1934), Richards (1953), Allison (1956), Lemos and Luts (1957), Hanks and Thorp (1956 & 1957) and others used modulus of rupture as an indication of the force exercised by seedlings in breaking soil crusts. Amongst laboratory methods, Richards (1953) as detailed in Richards (1954) is commonly used to calculate the modulus of rupture of the soils. The method described in this section incorporates some modifications made in the original equipment (Agrawal and Sharma, 1979).

The Equipment

The machine described in Richards (1954) comprises 2 parallel bars 5 cm apart to support the sample. The breaking force is supplied from a third overlying bar centrally located and parallel to the supporting bars. The bars are coated with a strip of soft rubber having a cross section of 0.16 cm^2. The breaking machine is mounted on the platform of a beam balance. The briquette sample is broken by an upward motion of the two lower bars, which are supported by the platform of the balance. Upward motion of the upper bar is constrained by a cross frame above the balance. The breaking force is supplied by water accumulating in a vessel hung from the end of the balance beam. The locally made modified equipment (Fig. 12.13) also consists of knife-edge assembly, simple two pan balance (5 kg capacity), briquette support and source of water supply. The flow of water in the modified version is regulated @ 1000 mL/min for greater sensitivity instead of 2000 mL/min used by Richards (1954). The beam motion occurring at the time of breaking of the sample is used to automatically stop the water flow to the vessel.

Preparation of Briquette

Sieve the air-dry soil through a 2-mm round hole sieve. Take samples slightly larger than that will fill the briquette mould (7.0 cm × 3.5 cm × 0.95 cm). A thin layer of Vaseline is applied inside the mould to prevent the soil from sticking the walls. Place the screen-bottomed tray in the pan. Place the moulds on the blotting paper on the screen. Rest the tremie (a long necked funnel) on the blotting paper at one end of the mould. Dump whole of the soil sample into the tremie. Move the funnel around inside the mould while raising it continuously so as to give a uniform smooth filling of the mould. Strike off excess soil level with the help of a spatula. Ensure that it in no way compacts the soil. Add water to the pan until free water surrounds every mould. Allow samples to stand for 1 hr after all the soil samples are wet. Raise the screen very carefully so as not to jar the samples. Transfer to a forced draft oven at 50° C to dry to a constant weight. Soil briquettes are removed from the moulds and their dimensions are measured using a set of micro calipers. These briquettes are now ready for use to determine the breaking strength. Raise the temperature to 100 -105°C for few briquettes moulds to measure oven-dry coefficient of linear extensibility (COLE) value (See Chapter 14).

Procedure

1. Balance the pan against a weight of 1000 g
2. Place the briquette in such a way that it is fully centered on the briquette assembly
3. Adjust the top of the knife of knife-edge assembly to just touch the briquette
4. Apply the water at a constant volume at the other end of the beam balance
5. As soon as the briquette is broken, the water flow will stop. Note the volume of water or its weight at this time
6. Conduct at least 6 tests and take the average of these values as a measure of the modulus of rupture.

Calculations

The modulus of rupture is calculated with the following formula.

$$S = 3FL/(2bd^2) \qquad ...(12.27)$$

Here S is the modulus of rupture (dynes/cm^2), F is the breaking force in dynes (the breaking force in gm wt. × 980); L is the distance between the lower two supporting bars, b is the width of the briquette, and d is the depth or thickness of the briquette, all expressed in cm. F in this set-up is calculated by the following relation.

$$F = 980 \times 1000 \times W_2/W_1 \qquad ...(12.28)$$

W_1 is the counterweight required to balance out 1000 g on the weighing platform (g) and W_2 is the grams of water added to the vessel to break the sample (g).

The modulus of rupture equation after substituting the above values is given as:

$$S = 1.47 \times 10^6 \times W_2 \times L/(W_1 \times b \times d^2) \qquad ...(12.29)$$

Note 1 bar = 10^6 dynes/cm^2 = 10^2 kPa = 1023 cm H_2O

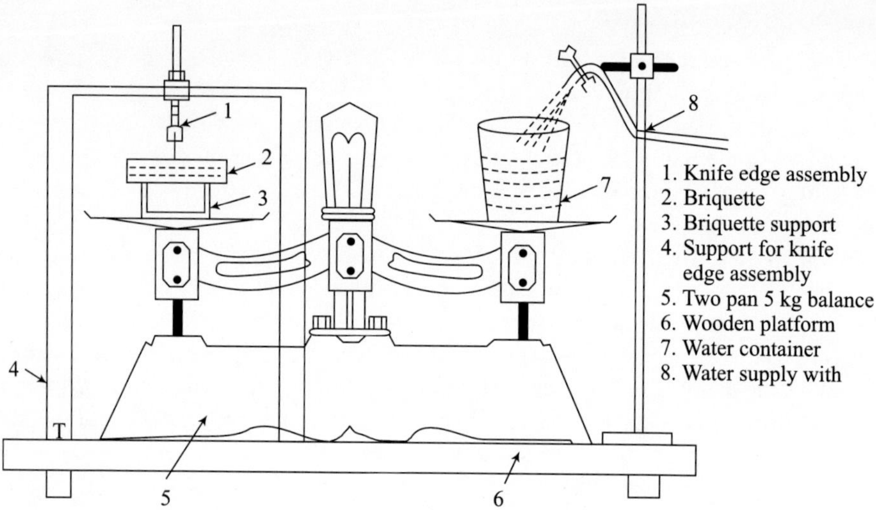

Fig. 12.13. *Schematic diagram of Richards' modulus of rupture apparatus (modified)*

COLE Index

The coefficient of linear extensibility is determined from (Grevers, 1991):

$$\text{COLE} = (L_m - L_d)/L_d \quad \text{...(12.30)}$$

Here L_m is the length of moist sample, and L_d is length of dry sample (usually air-dry or oven-dry).

Note: The modulus of rupture test is performed only for soils showing soil crusting problem.

ANGLE OF REPOSE

The angle of repose is defined as the steepest slope of the unconfined material, measured from the horizontal plane on which the material can be heaped without collapsing. The angle of repose varies from 0° to 90°, but for sands, it is only in the range of 25° to 35°. The angle of repose is related to the density, surface area, moisture content, shapes of the particles, and the coefficient of friction of the material under consideration. Because of some conflicting and inconsistent definitions that led to confusion as to its meaning, the term has now been replaced by the maximum allowable slope. The angle of repose is primarily applied to determine slope stability and in designing retaining structures. In addition, the angle of repose is also utilized in the design of barriers or rock fall collectors. In slope stability problems, the slope of the soil or granular material becomes unstable if its slope angle exceeds the angle of repose. In certain cases such as backfilling, if the slope angle of the soil exceeds the angle of repose, a retaining structure is required to maintain the slope stability. As such, the angle of repose is used to justify whether a retaining wall or structure is required. Angle of repose is widely used in food processing and pharmaceutical industries. A good review of the angle of repose is presented by Al-Hashemi and Al-Amoudi (2018). Although numerous methods are designed to measure the angle of repose, the most commonly used methods continue to be the funnel and hollow cylinder methods.

Funnel Method

The angle of repose can be measured in the laboratory using the funnel test. This test forms part of the standard tests backed by the American Society of Testing and Materials (ASTM) (Method No. C1444-00; ASTM, 2001).

Procedure

1. Prepare a smooth surface on which the pile is formed. It could be smooth or rough surface, say sand paper of appropriate roughness
2. Take the funnel and close its stem orifice with the help of thumb. Transfer the weighed amount of granular material into the funnel
3. Adjust the height of the funnel from the surface and allow the material to fall so that it is slowly deposited on the horizontal surface (Fig. 12.14)
4. Simultaneously raise the funnel so that the distance between the funnel bottom and the peak of the material is minimal. It is achieved with experience. The granular material moves over the face of the accumulated triangular pile
5. Once the funnel is emptied, measure the angle of repose by measuring the height, h of the pile. Two rulers method is commonly used for this measurement. The pouring of the material can even be stopped when the heap reaches a predetermined height or width
6. Measure the radius of the pile, $r = d/2$, such that d is the diameter
7. Calculate the angle of repose, θ from the radius and height as follows

   $$\theta = \tan^{-1}(2h/d) \qquad ...(12.31)$$

   ASTM (2001) however, suggests the following formula to calculate the angle of repose.

   $$\theta = \tan^{-1}[2h/(d - d_1)] \qquad ...(12.32)$$

   Here d_1 is the internal diameter of the funnel nozzle.

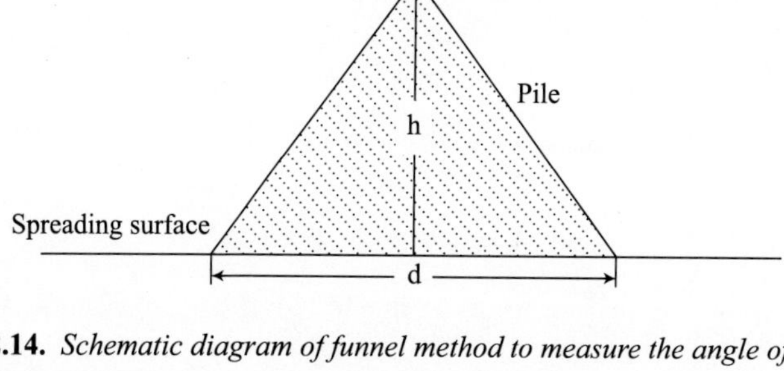

Fig. 12.14. *Schematic diagram of funnel method to measure the angle of repose*

The angle of repose in funnel method is affected by the roughness of the base at which the material is heaped, segregation of the material, the mass of the material, and pouring height. Miura et al. (1997) introduced a funnel-type device to determine the angle of repose in which the pile of soil was formed on a cylindrical pedestal with a depression. The method is not included for brevity as no major benefit has been projected.

Hollow Cylinder Method

The hollow cylinder method is employed to determine the static angle of repose of cohesion less materials (Liu, 2011). The test material is placed into a hollow cylinder of a certain diameter and height atop a selected base of known roughness properties (Fig. 12.15). The cylinder is then carefully pulled off of the base at a particular velocity, like in a concrete slump test. The material is allowed to flow and form a conical shape. The angle of repose is then measured by the (arctan) rule as in the funnel test. The angle of repose in this case is affected by the stratification, interface friction angle (which is the friction angle between the base and the granular material), lifting velocity, cylinder size, granular material mass and height of the sand in the cylinder.

Procedure

1. Prepare a smooth surface on which pile is to be formed. It could be smooth or rough surface, say sand paper of appropriate roughness
2. Partially fill the granular material in a hollow cylinder (Fig. 12.15)
3. Place the hollow cylinder on the surface by closing its lower end with a suitable plate and remove the plate thereafter
4. Raise the hollow cylinder at a constant speed. One can do it after some experience
5. Once the hollow cylinder is fully lifted, the material forms a pile
6. Measure the angle of repose by measuring the height, h of the pile using two rulers method
7. Measure the radius of the pile, r
8. Calculate the angle of repose, θ from the radius and height using eq. (12.31).

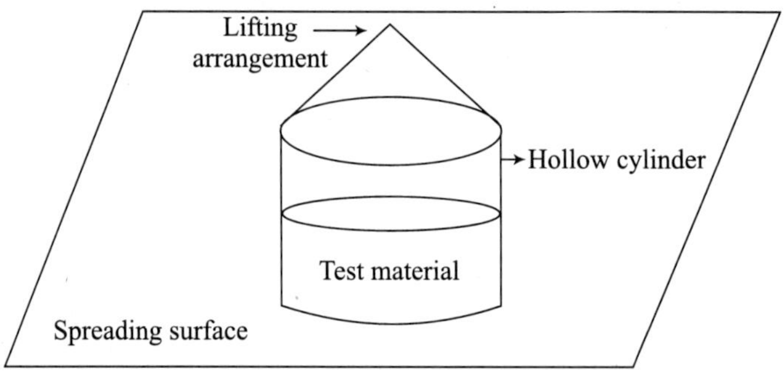

Fig. 12.15. *Hollow cylinder technique to measure the angle of repose*

Several other methods such as tilting box method for fine-grained non-cohesive materials and revolving cyclinder method for dynamic angle of repose are used. Since dynamic angle of repose may vary from static angle of repose, it is always desirable to mention the method of determination while reporting angle of repose. The angle of repose for various soil types are reported in Table 12.4 as a reference (structx.com, 2020)

Table 12.4. *Angle of repose of various materials for varying degrees of moisture*

Soil type	Degree of moisture		
	Dry	Moist	Wet
Top soil, loose	35-40	–	45
Loam, loose	40-45	–	20-25
Peat, loose	15	45	–
Clay/silt, solid	–	40-50	–
Clay/silt, firm	–	17-19	–
Clay/silt, loose	–	20-25	–
Puddle clay	–	–	15-19
Silt	–	19	–
Sandy clay	–	15	–
Sand, compact	–	35-40	–
Sand, loose	30-35	–	25
Sandy gravel, compact	–	40-45	–
Sandy gravel, loose	–	35-45	–
Sandy gravel, natural	–	25-30	–
Gravel, medium to coarse	25-30	–	25-30
Shingle, loose	–	40	–
Shale, hard	–	19-22	–
Broken rock	35	–	45

LAND SLOPE

Slope is the incline or gradient of a surface commonly expressed in percent or degrees. The percent slope is determined by measuring the difference in vertical elevation in m over 100 m of horizontal distance. For example, a 5% slope means a rise or fall of 5 m per 100 m of horizontal distance. Slope has been an important element in soil formation because of its influence on runoff and soil erosion. It also plays a great role in soil management as it governs soil drainage, use of machinery and choice of crops. The topography of a field greatly determines the susceptibility of the soil to erosion by water. In general, the steeper and longer the slopes, the greater is the soil erosion potential. Slopes are described as simple or complex. Simple slopes have smooth appearances, their surfaces extending only in one or at the most two directions. For example, slopes on alluvial fans and foot slopes of river valleys are regarded as simple. Complex areas have short slopes that extend in several directions and consist of convex and concave slopes.

Clinometer

Abney or any other optical clinometers can be used for this purpose. An Abney clinometer has a sighting tube with an angle scale of -90° to +90°. It has a spirit level with a Vernier index that can be moved along the scale while the user looks through the sighting tube (Fig. 12.16 left). A small mirror and lens is used so that level bubble is visible in the field of view. When the object is aligned with the cross-hair in the sighting tube, the spirit level is rotated so that the bubble is bisected by the cross-hair. The elevation of the line of sight is read on the scale. The scale on the left gives per cent slope. The scale on the right measures topographic slope. For percent slope, the minus (–) symbol means down the slope, and a plus (+) symbol means up slope (Fig. 12.16 right).

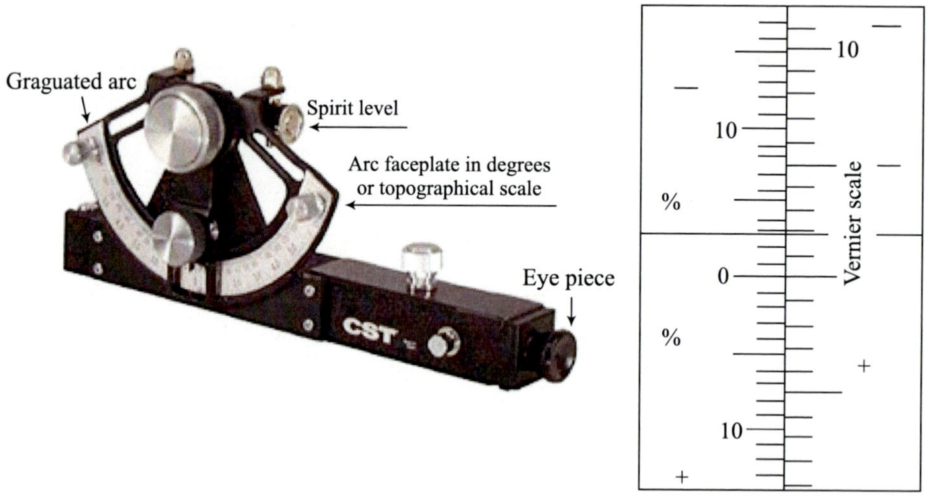

Fig. 12.16. *Abney clinometers (left) and vernier reading scale (right)*

Procedure

1. The first step is to calibrate the clinometer at sight level. To calibrate the instrument, one can either use a level rod or the level of one's vision height looking at another person. This level in the clinometer is read wherever 0-0 hits the rod or the body of the other person. This initial calibration must be done at a completely levelled site and is the reference point for the rest of the measurements
2. To determine the slope, one partner walks up or down the slope to points where observations are proposed to be taken. The observation is taken and recorded to the nearest percentage.

Use of Topographic Maps

Another useful way to calculate field length and steepness is to use information available on detailed field contour or topographic maps. The contour maps are available from a variety of data sources. Determining the average slope of the land using a topographic map is fairly simple.

1. Decide on an area for which the slope is to be measured (Note that it should be an area where the slope direction does not change; it should not cross the top of a hill or the bottom of a valley)

2. Once an area of interest is deciphered, draw a straight line perpendicular to the contours on the slope. For high accuracy, start and end the line on, rather than between contours on the map
3. Measure the actual length of the line using the scale of the map
4. Determine the total elevation difference along the line. It is obtained by subtracting the elevation of the lowest contour from the elevation of the highest contour used. No conversions are required for this measurement, as it is a real-world elevation change
5. To calculate percent slope, simply divide the elevation change in m by the distance of the line measured in step 3. Multiply the resulting number by 100 to get a percentage value equal to the percent slope of the area
6. To calculate the angle of the slope, divide the elevation change by the distance of the line. It gives the tangent of the angle of the slope. Applying an arctangent function, it is possible to calculate the angle of the slope. The angle calculated is the angle between a horizontal plane and the surface of the area. One can also use the conversion table to determine the angle (Table 12.5). Once the slope is known, the qualitative slope can be described as per the guidelines given in Table 12.6.

Table 12.5. *Conversion of slope (%) in degrees*

Slope (%)	Degrees-min-s (°– '–')
0.5	0°17'10"
1	0°35'
2	1°08'40"
5	2°51'40"
10	5°42'40"
20	11°18'36"
30	16°42'
40	21°48'05"
50	26°33'55"
100	45°

Table 12.6. *Description of slope classes for different ranges of the slope*

Class	Slope (%)	Description
A	0 – 0.5	Level
B	0.5 – 2.0	Nearly level
C	2 – 5	Very gently sloping
D	5 – 9	Gently sloping
E	9 – 15	Moderately sloping
F	15 – 30	Strongly sloping
G	30 – 45	Very strongly sloping

Illustration

To illustrate the application of a topographic map to determine the slope, we use the topographic map of the area shown in Fig. 12.17. Let the area A be the identified spot where we wish to determine the slope.

Highest contour = 7000 m

Lowest contour = 1000 m

Difference = 6000 m

Distance between the two selected points = 2.75 km = 27500 m

Slope = 6000/27500 = 0.218 or 21.8%

Slope angle = 12.3^0 or 12^{0}18'

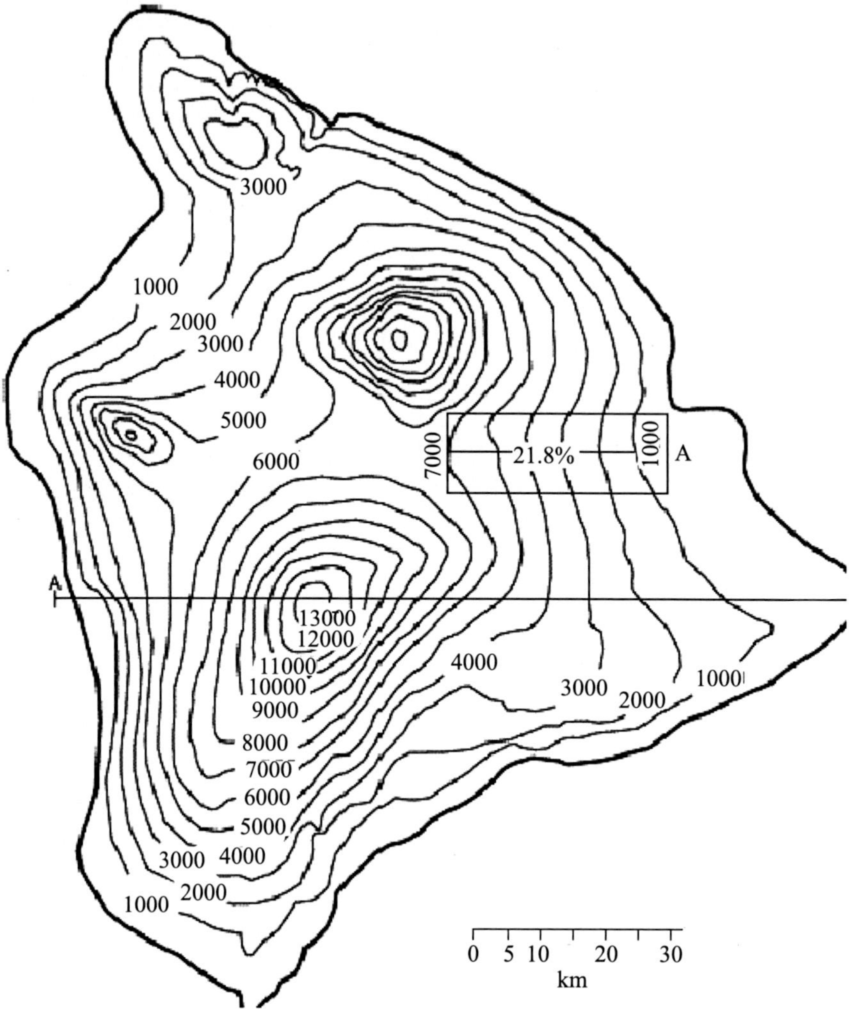

Fig. 12.17. *An illustration of the determination of the slope using a topographic map*

Chapter **13**

Thermal Properties

The thermal properties of soils, an integral component of soil physics, have applications in agriculture, engineering, and climatology. While most people immediately relate themselves to soil temperature, thermal properties go beyond temperature, being associated with the transfer of heat through the soil profile. Soil thermal properties have been extensively studied since the 1940s, when Kersten (1948) published the thermal conductivity measurements under frozen and unfrozen conditions for a wide range of soil types. Some of the soil thermal properties of interest are: thermal conductivity, specific heat, volumetric heat capacity, thermal diffusivity etc. This chapter looks at the methods of determining some of these soil properties.

TEMPERATURE

The measurement of air/soil temperature has great implications in meteorological and agricultural studies. It is one of the important environmental factors for optimum plant growth and development in IMD observatories in India. Following temperature measurements are routinely made.

- Dry and wet bulb temperatures
- Maximum and minimum temperatures
- Soil temperature at various depths
- Grass minimum temperature

While the first two are observed at a height of 1.25–1.40 m above the ground surface in a Stevenson screen, the last two are observed on or below the soil surface and near the ground surface (about 2.5 cm above the ground) respectively. The following sections discuss the soil and grass minimum temperature.

SOIL TEMPERATURE

The soil temperature is known to affect root growth, water and nutrient uptake, root metabolism, microbial activity, decomposition of organic matter, soil chemical processes, and soil moisture distribution. A liquid-in-glass thermometer or a thermistor temperature sensor is commonly used to measure soil temperature. A thermometer consists of a temperature sensor and some type of calibrated scale or readout device. In India, the unit of measurement is in degree Celsius (°C) but it is also measured in degree Fahrenheit (°F). The soil temperature measurements are made with simple or specialized soil thermometer with their bulbs exposed to the soil surface or to various depths going up to 100 cm. The standard depths at which soil temperature is measured are 5, 10, 20, 50 and 100 cm, the former 3 being part of the standard measurements at most agricultural observatories in India (Fig. 13.1). The measurement range of a soil thermometer is –20°C to +60°C. The temperature in observatories is measured daily at 07:18 and 14:18 hrs.

Permanent Installation

Each thermometer is installed at required depth almost permanently. The specialized soil thermometer has bulb bent at 120^0 (BIS 1972) while the rest of the stem is straight (Fig. 13.2). Iron stands fixed in an inclined position, making an angle of 60^0 with ground surface, hold these thermometer in position (Yadav et al., 2012). Soil thermometers, for 3 vertical depths of 5, 10 and 20 cm, are placed 45 cm apart at inclined depths of 5.8, 11.6 and 23.2 cm. These thermometers with increasing depth are fixed from west to east, soil thermometer at 5 cm depth being in the west followed by the thermometers at depths of 10 cm and 20 cm towards the east.

Fig. 13.1. *The soil thermometers buried at various depths in a meteorological laboratory*

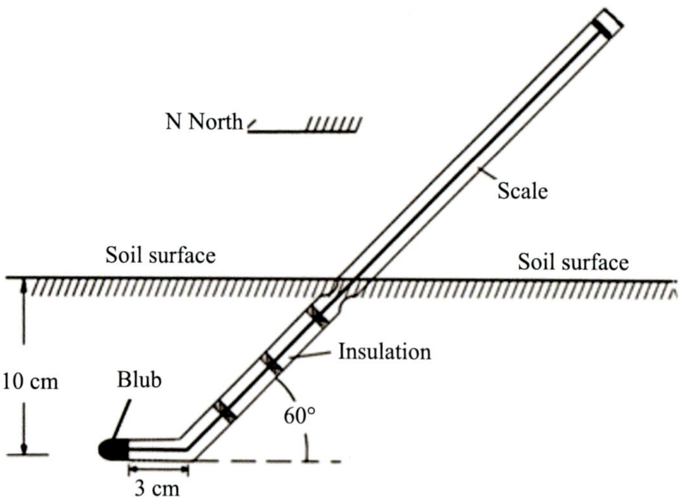

Fig. 13.2. *A bent bulb thermometer*

A small pit is made to install a bent stem soil thermometer. One of the walls of this pit is left undisturbed. A rod is pushed into the undisturbed vertical wall of the pit to make a horizontal hole at the desired depth below the soil surface. the rod used for this purpose is of slightly smaller diameter than the diameter of the probe to ensure that the thermometer probe makes a good contact with the undisturbed soil. After inserting the thermometer, the soil pit is backfilled and the thermometer is fastened to the stand (Fig. 13.1). For measuring temperature at soil depths of over 20 cm, mercury-in-glass thermometers are used. Their bulbs are embedded in wax or metallic paint. These are mounted on wooden, glass or plastic tubes. The thermometer-tube assemblies are then suspended or slipped in thin-walled metal or plastic tubes sunk into the ground to the required depth. For greater depths, lagged thermometers are used, which are suspended in iron tubes sunk into the soil to the required depth (BIS, 1972). In cold climates, the tops of the outer tubes extend above the ground to a height higher than the expected depth of snow cover. Required information on soil temperatures can be obtained from nearby meteorological observatories. A sample data set is shown in Table 13.1 along with the air temperature measured in the same observatory.

Table 13.1. *Sample data set from soil thermometers installed at CSSRI observatory*

Date	Soil temperature at depths (°C)						Air temperature (°C)	
	5 cm		10 cm		15 cm		Max	Min
1.1.2019	7.5	14.0	8.5	13.0	9.0	12.0	19.4	3.4
11.1.2019	10.0	16.5	10.5	15.5	11.0	14.0	19.9	6.6
21.1.2019	12.0	17.5	13.0	16.5	14.0	16.0	24.4	12.0
31.1.2019	8.5	13.0	9.0	12.5	10.0	12.0	16.4	7.0
11.2.2019	10.0	17.0	10.5	16.0	11.0	15.0	19.4	7.4
21.2.2019	13.0	20.0	14.0	19.0	15.0	18.0	18.0	13.4
1.3.2019	9.0	19.0	10.0	18.0	11.0	17.0	19.0	5.0
11.3.2019	14.5	20.0	15.0	19.5	16.0	18.5	24.4	10.0
21.3.2019	18.0	26.0	19.0	25.5	20.0	24.0	27.0	14.8
31.3.2019	20.5	30.5	21.5	29.5	*	*	34.0	15.0

*Out of order; max and min are the maximum and minimum temperatures

Spot Measurements

Many times, one is interested in observing the spot soil temperature say before the crop sowing or in an experimental set up to compare soil temperature under various treatments. Soil temperature under this situation is observed as per the following procedure.

Procedure

As far as possible, measurements are made in the middle of a sunny day. In order to have comparable observations, all measurements at various sampling points are completed within the shortest possible time.

1. Select the appropriate sampling point in the field under investigation. With the 15 meter mark at the center of this point, put a 30-m measuring tape on an axis perpendicular to the slope of the field. Starting at the 1 meter mark, identify points on both sides of the sampling point

2. At each location use a straight bulb thermometer to measure the temperature at the soil surface, and at 5 and 15 cm depths below the soil surface. If seeds are to be sown and/or new plants are to be planted, one measurement at the recommended sowing/planting depth may be sufficient
3. The soil surface for the purpose of temperature measurements is the surface of the soil proper. Therefore, in mulched fields, the mulch covering or organic litter must be carefully pushed aside before observing the temperature of the soil surface. All depths must be measured with respect to this depth i.e. without the surface mulch

Following guidelines may help to take the appropriate reading(s).

1. Refer to thermometer package for specific instructions
2. For taking soil surface temperature, ensure that the probe is in full contact with the soil surface, but not buried. The best way is to hold the probe nearly horizontal with the probe tip touching the soil. Practice with some researchers is to put the probe within 1 cm of the soil surface, which is taken as the soil surface temperature
3. For below surface measurements, mark the appropriate depth of insertion using a permanent marker or a piece of tape
4. Use a screw driver or appropriate sized nail to make a hole to the desired depth. If some resistance is encountered, shift the spot and make a new hole few centimeters away from the identified spot. Do not put excessive pressure while inserting the probe into the soil, lest it may get damaged
5. Allow enough time for the thermometer to reach equilibrium before recording the observation
6. Shade the probe with both hands, if direct bright sunshine is falling on the probe
7. Record the soil temperature on the observation sheet (Table 13.2). Same sheet can be used to record diurnal changes in soil temperature

Table 13.2. *Soil temperature data sheet*

Study Site: ____

Name of collector/analyst/recorder: ________________________

Date: _______ Soil thermometer type: Dial / Digital / Other _____

Has there been precipitation within the last 24 hr? Yes _____ No _____

Daily/Weekly measurements:_________

No.	Sample Time		Temperature (°c)		
	Hour	Minute	5 cm	10 cm	Air
1					
2					
3					
4					
5					

Comments, if any

____

____

____

GRASS MINIMUM TEMPERATURE

The occurrence of frost during the winter season is a common phenomenon in many parts of India notably the north-western parts. Some information on frost is of great importance to the farmers, gardeners and even auto-drivers. The grass minimum temperature provides some indication of the occurrence of frost. Grass minimum thermometer measures the temperature attained by the grass/vegetation during night time, giving an indication of the conditions through which crops have undergone during night hours. A grass minimum temperature of 0^0C or less indicates that water vapours can directly change from liquid to solid state. The air temperature simultaneously recorded at the same place may be 4 to 5^0C higher than the grass minimum thermometer.

Construction and Working

The grass minimum thermometer is a liquid-in-glass thermometer filled with spirit or alcohol (Fig. 13.3). The tube is attached to a graduated capillary tube having a dumbbell shaped metallic index of about 2.5 cm long. Its construction and working is similar to that of minimum thermometer except that its bulb is spherical and the stem is encased in a glass jacket to protect the thermometer seal marks on the stem from dew and/or rain. Alcohol in the capillary tube expands or contracts as it experiences any change in the temperature of the surrounding air. The metallic index also moves down with decreasing temperature. If the temperature rises before the observation is taken, the alcohol expands and flows around the index leaving it at the lowest temperature. Therefore, the index does not move away from the lowest position it has reached. To set the thermometer for the next day, it is removed from its supports and tilted slowly, bulb upwards, until the index touches the end of the spirit column. Never give jerks to the thermometer to move its index. The temperature of the ambient air can be noted on the graduated capillary tube.

Fig. 13.3. *A view of grass minimum thermometer in an observatory*

Placement

The grass minimum thermometer is placed in the observatory in the evening hours and removed after taking the observation in the morning hours. The thermometer is placed at a height of 2.5 to 5.0 cm above the ground at an inclination of 2° to the horizontal on two Y-shaped pegs (Fig. 13.3). The level of the bulb is lower than that of the stem. The bulb is always in contact with the tips of the well maintained grass

Notes:

- The most open place at the site should be selected for installing the thermometer

- The grass should ideally be kept short so that its condition resembles that of a lawn. Apply light irrigation whenever required to maintain the greenness in the measurement plot
- Do not leave the instrument exposed to the sun during the daytime to avoid condensation of spirit. The instrument should be shifted indoors after taking the observation. Direct exposure to solar radiation should be avoided.

THERMAL CONDUCTIVITY

Three key thermal properties of soils are thermal conductivity, heat capacity and thermal diffusivity. These properties are related as per the following relation:

$$S = K/(\rho c) \qquad ...(13.1)$$

Here S is the thermal diffusivity (m^2/s), K is the thermal conductivity (W/ m-K), c is the specific heat (J/kg-K) and ρ is the soil density (kg/m^3). Note that 1 W = 1 J/s.

The ability of a soil to conduct heat is governed by its thermal conductivity. Thermal conductivity determines how fast the soil's temperature changes during a day or between seasons. The thermal conductivity of the soil essentially depends on the thermal properties of its constituent i.e. mineral particles, organic matter, water and air contained in its pores. Thus, the thermal conductivity of soils has been found to be a function of several parameters notably: dry density, moisture content, mineralogy, temperature, particle size and shape and volumetric proportions of various soil constituents (Nusier and Abu-Hamdeh, 2003). In fact, the volumetric heat capacity or specific heat capacity can be determined to a reasonable accuracy based on the fractions of the soil constituents.

Several methods are used to measure soil thermal conductivity (Mitchell and Kao, 1978). The methods either use steady or transient state theories. In the steady state set-up, steady one directional heat flow is used to pass through a soil specimen. Once the steady state has reached, power input and the temperature difference across the specimen are noted. Fourier's law of heat conduction is then used to calculate the thermal conductivity. In the transient methods, heat is applied to the specimen. The temperature change over time is recorded. This temperature variation over time is used to determine the thermal conductivity. This section describes steady-state thermal cells and needle probe apparatus under steady and transient state methods.

Steady State Approach

To measure the thermal conductivity, two set-ups for the steady-state approach are shown in Fig. 13.4 (Clarke et al., 2008 and Low et al., 2017). It is possible to locally prepare both these set-ups.

Theory

For the first case shown in Fig. (13.4 left), Fourier law is used to calculate the thermal conductivity, K.

$$Q = KA\,(T_{top} - T_{base})/L \qquad ...(13.2)$$

Here Q is the power input, A is the cross-sectional area, T_{top} and T_{base} are the temperatures of the top and lower thermistor and are used to calculate the temperature gradient, ΔT, across the length of the specimen, and L is the length of the specimen. In eq. (13.2), Q and K are the two unknowns. To determine Q, T_{top} versus t curve of the top thermistor is plotted after switching off the heater. Theoretically, it has been proved that the temperature of the top thermistor at any time can be calculated by the following equation.

$$T = T_{air} + (T_0 - T_{air}) \exp(-hAt/mC_s) \qquad ...(13.3)$$

Here T_0 is the temperature of the soil at time $t = 0$, subscript air refers to the temperature of the ambient air, h is the convective heat transfer coefficient, m is the total mass of soil, and C_s is the soil specific heat capacity. The specific heat capacity is estimated from the mass-weighted properties of the soil constituents:

$$mC_s = m_s C_{ds} + m_w C_w \qquad ...(13.4)$$

Such that m is the weight of the soil, m_s the weight of the dry soil, m_w is the weight of water in the soil, C_{ds} is the heat capacity of the dry soil and C_w the heat capacity of water. Knowing mC_s and assuming various values of h, the value of h that best describes the cooling curve is selected. Following relation is then used to calculate, Q.

$$Q = hA(T_{top} - T_{air}) \qquad ...(13.5)$$

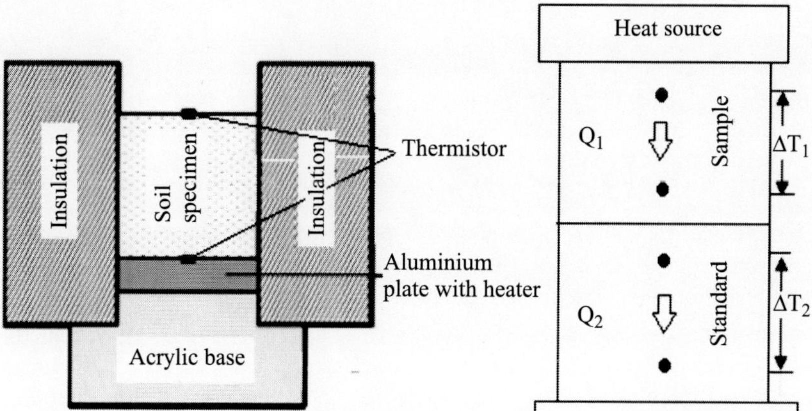

Fig. 13.4. *Two set-ups of the thermal cells used to determine thermal conductivity*

Comparing eq. (13.2) and eq. (13.5), we have

$$KA(T_2 - T_1)/L = hA(T_{top} - T_{amb}) \qquad ...(13.6)$$

$$K = hL(T_{top} - T_{amb})/(T_2 - T_1) \qquad ...(13.7)$$

Procedure

1. Fix the soil specimen in the set-up shown in Fig. (13.4 left). Note down the temperatures of the two thermistor
2. Switch on the heater in the aluminium plate. Continue to heat until steady state is reached. Note down the temperatures

3. Switch off the heater. Begin monitoring the temperature of the top thermistor
4. Plot T_{top} versus t curve and fit the curve using eq. (13.3). Select the best fit value of *h*
5. Use eq. (13.7) to calculate thermal conductivity, *K*.

Note: Several errors may be introduced during the estimation of the specific heat capacity from soil constituents whose properties are not accurately known. Some errors might also be introduced during curve fitting.

Improved Steady State Approach

An improved version of this set-up is shown in Fig. (13.4 right). In this method, a standard material, the thermal conductivity of which is known, is used along with the unknown sample, (Blázquez et al., 2017).

Theory

The heat flow through the test sample is given by the relation

$$Q_1 = K_1 A_1 \Delta T_1 / L_1 \qquad ...(13.8)$$

The heat flow through the known standard is given by the relation

$$Q_2 = K_2 A_2 \Delta T_2 / L_2 \qquad ...(13.9)$$

At steady state, $Q_1 = Q_2$ and therefore

$$K_1 A_1 \Delta T_1 / L_1 = K_2 A_2 \Delta T_2 / L_2 \qquad ...(13.10)$$

$$K_1 = K_2 A_2 \Delta T_2 L_1 / (A_1 \Delta T_1 L_2) \qquad ...(13.11)$$

Here *Q* is the heat transfer, *K* is the thermal conductivity, *A* is the area, *L* is the length and ΔT is the temperature difference. Subscripts 1 and 2 indicate the respective parameters for the unknown sample and the known standard respectively.

Procedure

The procedure essentially remains similar to the first approach discussed in the previous section except the cooling test, which is not required in this set-up. Although the set-up in the Lee's method is slightly different, yet the basic principle is same. As such, the Lee method described in a detailed manner will further clarify the procedure.

Lee's Method (Lee and Chorlton, 1896)

Apparatus: Lee's apparatus, Soil specimen in the form of a disc, Thermometers, Stop watch, Weighing balance, Stand, Boiler and Heater

Construction

The Lee's apparatus shown in Fig. 13.5(A) consists of two parts. A steam chamber, and a circular metal plate marked as O is one part. The steam chamber is hollow in which steam enters and exits from the respective sides. The lower part of the Lee's apparatus marked as M is also a circular metal disc of the same metal and diameter as O. Both the discs M and O have radial holes for inserting the thermometers shown as T_1 and T_2 respectively. The soil specimen *N*, having uniform thickness and of the same diameter as *M*, is sandwiched between *M* and *O*. Three hooks attached to *M* are used to suspend the whole set-up in the air.

Soil Specimen

The soil specimen is prepared from the air-dry soil, which has been passed through 2-mm sieve. The soil sample after mixing with the appropriate quantity of water is moulded into the desired shape using disc shaped plastic rings. The disc so prepared has the same density as of the field soil. The moulds are dried on the ground or floor to solidify in the form of discs. Numbers of such specimens are prepared.

Theory

When steam is allowed to flow, the temperatures of the thermometers begins to rise and gradually attain steady-state temperatures say T_1 and $T_2 (T_2 > T_1)$. Hence amount of heat flowing per second through *N*, *Q* is given by the Fourier's law as follows.

$$Q = KA(T_2 - T_1)/X \qquad ...(13.12)$$

Here *A* is the surface area of the specimen, *X* its thickness and *K* the thermal conductivity of the material. All parameters except *Q* in the above equation are known.

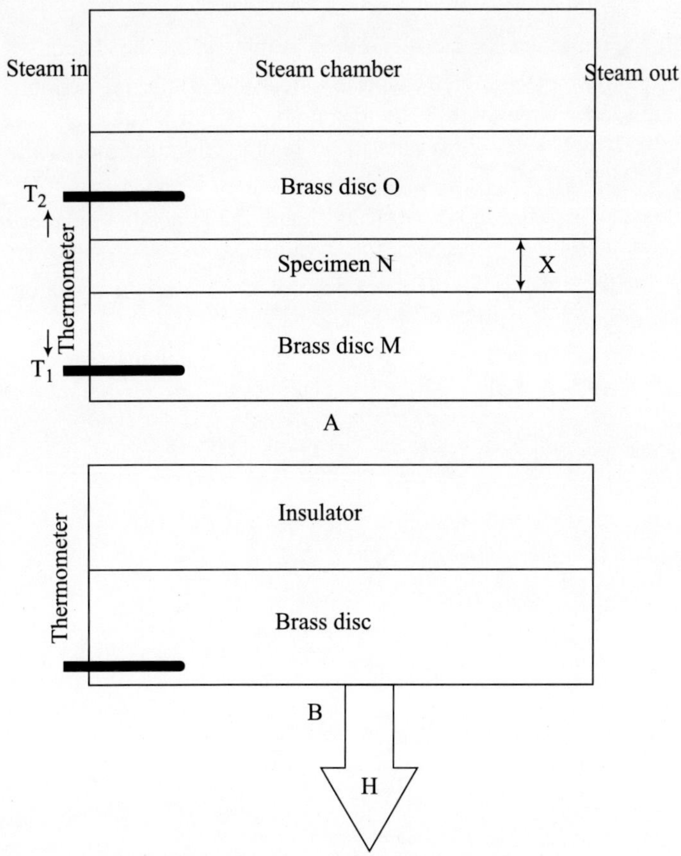

Fig. 13.5. *A schematic view of the Lee's apparatus to determine thermal conductivity*

At the steady state, the rate of heat conduction into the brass disc M is equal to the rate of heat loss from the bottom of it. The rate of heat loss is determined by measuring how fast the disc *M* cools at the previous (steady state) temperature T_1 (with the top of the brass disk covered with insulation). The amount of heat radiated per second is calculated from the known mass and specific heat of the lower disc, say *m* and *S*, respectively and the rate of cooling, dT/dt at T_1 as per the following equation:

$$Q = mSdT/dt \quad ...(13.13)$$

Equating (13.12) and (13.13) and simplifying, *K* can be determined as:

$$K = mS\,(dT/dt)X/[A(T_2 - T_1)] \quad ...(13.14)$$

Here dT/dt can be replaced by $\Delta T/\Delta t$ (Fig. 13.6).

Procedure

1. Fill the boiler with water to produce steam
2. Note down the specific heat of the metal of disc M (from Tables on specific heat) and its weight
3. Measure the diameter and average thickness of the specimen. Calculate the surface area of the specimen, $A = \pi r^2$
4. Assemble the equipment and suspend the assembly using the clamps
5. Insert the thermometers in their respective positions. Both the thermometers should display the room temperature. If the display is not same, note down the difference. Let it be θ. It will be used to take into account this difference while calculating $(T_2 - T_1)$
6. Allow the steam to pass through the steam chamber. Temperatures in the thermometers begin to rise, settling at some steady state values
7. Wait for another few minutes before noting down the steady temperatures say T_1 and T_2.
8. Stop the steam inflow and remove the steam chamber and the specimen *N*. *M* is still suspended. Heat *M* directly by the steam chamber till its temperature is about $T_1 + 7°$

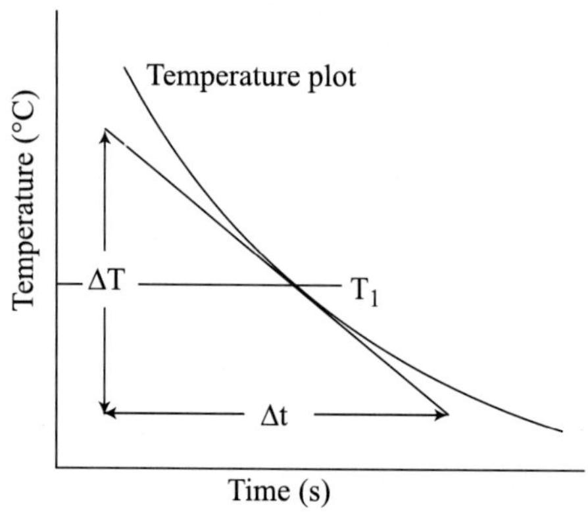

Fig. 13.6. *Cooling curve of disc M*

9. Remove the steam chamber and allow few minutes for the heat to get uniformly distributed over the disc M
10. Place some insulating material on *M* (Fig. 13.5 B). Record the temperature of the thermometer
11. Plot the cooling curve and determine its slope $dT/dt = \Delta T/\Delta t$ at the steady temperature T_1 (Fig. 13.6).

Precautions

- Record T_1 and T_2 only after they have attained a steady state for about 10 min
- The heat loss due to the surface of the soil specimen *N* in contact with *M* is taken into account. There may be some loss at the edges as well. Therefore, the specimen should always be a thin disc with a large cross sectional area ($A = \pi r^2$) compared to the area exposed at the edges ($a = 2\pi rX$). It will minimize energy loss from the edges
- Perform the test in the shortest possible time. Room temperature during the test may change, if undue large time is taken to complete the test
- Since the thermal conductivity increases with increasing moisture content, it is customary to report the moisture content at which measurements are made.

Illustration. Following illustration explains the use of eq. (13.14) to calculate the thermal conductivity.

Diameter of soil sample = 6.73 cm =0.0673 m

The area of the sample $A = 3.56 \times 10^{-3}$ m²

Thickness of soil sample = 1.75 cm = 0.0175 m

Mass of tin disc = 4.9 g = 4.9×10^{-3} kg

Specific heat capacity of tin (from tables) =217 J/kg K

At steady state

$T_1 = 52$ °C

$T_2 = 64$ °C

Slope = 0.022 °C/s

$K = mS\,(dT/dt)X/[A(T_2 - T_1]$

$K = 4.9 \times 10^{-3} \times 217 \times 0.022 \times 0.0175/[3.56 \times 10^{-3} \times (64 - 52)]$

$K = 0.01$ W/m-°C

Needle Probe Method

Needle probes are suitable for soft materials e.g. unconsolidated soil samples, flexible materials and pastes where the needle of the probe can be easily inserted without pre-drilling. They can also be adapted to somewhat harder samples where a drill hole matching the probe dimensions can be made, e.g. soft rock (sediments), construction materials and plastics. In such cases, a high thermal conductivity gel is filled in the hole before inserting the needle. It ensures full contact between the material and the needle (ASTM, 2000).

The needle probe method has been designed on the premise that the rate of temperature rise/fall of a line heat source depends upon the thermal constants of the medium in which it is placed. Based on the construction and procedures used in the data analysis, several alternative options are available to perform the tests (Kukkonen and Suppala, 1999). On the basis of construction, we have

- Single needle probe
- Dual needle probe

Dual probe is a modification of the single needle probe method, the modification being in the use of a second probe recording temperature at a finite distance from the conductivity probe. The distance between the two probes is approximately 20–25 mm. Similarly, on the basis of data collection and analysis, we have the following options.

- Analysis of data for heating cycle alone
- Analysis of data for heating and cooling cycles

In the latter case, thermal conductivity is determined for both the cycles. The average of the heating and cooling cycle is taken as the final value of the thermal conductivity. Analysis of both kinds of data can be made by simplified equations or by fitting the curves to the data sets using highly elegant theoretical equations. This section only describes a single needle probe with heating cycle using the approximate equation.

Apparatus: A thermal needle probe consists of a device that creates a linear heat source and incorporates a temperature measuring element (Fig. 13.7), Constant current source, Thermal readout unit: a device to produce a digital readout of temperature in °C, Voltage-Ohm-Meter (VOM): A device to read the voltage and current to the nearest 0.01 V and ampere, Timer, stopwatch or similar time measuring instrument, Drilling equipment: only required for hard materials. Drilling unit should be capable of drilling a straight vertical hole of diameter close to the diameter of the probe to a depth equal to the length of the needle

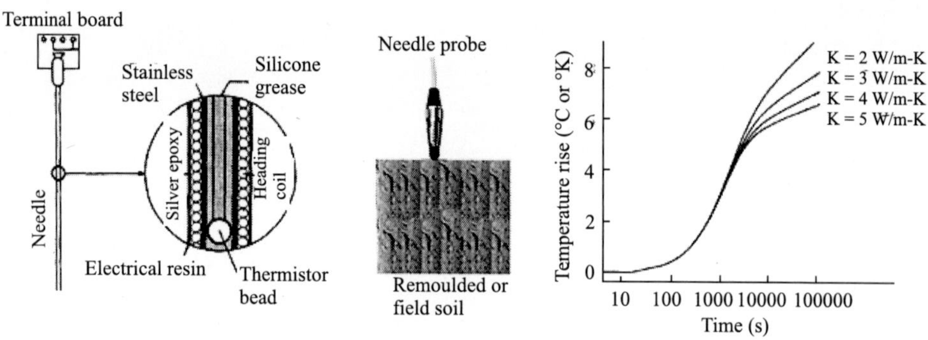

Fig. 13.7. *A single needle probe: construction details (left), a commercial probe in remoulded soil core (middle) and characteristic temperature curve*

Preparation of Specimen

Cut a 200 ± 30 mm long section from a core sampling tube containing an undisturbed soil specimen. The minimum diameter of tube section should be 51 mm. For moulding disturbed samples, use a thin-walled metal or plastic tube. The minimum diameter of the tube should be 51 mm. Its length shall be 200 ± 30 mm. Compact the specimen in an appropriate manner to achieve the desired density and water content. Determine moisture content and dry density of the representative samples at the time of carrying out the test, which should be reported along with the test results.

Calibration

Calibrate the instrument using one or more materials of known thermal conductivities covering the range of the thermal conductivities of the materials being tested (typically $0.2 < 1 < 5$ W/m-K).

Procedure

1. In laboratory tests, allow the specimen to come to equilibrium with the room temperature. Connect the heater wire of the probe to the constant current source and the leads of temperature measuring element to the readout unit. Apply a known constant current to the heater wire such that the temperature change is less than 10 K in 1000 s
2. Record the temperature readings at 0, 5, 10, 15, 30, 45, and 60 s. Take further readings at 30 s time intervals for a minimum of 1000 s or more
3. Turn off the constant current source
4. Plot the temperature data as a function of time on a semi-log graph. Use the approximately linear portion of the quasi-steady-state temperature-time-response portion of the curve to obtain the thermal conductivity of the material. For this purpose, select times t_1 and t_2 at appropriate points on the line and read the corresponding temperatures T_1 and T_2

Calculations

Following relation is used to calculate thermal conductivity.

$$K = Q\,\mathrm{Ln}(t_2/t_1)/[4\pi(T_2 - T_1)] \quad \text{...(13.15)}$$

$$Q = VI/L \quad \text{...(13.16)}$$

Here V is the measured voltage (V), I is the current flowing through heater wire (A), L is the length of heater wire, m, K is the thermal conductivity, $W/(m\text{-}K)$, Q is the power consumption of the heater wire in watts per unit length, T_1 is the initial temperature, K, t_1 is the initial time, s, T_2 is the final temperature, K, t_2 is the final time, s.

Note: A commercial meter comes with the standard 100 mm sensor (50 mm sensor is optional) for testing soils and soft and easy to drill materials. After the needle sensor is fully inserted into an isothermal sample, the measurements can be made with the push of a button.

Guidelines

Thermal conductivity of soils and soil formations depends on temperature, pressure, porosity, composition, and properties of pore-filling fluids and gases. Nonetheless, thermal conductivities of some of the soil constituents are given in Table 13.3 (Poelchau et al., 1997, Bristow 2002) and for some soil types in Table 13.4. These values may be used as guidelines to compare the field or laboratory measured values.

Table 13.3. *Thermal conductivities of soil-related materials*

Material	Thermal conductivities (W/m-K)
Earth's crust	2.0-2.5
Basalt	2.2
Granite	2.0
Sandstone	2.5
Limestone	2.5-3.0
Quartz	8.8
Clay minerals	2.9
Organic matter	0.25
Water	$0.552 + 2.34 \times 10^{-3}$ T – 1.1×10^{-5} T^2
Air	0.0237 + 0.000064 T
Ice (0°C)	2.18

*T in °C

Table 13.4. *Typical values of apparent thermal conductivity of soils (W/m- K)*

Soil	Normal range[1]	Recommended values	
		Low[2]	High[3]
Sands	0.60 - 2.50	0.75	2.25
Silts	0.85 - 2.50	1.65	2.25
Clays	0.85 - 1.65	1.10	1.55
Loams	0.85 - 2.50	0.95	2.25

Source: Sterling R. L. 1. Reasonable values for use when no site or soil specific data 2. Moderately conservative values for minimum heat loss 3. Moderately conservative values of maximum heat loss

SPECIFIC HEAT

The quantity of heat (Q) received by a body is proportional to its mass (M) and temperature change ($T_2 - T_1$). It also depends on the type of the material of the body. Mathematically, it is expressed as:

$$Q = MC(T_2 - T_1) \quad \text{...(13.17)}$$

$$C = Q/M(T_2 - T_1) \quad \text{...(13.18)}$$

Here Q is the quantity of heat, M is the mass of the sample (body), kg, ($T_2 - T_1$) is the temperature change, °C or K, C is the constant of proportionality or specific heat, which depends on the nature of the body. The unit of C is J/kg-K. Thus, specific heat, also known as specific heat capacity, is defined as the amount of heat required to raise the temperature of a unit weight (1 gm) of the substance by 1°C. The specific heat capacity of water is 4.18 J (or 1 calorie/gram °C), being much higher than most other substances. For example, the specific heat capacity of copper is only 0.39 J. Similarly, the specific heat of dry soil (0.2 cal/g°C) is only about one-fifth that of water. Thus, water is exceptionally good to regulate temperature. It is why; moist soils are cooler than dry ones. It is mainly attributed to their high specific heat besides having to spent heat energy in evaporating the soil moisture. Three amongst several methods to determine the specific heat are described in the following sections.

Constituents Method

Relative proportions of various soil constituents having different specific heat vary from one soil to another. Quartz is known to have the lowest specific heat amongst the major soil constituents while humus the highest after water. As a result, the specific heats of the soils vary. It is possible to calculate the specific heat of a soil, C_s from the summation of the specific heat times the mass of the individual constituents expressed as a fraction of the soil mass.

$$C_s\,(\text{cal/g-°C}) = C_1M_1 + C_2M_2 + C_3M_3 + C_4M_4 + \ldots\ C_nM_n \qquad \ldots(13.19)$$

Here C_s is specific heat of soil, M_1, M_2, M_3, M_4 ... M_n are the masses in fraction of the total soil mass and C_1, C_2 ... C_n are the specific heats of respective constituents. Eq. (13.19) can be simplified by including only major constituents and neglecting those contributing very little. Thus, eq. (13.19) reduces to

$$C_s = C_1M_1 + C_2M_2 + C_3M_3 + C_4M_4 \qquad \ldots(13.20)$$

Here subscript 1,2,3,4 denotes the respective values for soil minerals, soil water, soil organic matter and air. Even air can be neglected having little contribution due to its low specific heat and weight. Upon incorporating specific heat values of various constituents (Table 13.5), we have

$$C_s = 0.175M_1 + 1.00M_2 + 0.46M_3 \qquad \ldots(13.21)$$

Table 13.5. *Specific heat and volumetric heat capacities of various soil constituents*

Material	Density (g/cm³)	Specific heat (cal/g-°C)	Volumetric heat capacity (cal/cm³-°C)
Quartz	2.65	0.175	0.46
Many soil minerals*	2.65	0.175	0.46
Soil organic matter*	1.3	0.46	0.60
Water	1.00	1.00	1.00
Air	0.0012	0.00029	3.5×10^{-7}

*Approximate average values

Example 13.1. If the moisture content of the soil is 28%, calculate the specific heat of the soil.

If we consider 1 g soil, it has 0.72 g of soil minerals and 0.28 g of water. Applying eq. (13.21), we have

$$C_s = 0.175 \times 0.72 + 1.00 \times 0.28 = 0.126 + 0.28 = 0.41\ \text{cal/g-°C}.$$

The Radiation Method

The basic principle of this method is the Newton's law of cooling, which states that the cooling rate of a body is directly proportional to the temperature difference between the body and its surroundings, provided the temperature difference is small (Bowers and Hanks, 1962). Customarily, a substance with known specific heat and another of unknown value are heated to the same temperature and then placed in a radiation calorimeter to cool. By comparing the cooling curves of the two, it should be possible to compute the specific heat (Weber, 1950).

Apparatus: Radiation calorimeter, Thermometers, Stop watch, Weighing balance, Heater

Procedure

1. Ground, screen and oven-dry the soil at 50°C
2. Put the soil sample in one of the tubes of the radiation calorimeter. Fill the other tube with distilled water. Fill the soil tube also with water ensuring that the water level in both the tubes is nearly same. Normally, the tubes are filled to 4/5th of their volumes. Allow sufficient time for the soil to saturate
3. Insert both the tubes in boiling water and allow the temperature to rise to about 50°C
4. Remove the tubes, clean the water on the outer surfaces of the tubes and immediately place the tubes in the calorimeter (Fig. 13.8). Fix the stoppers and thermometers
5. Take the thermometer readings at 1 min intervals for about an hour

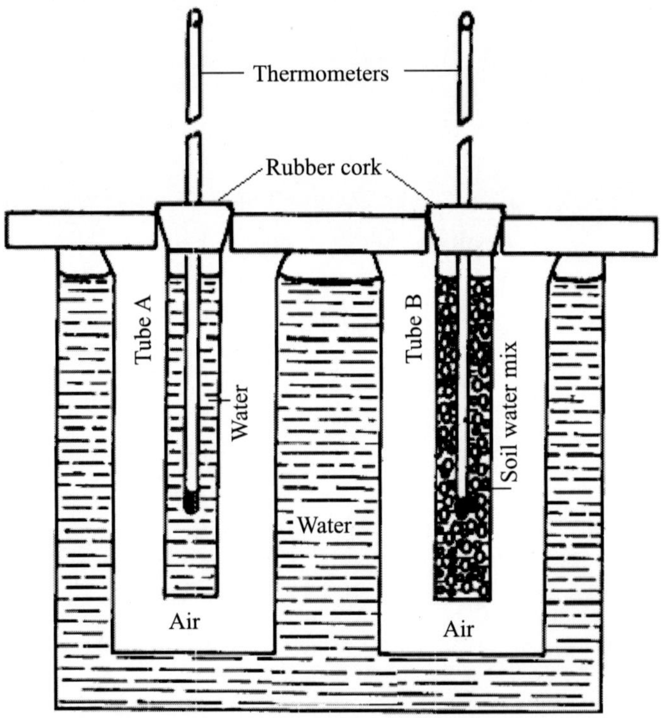

Fig. 13.8. *Radiation calorimeter for the cooling of the water and soil-water mixture*

6. Plot the cooling curve for both the tubes (Fig. 13.9). From this curve note down the temperatures and time in the range of 25 to 40°C, say 30 and 40 degrees.

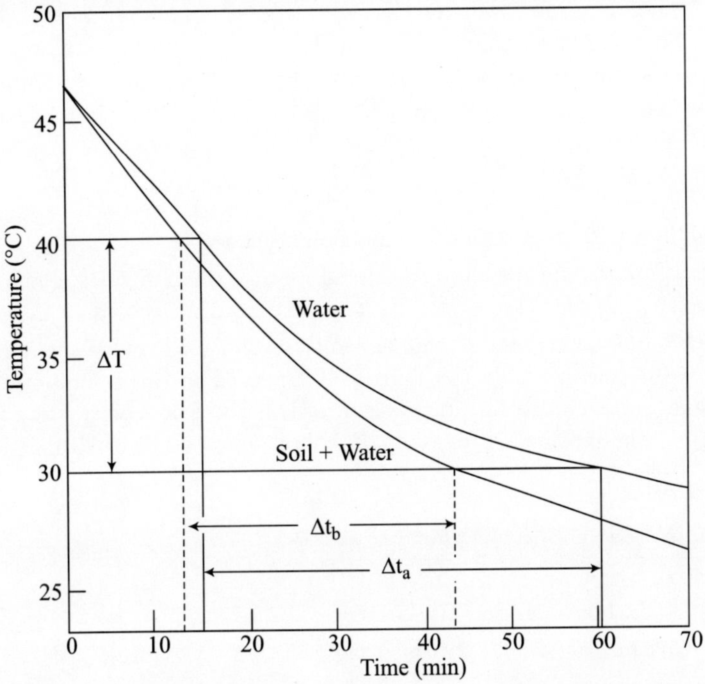

Fig. 13.9. *Cooling curves for selection of time for the temperature drop of* ΔT

Calculations

$$\Delta Q_a/\Delta t_a = (C_a + C_w\, m_{wa})\Delta T/\Delta t_a \quad ...(13.22)$$

$$\Delta Q_b/\Delta t_b = (C_b + C_s m_s + C_w m_{wa})\Delta T/\Delta t_b \quad ....(13.23)$$

When the two tubes have the same temperatures (not necessarily simultaneously) their net rates of energy loss are equal. Thus, for the same ΔT, comparing eq. (13.22) and eq. (13.23), we have

$$(C_a + C_w\, m_{wa})\Delta T/\Delta t_a = (C_b + C_s m_s + C_w m_{wb})\Delta T/\Delta t_b \quad ...(13.24)$$

$$C_s = [(C_a + C_w m_{wa})\Delta t_b - (C_b + C_w m_{wb})\Delta t_a]/m_s\, \Delta t_a \quad ...(13.25)$$

In these equations, C_a and C_b are calculated with the help of the following equation.

$$C_a = C_b = V_{th} \times 0.46 + m_c \times 0.090 \text{ cal/}^0\text{C} \quad ...(13.26)$$

In the above equations, ΔT is the temperature drop in °C for tubes A or B, Δt_a is the time in minutes for tube A contents to drop by ΔT, Δt_b is time in minutes for cup B contents to drop by ΔT, m_{wa} is the mass of water in tube A, g, m_{wb} is the mass of water in tube B, g, m_s is the g of soil sample in the tube B, C_a is the thermal capacity of tube A and thermometer, cal/°C, C_b is the thermal capacity of tube B and thermometer, cal/°C, C_s is the specific heat of soil, cal /g°C, C_w is the specific heat of water, cal/g°C, m_c is the weight of the calorimeter, g. Besides, 0.46 is volumetric specific heat of the thermometers and 0.090 is the specific heat of the calorimeter.

Mixing Method

The procedure to determine the specific heat using mixing method is proposed by Ebong et al. (2016). It comprises mixing two substances, one at a lower temperature say T_1 and the other at a higher temperature say T_2. If the two substances are mixed and the resulting temperature is T_3 then heat gained and heat lost are used to calculate the specific heat.

Apparatus: Weighing balance, Thermometer, Calorimeter, Heat source, Beaker, Water, Stop watch

Procedure

1. Weigh the calorimeter with stirrer and note its mass, M_{ca}
2. Put the sample into the calorimeter and weigh, M_1. Half fill the calorimeter with water. The calorimeter and the mixture are weighed and record the mass, M_2
3. The initial temperature of the mixture is taken as T_1. A known weight of the boiling water with its temperature T_2 and mass M_h is poured into the calorimeter. The change in temperature with change in time is recorded. The final temperature of the mixture is say T_3. The specific heat capacity of the soil sample is then calculated as follows.

Calculations

Heat gained by the calorimeter + sample + cold water is given as:

$$\text{Heat gained} = M_{ca}C_c(T_3 - T_1) + M_sC_s(T_3 - T_1) + M_cC_w(T_3 - T_1) \quad ...(13.27)$$

$$\text{Heat lost by the hot water} = M_hC_w(T_2 - T_3) \quad ...(13.28)$$

Equating eqs. (13.27) and (13.28), we have

$$M_hC_w(T_2 - T_3) = M_{ca}C_c(T_3 - T_1) + M_sC_s(T_3 - T_1) + M_cC_w(T_3 - T_1) \quad ...(13.29)$$

The specific heat capacity is then calculated as:

$$C_s = [M_hC_w(T_2 - T_3) - M_{ca}C_c(T_3 - T_1) - M_cC_w(T_3 - T_1)]/[M_s(T_3 - T_1)] \quad ...(13.30)$$

Here, C_s is the specific heat capacity of the soil sample, C_w is the specific heat capacity of water, C_c is the specific heat capacity of the calorimeter, M_c is the mass of cold water $(M_2 - M_1)$, and M_s is the mass of the soil sample.

VOLUMETRIC HEAT CAPACITY

The volumetric heat capacity of a material is the heat capacity of a sample of the substance divided by its volume. The volumetric heat capacity can also be expressed as the specific heat capacity (heat capacity per unit of mass, in J/K-kg) times the density of the substance (kg/m³). Thus, the units of volumetric heat capacity are J/m³-K. The heat capacity of the soil per unit volume can be computed by the following equation.

$$C_{vs}\,[\text{cal/cm}^3\text{-°C})] = X_SC_S + X_WC_W + X_oC_o + X_aC_a \quad ...(13.31)$$

Here C_{vs} is the volumetric heat capacity of the soil, X_s, X_w, X_o and X_a are the volume fractions of the solid mass, water, organic matter and air respectively. C_s, C_w, C_o and C_a are the volumetric heat capacities of the respective constituents. Since the solid mass of the soil consists of mineral and organic matter whose heat capacities per unit volume are approximately 0.45 and 0.60 respectively (Table 13.5) and the value of the air component is too small to be significant, we have.

$$C_{vs} = 0.46X_s + 0.60X_o + X_w$$

Chapter 14

Miscellaneous Properties

Soil physical properties are evaluated for various purposes including the assessment of soil quality. Some of these properties are closely interlinked with each other. The others often lack such interlinks with other properties. Nonetheless, these are important entities to describe many physical and chemical processes. Some such properties like specific surface area, soil repellency, capillary rise, solid-liquid contact angle, free-swell characteristics, cracking and coefficient of linear extensibility have been put together in this chapter.

SPECIFIC SURFACE AREA

Specific surface area (SSA) of a soil is its intrinsic property governed by grain-size distribution and clay mineralogy. The specific surface area of a soil is the total surface area contained in a unit mass or unit volume of soil usually expressed in units of cm^2/g or cm^2/cm^3. Specific surface area increases with decreasing particle size. For example, if the size of each side of a cubical particle is 1 cm, then specific surface area will be 6 cm^2/cm^3. If the same cube is divided into 10 cubes, each of 0.1 cm (1 mm) sides, its SSA will increase to 60 cm^2/cm^3. In practice, SSA of sand particle 1 mm size and of cubical shape would be 60 cm^2/cm^3, whereas it will be 60,000 cm^2/cm^3 for clay particle of 0.001 mm i.e. around 1000 times more than the sand particle. Besides size, shape and type of clay minerals also affect specific surface area. Plate shaped clay particles exhibit higher SSA than cube shaped particles. Swelling clays such as montmorillonite have SSA ranging from 600-800 m^2/g compared to non-expanding clays such as kaolinite, typically having SSA ranging from 30 to 50 m^2/g (Table 14.1, Phogat et al., 2015).Thus, the type of clay minerals present in the soil have a major influence on SSA and consequently other soil properties. SSA is critical in many chemical and physical processes like cation exchange capacity, availability of nutrients, adsorption of water, swelling, shrinkage, plasticity, and soil strength etc. Soils with high specific surface areas have high water holding capacities, more adsorption of contaminants, and greater swell potential. The SSA is also a dominant parameter in modelling water and solute transport especially chemical species that adsorb on the mineral surfaces.

SSA has been physically divided into two categories namely external or particle surface area, and internal or intra-crystalline surface area. Both of them govern physical, chemical, and biological properties of soils in different ways. For example, swelling behavior of a soil is linked to internal SSA while adsorption of non-polar materials to the external SSA.

Table 14.1. *Specific surface areas of soil separates, soils of different textures and clay minerals*

Soil separates/ soil	Diameter (mm)	Specific surface area (cm^2/g)	Clay mineral	Specific surface area (m^2/g)
Coarse sand	2.0-0.2	45	Kaolinite	37-45
Fine sand	0.2-0.02	446	Illite	120-170
Silt	0.02-0.002	4458	Chlorite	130-180
Clay	10^{-4} - 10^{-6}	1000×10^4	Montmorillonite	580-750
Sandy loam		$10 \times 10^4 - 40 \times 10^4$	Vermiculite	780-900
Loam		$50 \times 10^4 - 100 \times 10^4$		
Clay		$150 \times 10^4 - 250 \times 10^4$		

EGME Method

In most approaches used to measure SSA, some liquids or gases of known characteristics are allowed to adsorb on the soil surface. The amount adsorbed is then used to calculate the SSA. The most common method uses ethylene glycol monoethyl ether (EGME) for this purpose (Carter et al., 1986). The process comprises equilibrating the soil samples in a vacuum over a $CaCl_2$- EGME solvate. The specific surface is then determined from the mass of EGME retained in comparison to the amount retained by pure montmorillonite clay, which is assumed to have a SSA of 810 m^2/g (Carter et al., 1986). An easier approach to calculate the specific surface area is to use the standard value of 0.000286 g of EGME forming a monomolecular layer on 1 m^2 surface area. It obviates the need to conduct a separate test with pure montmorillonite clay. The typical measurement consists of 3 steps and may take around 2 days to complete the test.

Pre-drying of Soil Samples

Weigh and record the weight of each oven-dried empty weighing bottle. Weigh ~ 1.1 g soil sample passing a #40 sieve in weighed weighing bottle. Spread the sample evenly over its bottom. Place the weighed bottles with samples in a vacuum desiccator over about 250 g of P_2O_5. Evacuate the desiccator using a vacuum pump for one hr, close the stopcock, and dry the sample to constant weight. Record the weight of P_2O_5-dried sample.

Preparation of $CaCl_2$-EGME Solvate

Weigh ~ 120 g of $CaCl_2$ passing through 40-mesh into a 1 L beaker and dry in an oven at 210°C for one hr or more to remove all traces of water. Take 21.5 mL or 20 g of EGME and pour it in a 400 mL beaker. Remove the $CaCl_2$ from the oven, weigh 100 g without cooling, and add it to the beaker containing the EGME. Mix immediately and thoroughly with a spatula. After cooling the $CaCl_2$-EGME solvate, transfer it to a glass vessel. Keep the vessel with its contents in sealed desiccator.

Sorption of EGME by Soil

Wet the P_2O_5-dried soil sample with approximately 3 mL of EGME solvate to form soil-adsorbate slurry. Place the weighing bottles containing the sample-adsorbate slurry into a vacuum desiccator containing anhydrous $CaCl_2$ at its bottom. Transfer the vessel containing $CaCl_2$-EGME solvate into the same vacuum desiccator. It takes about 30 min for sample–solvate slurry to equilibrate. Evacuate the desiccator with a vacuum pump for about 1 hr. Allow the desiccator to stand at room temperature for 4 to 6 hr. Release the vacuum, open the desiccator and close each weighing bottle with its cap to prevent the sample adsorbing moisture from the atmosphere. Take the weight of each weighing bottle containing the sample and record their weight. Each weighing bottle with their caps removed is then returned to vacuum desiccator. Repeat steps beginning with evacuate the ---- desiccator onwards in this paragraph until the sample attains a constant weight (Cerato and Lutenegger, 2002).

Calculation

The total specific surface area is calculated by the equation:

$$A = W_a / (W_s \times 0.000286) \qquad ...(14.1)$$

Here A is the total specific surface area in m^2/g, W_a is the weight of EGME retained by the sample in g, and W_s is the weight of P_2O_5-dried sample in g. As already stated 0.000286 is the weight of EGME required to form a monomolecular layer on a m^2 surface area.

Example Calculations

Initial tare weight: 30.0485 g

Initial soil weight (W_s): 1.0030 g

18 h tare + slurry weight: 31.0634 g

18 h slurry weight: 31.0634 – 30.0485 = 1.0149 g

24 h tare + slurry weight: 31.0625 g

24 h slurry weight: 31.0625 – 30.0485 = 1.014 g

W_a = 1.0140 – 1.0030 = 0.011 g

Total SSA = 0.011 g/(1.0030 g × 0.000286) = 38 m^2/g

External Specific Surface Area

To measure the external specific surface area, soil sample is heated at 600°C for 2 hr. This sample instead of the air-dry soil is used to determine the specific surface area. The procedure remains the same as discussed in the previous section of this chapter.

Internal SSA = Total SSA – External SSA

Methylene Blue Absorption Technique

Methylene blue (MB), $C_{16}H_{18}ClN_3S$ has been used to measure the surface area of clay minerals for several decades (Kandhal and Parker, 1998 and Santamarina et al., 2002). Methylene blue in aqueous solution is a cationic dye, $C_{16}H_{18}N_3S^+$, having the molecular weight of 319.87 g/mol. It absorbs to negatively charged clay surfaces (Chen et al. 1999). Hence, the amount of absorbed methylene blue has a close relationship to the specific surface area of particles. The surface area covered by one methylene blue molecule is typically assumed to be 130 Å^2 (1 Å = 0.1 nm).

Apparatus and Reagents: Sheets of filter paper (Fisher brand P5, diameter 12.5 cm), 10 mL pipette with 1/10 readings, A small drop-pipette, Glass beaker, De-ionized water and Oven to dry the sample

Methylene Blue Solution: Mix 1g of dry methylene blue chloride powder with 200 mL of de-ionized water in a 500 mL glass beaker. Stir the solution until the powder dissolves

Preparation of Soil Solution

Mix 10 g of the oven-dry soil with 30 mL of de-ionized water. Stir the solution until the sediments are suspended.

Spot Test

Methylene blue solution is added to the soil solution in 0.5 mL increments using the pipette. For each addition of methylene blue, mix the soil suspension for 1 min by swirling the solution in the beaker. If available, an electric stirrer may be used. After each addition of methylene blue, remove a small drop of the suspension with the drop-pipette (does not have to be an exact amount), and drop it on Fisher brand filter paper P5. The end point has reached, once the unabsorbed methylene blue forms a permanent light blue halo around the wet soil aggregate spot on the filter paper. It means the methylene blue has replaced all cations in the double layer and coated all of the mineral surfaces. The amount of MB used in the process is known as a ballpark amount. Continue counting the number of additions of methylene blue. Determine the specific surface area from the amount of methylene blue (number of additions) required to reach the end-point. Store the leftover methylene blue solution in an appropriately labeled plastic bottle for further use.

Calculations

Calculate the specific surface area with the help of the following equation.

$$S_s,\ \mathrm{m^2/g} = [(1/319.87)(1/200)(1/10)](0.5\ \mathrm{N})]\ Av\ A_{mb}$$

Here N is the number of MB increments added to the soil suspension solution, Av is Avogadro's number (6.02×10^{23}/mol), and A_{mb} is the area covered by one MB molecule (typically assumed to be 130 Å^2). Simplifying, we have

$$S_s,\ \mathrm{m^2/g} = 0.78 \times 10^{-6}\ \mathrm{N}\ Av\ A_{mb} \qquad ...(14.2)$$

Note: This test takes a long time, especially for swelling clays. If one is aware that the sample contains a large amount of very fine clay particles or swelling clay minerals with a large surface area, a quick first run is performed using larger than 0.5 mL increments until the end point is reached. Then in the second run, increments of 0.5 mL are used well in advance of reaching that ballpark amount. One has to be very careful when MB is added in an amount larger than 0.5 mL at a time. It requires much longer time than the suggested 1 min of mixing/equilibration. It is also possible to improve the precision by means of analytical techniques notably the use of a spectrophotometer to determine the amount of absorbed MB (Santamarina et al., 2002).

Specific Surface Area: Soil Characteristics Curves

Quirk (1955) made a comprehensive review of measurement techniques for determining SSA. He concluded that water is not an appropriate liquid to calculate SSA. Water clusters around cation sites result in considerable measurement errors. Recently, modern methods to measure the energy state of water in the soil have come into being, and such methods have proved their potential to measure SSA using water. Campbell and Shiozawa (1992) for example, correlated specific surface of six soils by measuring the slope of a moisture release curve and found excellent correlation. The slope in this case (x-axis value) is the water content of the sample at a water potential of -123 MPa (Fig. 14.1). While six points relate to data of six soils, a point on the curve is for Ca-montmorillonite. The excellent correlation shows that water potential of -123 MPa can be used to measure the specific surface area of the soils.

A recent paper by Tuller and Or (2005) obtained the following equation relating the surface area and the moisture characteristic:

$$W = [k/(6\pi \rho w \Psi)]^{1/3} \rho_w S \quad ...(14.3)$$

Here W is water content (g/g), ρ_w is the density of water (1000 kg/m^3), Ψ is the water potential (J/kg), S is the specific surface (m^2/kg), and k is the Hamaker constant, which is taken as -6×10^{-20} J. They concluded that the psychrometric approach to determine the specific surface area should provide reliable values for natural soils with surface areas below 200 m^2/g. As a first approximation, they recommend the use of water content at –10 MPa or lower (drier) to predict SSA values.

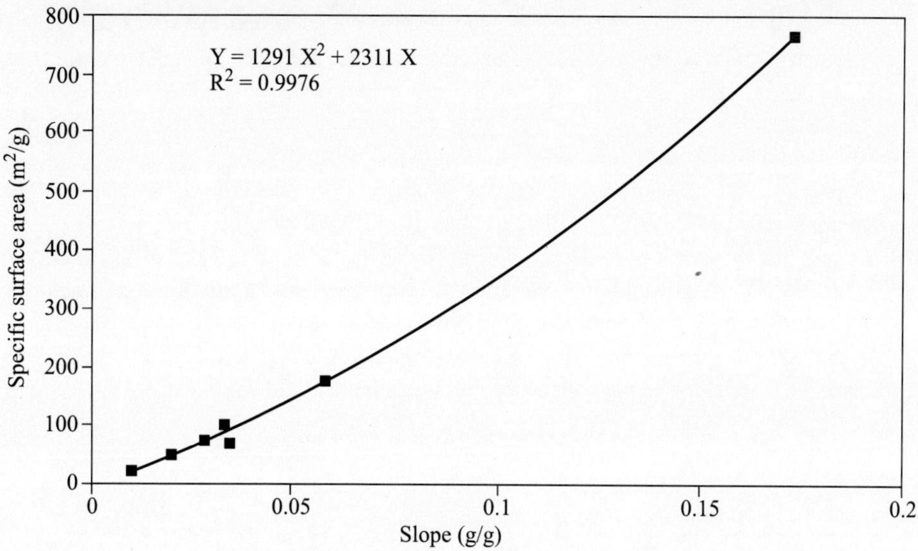

Fig. 14.1. *Correlation of water content of the soil at a water potential of –123 MPa to its specific surface area*

SOIL REPELLENCY

Soils that repel water are considered hydrophobic or water repellent. In severe cases, water sits as droplets on top of the dry soil and the soil fails to wet up. Sand grains in some sands to sandy clay loams may become water repellent as they get coated with organic residues of some plant materials or even following intense heating (King, 1981). Water repellency in soil can also be induced by long-term irrigation with treated sewage effluent. Water repellent soils do not permit normal infiltration of water. In such cases, water pools on the surface, or evaporates, or moves down through 'preferred pathways' leaving large volumes of soil dry (Dekker and Ritsema, 1995). Farming practices such as zero/no tillage and stubble retention can be adopted to manage water repellent soils.

Water Drop Penetration Time Method

The method described herein is after USDA–NRCS (2000b), with some modifications related to water drop penetration time (WDPT) proposed by Letey (1969), Wallach et al. (2005), Schoeneberger et al. (2012), and Department of Sustainable Natural Resources (2015). The procedure, semi-quantitative in nature, measures the rate of disappearance of water drops on a soil surface to characterize water repellency. The method is applicable to undisturbed as well as disturbed soil samples (Steenhuis et al., 2001).

Apparatus: Knife or some other tool to scrape and excavate soil, Eye dropper, Deionized (distilled) water

Procedure

1. Scrape away the top layer to expose mineral soil surface. Prepare a clean, level, 15×15 cm^2 horizontal area at the desired depth of investigation
2. Put a drop of deionized water (approximately 6 mm diameter) from a height of about 1.0 cm so that no crater is formed on the soil surface. Put 5 such drops one by one and note down the time taken by the soil to absorb each drop of water
3. If the drop attains a spherical shape on the soil surface, water repellency is indicated. If the bead remains even after 1 min, the soil is hydrophobic (USDA–NRCS, 2000). It indicates that the surface tension of the soil surface is less than that of the droplet within a particular time interval
4. Average out the time taken by the drops for absorption. Assess the class based on the time taken as per Table 14.2 (Bisdom et al., 1993, Wallach et al., 2005)
6. To find out the depth of hydrophobic soil layer, test is repeated at various depths to identify the first hydrophobic layer and then continued until a non-hydrophobic layer is encountered. The difference in depth gives the hydrophobic soil layer
7. In the laboratory tests, 5 to 30 g of air-dry soil sample is taken. It is levelled with a spatula. Other steps of the procedure remain same as in the field test.

Table 14.2. *Categorization of soil for repellency on the basis of time taken for the water drop to infiltrate*

Class	Name	Time taken (s)
Class I	Wettable, not water repellent	< 5
Class II	Slightly water repellent	> 5 to <60
Class III	Strongly water repellent	> 60 to < 600
Class IV	Severely water repellent	> 600 to < 3600
Class V*	Extremely water repellent	> 3600 to < 3 hr
Class VI	Extremely water repellent	3-6 hr
Class VII	Extremely water repellent	>6 hr

*Class V modified and class VI and VII added by Dekker et al. (2009)

Report: Report positive or negative results for hydrophobicity. If positive, report depth to layer (cm) and thickness of layer (cm).

Soil Repellency: Ethanol Droplet Test

The degree of soil repellency to water can be determined using the Molarity of an Ethanol Droplet (MED) test (Kořenková et al., 2015). In principal, the MED test is an indirect measure of the surface tension of the soil surface, utilizing standardized solutions of ethanol in water of known surface tensions. The surface tension of the solution increases with the decreasing concentration (molarity) of ethanol. Drops of ethanol dilutions are applied onto a smoothed soil surface in a similar manner as for water droplet test. A droplet of lower surface tension (higher ethanol concentration) infiltrates into the soil quite rapidly than a droplet with a higher surface tension (lower ethanol concentration). Droplets are usually applied in order of increasing surface tension until a droplet resists infiltration. A completely wettable soil will be readily wet at zero molarity of ethanol. Soil is categorized as per the guidelines reported in Table 14.3 following an observation made in 3 seconds (Doerr, 1998). Surface tension γ_{ND} of an ethanol droplet, which wets the soil at 90° (Carrillo et al., 1999) and the surface tension of a solid/air interface γ_{sa} are calculated to assess the degree of repellency. These calculations are beyond the scope of this book.

Table 14.3. *Guidelines on soil repellency based on absorption of ethanol of varying molarities*

Molarity of ethanol (M = mol/L)	**Soil category**
0–0.85	Wettable soil
0.85–1.45	Slightly water repellent (WR) soil
1.45–2.22	Moderately WR soil
2.22–3.07	Strongly WR soil
3.07–6.14	Very strongly WR soil
> 6.14	Extremely WR soil

Infiltration Ratio Test

Soil water repellency adversely affects water infiltration into the soil and consequently several soil functions, such as rainfall-runoff relations and soil erosion (Lowe, 2019). The ratio of infiltration rates using water and an aqueous solution of soil wetting agent has been used as a guide to the degree of water repellency (Murphy, 1986).

Apparatus and Reagents: Same as in double ring infiltrometer test (Chapter 8)

- **Ethanol 1% aqueous solution**

Procedure

Conduct the infiltration test with water and 1% aqueous solution of ethanol as described in Chapter 8.

Calculations

The infiltration ratio, IR is given by the following equation:

$$IR = I_a/I_w$$

Here I_a is the infiltration rate with the wetting agent/surfactant (1% solution), and I_w is the infiltration rate with water alone. Following guidelines are used to rate the soil repellency based on IR values (Murphy, 1986). Beatty and Smith (2014) observed that comparison of the late-time infiltration rates of aqueous solutions of varying ethanol concentrations proved a useful indicator of repellency and fractional wettability effects.

Infiltration ratio	Rating
<2	Not significant to very low
2–5	Low
5–15	Moderate
15–25	Severe
>25*	Very severe

* Values up to 1000 have been recorded

PORE SIZE: CAPILLARY RISE TEST

The phenomenon of capillary rise of water in soils is of great significance. This phenomenon is quite significant in saline environments owing to its potential to move the salts from below layers into the root zone whether present within the soil or in irrigation water. In drainage designs, it helps in deciding the depth to water table, which should be achieved to maintain favourable air-water regime for crop growth. Besides, it has applications in several engineering applications. The capillary rise test is a very quick method to characterize the pore size of the soils.

Principle

Water rises in a glass tube as soon as the tube is dipped in water. It is because of surface tension and adhesive properties of water. While the adhesion between water molecules and the glass surface tends to make the water move upward, surface of water in the capillary concaves as the surface tension tends to flatten the surface in order to minimize the surface area (Fig. 14.2 left). The surface tension acts all along the circumference ($2\pi r$), tangential to the surface and is inclined at an angle (θ). Thus, upward force supporting the column of water is surface tension while the weight of the water column acts as a force that tends to pull the water column down (Means and Parcher, 1965). At equilibrium, both the forces are equal.

$$\text{Upward force} = 2\pi r\ \sigma \cos\theta \quad ...(14.4)$$

$$\text{Downward force} = \pi r^2 \rho g h \quad ...(14.5)$$

Here r is the radius of the capillary, cm, σ is the surface tension (dynes/cm), θ is the contact angle between glass and water, h is the height to which the water rise in the capillary, cm, ρ is the density of water, g/cm^3, and g is the acceleration due to gravity, cm/s^2. At equilibrium, it is possible to equate the two equations as follows:

$$2\pi r\ \sigma \cos\theta = \pi r^2 \rho g h$$

$$h = 2\sigma \cos\theta/(\rho g r) \quad ...(14.6)$$

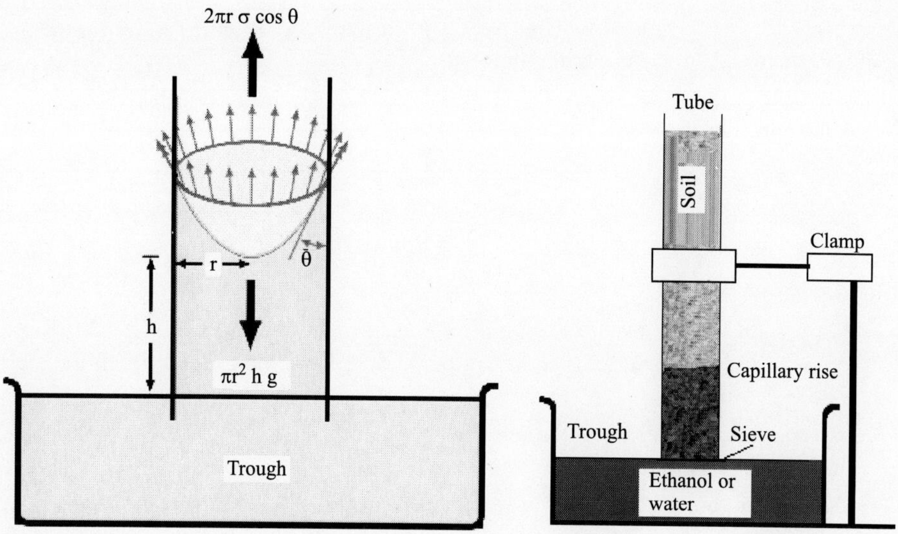

Fig. 14.2. *Schematic view of capillary rise phenomenon (left) and the experimental set-up (right)*

Considering the contact angle between the glass surface and water equal to zero, eq. (14.6) becomes

$$h = 2\sigma/(\rho g r) \qquad ...(14.7)$$

Equation (14.7) is used to calculate the height of water to which it rises in a capillary tube at steady state. By implication the same equation applies to soils where individual soil pore channels are treated as capillaries. This equation can be used to calculate the size of the soil pores (see next section on contact angle) from known heights of water rise.

Apparatus: Glass tube of 2-3 cm diameter, 70-80 cm long with paper strips graduated in cm pasted over them, A trough of water, Stands for holding the tubes, A spoon and rubber hammer to fill soil in glass tubes, Cheese cloth and string.

Procedure

1. Take dry soil samples and sieve them through 2-mm sieve
2. Place two folds of cheese cloth on one end of each glass tube and tie it firmly with the string
3. Pack the soil samples in tubes, spoon by spoon while gently tapping the sides by the rubber hammer
4. Place a small piece of filter paper at the top of tubes and place all the tubes in water at the same time
5. Record the height of the water rise in the tube after the lapse of 10 min, 30 min, one hr, 3-4 hr, two days, etc.
6. Observe and note down the time and height of water rise

Time of observation	Cumulative time from the beginning (min)	Cumulative height of water rise (cm)			
		I	II	III	IV

Calculations

- Plot heights of water rise versus time for each case. If the soils in all tubes are the same, average time versus height can be plotted. Note the time at which the steady state is reached
- Use eq. (14.7) to calculate the minimum pore size from the height of water rise in columns
- Make histogram of the height of water rise for each soil type for comparison.

LIQUID-SOLID CONTACT ANGLE

If we place a drop of liquid upon a dry solid surface, in an attempt to spread over the surface it displaces the gas covering the solid surface. When it's spreading ceases, the edges of the drop come to rest, its interface with the gas forms a typical angle at its interface with the solid. This angle is termed as the contact angle (Fig. 14.3). The contact angle of a wetting liquid is smaller than 90°, whereas the contact angle of a non-wetting liquid is in the range of 90 and 180°. The contact angle of a given liquid on a given solid under the given physical conditions is generally constant. The wetting angle of pure water upon a clean and smooth inorganic surface is generally zero, but it can be considerably greater than zero, if the surface is rough or coated with adsorbed surfactant of hydrophobic nature. The liquid-solid contact angle is used as an index to the wettability or hydrophobic: hydrophilic properties of the soil. The capillary rise of water also depends upon the contact angle, increasing with decreasing contact angle. Besides, the contact angle impacts soil infiltration, capillary rise, moisture retention, saturated hydraulic conductivity, and evaporation (Letey et al., 1962).

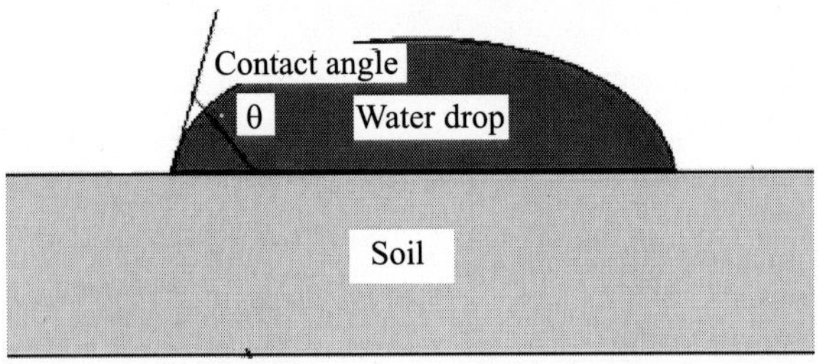

Fig. 14.3. *A schematic view of the contact angle for a wetting liquid*

Theory

The contact angle, θ is normally determined using capillary rise method (Liu et al., 2014) discussed in the previous section of this Chapter. The capillary rise can be simply calculated by means of the Laplace formula given by eq. (14.6). Since, it takes time to reach equilibrium capillary height (may be 4-6 days), we can also use Lucas-Washburn's equation (Washburn, 1921, Leelamanie et al., 2008), which is derived from Poiseuille's law of liquid flow in capillary rise. It states that a liquid will penetrate a distance, h into the capillary whose pore radius is r as per the following relationship:

$$h^2 = rc\,(\sigma \cos\theta / 2\eta)t \qquad \text{...(14.8)}$$

Here h is the height of the rising liquid in time t, m, σ is the surface tension of the test liquid, N/m, ρ is the density of the test liquid, kg/m^3, θ is the contact angle of the liquid on the solid, degrees, r is the mean static radius of the pores, m, η is the liquid viscosity, Pa-s, and c is the geometric factor, m^5. Since, this equation has three unknowns, we can combine r and c as a single variable and denote it as K, such that

$$h^2 = K\,(\sigma \cos\theta / 2\eta)t \qquad \text{...(14.9)}$$

These relations show that capillary penetration and fluid transport through porous structures exhibit diffusive behaviour akin to that which occurs in numerous physical and chemical systems. The other form of this equation in terms of mass is given as:

$$m^2 = C(\sigma \cos\theta / 2\eta)t \qquad \text{...(14.10)}$$

Here C is the capillary constant of the soil and includes the number of micro-capillaries and their mean radius. It depends on the nature of the soil and of the measuring tube. The parameter m is the increase in mass of the tube with respect to time. The three equations namely eq. (14.6), eq. (14.9) and eq. (14.10) each have two unknowns, θ and r, θ and K and θ and C (if m is measured) respectively. Therefore, capillary rise study with a reference liquid for which θ is known is conducted simultaneously. The reference liquid is usually an organic solvent with low surface energy, such as ethanol, hexane or pentane. Ethanol is proposed in this write-up. Considering the comparatively high surface energy of soil particles and low surface tension of the reference liquid, the contact angle between the reference liquid and soil is approximately zero such that $\cos\theta = 1$. Other values for the reference liquid and water are as follows.

σ for ethanol = 22.39 dynes/cm, ρ for ethanol = 0.810 g/cm, Viscosity of ethanol (Pa-s) = 0.001at 25^{0}C, $\cos\theta$ for ethanol = 1

σ for water = 72.75 dynes/cm, ρ for water = 0.998 g/cm, viscosity of water = 8.90×10^{-4} Pa-s at 25^{0}C, $\cos\theta$ = ?

Apparatus and Reagent: Trough, Capillary tubes, Cheese cloth or circular metallic sieves about 1 cm higher diameter than the external diameter of the tubes, Glue if sieves are used, Scale, Stop watch, Mass measurement arrangement

- **Ethanol**

Procedure

1. All tubes and vessels are rinsed with acetone and then dried before the tests are conducted
2. Take appropriate amount of soil sample passing through 2-mm sieve
3. Take two tubes of 2.25 cm internal diameter 60 cm long (Higher internal diameters tubes up to 5.75 mm have been used but narrower the better). Put cheese cloth or sieve pre-treated with paraffin dissolved in xylene on one end of the tube. The treatment with these chemicals helps to prevent the tube from becoming wetter than soil

4. Pack the soil in each tube to a desired bulk density
5. Place the tubes in troughs containing water and ethanol in such a way that the end with cheese cloth is immersed in the respective fluids (Fig. 14.2 right)
6. The height of capillary rise in the soil columns of both water and ethanol is checked periodically and recorded until about 120 h or more
7. If eq. (14.10) is to be used then suspend the tube from a force sensor to determine its mass with respect to time of measurement

Time	Capillary rise of water, h (cm)	h^2	Mass (kg)	Capillary rise of ethanol, h (cm)	h^2	Mass (kg)
6			$m_6 = M_6 - M_0$			
9			$m_9 = M_9 - M_0$			
12						
15						
--						
--						
150			$m_{150} = M_{150} - M_0$			

M_0, M_6....M_{150} are the respective values at respective times and m_6 and m_{150} are the increase in weight at respective times

Calculations

Effective pore radii or K from ethanol test

For steady state value of h for ethanol ($\theta = 0$), calculate the value of r from eq. (14.6):

$$r = 2\sigma/\rho gh$$

Similarly, for the value of h^2 for ethanol ($\theta = 0$), calculate the value of K from eq. (14.9). Note that in this case h should lie on the linear portion of the h^2 versus t plot:

$$K = 2\eta h^2/t\sigma$$

Cos θ for Water

Knowing the effective value of r, use the value of h for water in eq. (14.6) to calculate cos θ.

$$\cos\theta = h\rho gr/2\sigma = \text{say } A$$

$$\theta = \cos^{-1} A$$

Knowing the effective value of K, use the value of h^2 for water in eq. (14.9) to calculate cos θ.

$$\cos\theta = 2\eta h^2/\sigma Kt = \text{say } A$$

$$\theta = \cos^{-1} A$$

Higher values of the liquid-solid contact angle indicate that either the soil has poor water transmission characteristics or it is water repellent in nature.

If masses are noted, eq. (14.10) can be used in the same manner as eq. (14.9) by first calculating the value of C and then θ.

Note: Letey et al. (1962) observed that value of θ decreases with increasing h in eq. (14.6). Therefore, it is utmost necessary to use the steady state value of h in the equation. Letey et al. (1962) observed that such a state in soils reaches at about 120 hr.

Alternate Approach

A better approach to apply eq. (14.9) is to plot the data on h^2 versus t of both the ethanol and water on an arithmetic paper. The slopes of the straight line portions Y and Y_1 respectively are then used to calculate K and θ as follows.

$$K = 2\eta\, Y/\sigma \qquad \text{...(14.11)}$$

$$\text{and } \cos\theta = 2Y_1\eta/\sigma K = \text{say } A \qquad \text{...(14.12)}$$

$$\theta = \cos^{-1} A$$

Similar analysis procedure is used for the change in mass and time data sets for eq. (14.10).

FREE SWELL TEST

Free swell index of a soil has been defined as the increase in the volume of soil upon submergence in water with no external constraints (BIS, 1977, Sridharan and Prakash, 2000). Earlier Keen and Rackzowski (1921) method was described to determine the swell volume of the soils (See chapter 6). The objective of this section is to outline a procedure to determine the swell potential of fine-grained soil and assess its dominant mineralogical clay composition.

Apparatus and Reagents: 425 μm IS sieve, Measuring jars of capacity 100 mL

Kerosene or carbon tetrachloride (CCl_4) (Sridharan et al., 1990)

- Distilled water

Procedure

1. Take two separate samples 10 g each of oven-dry soil passing 425 μm IS sieve in two 100 mL measuring jars
2. Add distilled water to one of the jars to make up the volume to 100 mL. Add kerosene to the other jar to make up the volume 100 mL
3. Mix the contents of each jar with glass rods to expel any air entrapped in the mixture. Keep them undisturbed for about 24 hr so that soil samples settle freely
4. After 24 hr, note down the equilibrium sediment volumes of the soils in the jars containing distilled water (V_d) and kerosene (V_k).
5. Calculate the free swell index and free swell ratio of the fine-grained soil. Classify the swell potential of the soil and predict its dominant clay mineralogical composition as follows.

Observations and Calculations

Soil type:

The dry weight of the soil taken

Equilibrium sediment volume in distilled water = V_d

Equilibrium sediment volume in kerosene = V_k

Free swell index, % FSI = $100(V_d - V_k)/V_k$ ...(14.13)

The free swell index is reported to the nearest whole number. Using the same data sets, free swell ratio is calculated as follows:

Free swell ratio = (FSR) = V_d/V_k

Assess the swell potential as per Table 14.4 (BIS, 1980a, IS 2911) and Table 14.5 (BIS, 1970) and mineralogy as per Table 14.6 (Prakash and Sridharan, 2004).

Table 14.4. *The degree of expansiveness and consequent damage to the structures with light loading*

The degree of expansiveness	Free swell index (%)
Low	Less than 20
Moderate	20 to 35
High	35 to 50
Very high	Greater than 50

Table 14.5 *Expansive soil classification based on IS: 1498 (BIS, 1970)*

The degree of expansion/ swell potential	Liquid limit	Plasticity index	Shrinkage index	FSI (%)	Degree of severity
Low	20-35	< 12	< 15	< 50	Non-critical
Medium/Marginal	35-50	12-23	15-30	50 – 100	Marginal
High	50-70	23-32	30-60	100 – 200	Critical
Very High	70-90	> 32	> 60	> 200	Severe

Table 14.6. *Expansive soil classification based on FSR (Prakash and Sridharan, 2004)*

Free swell ratio	Clay type	The degree of soil expansiveness	Dominant clay mineral type
≤ 1.0	Non-swelling	Negligible	Kaolinitic
1.0 – 1.5	Mixture of swelling and non-swelling	Low	Mixture of Kaolinitic and Montmorillonitic
1.5 – 2.0	Swelling	Moderate	Montmorillonitic
2.0 – 4.0	Swelling	High	Montmorillonitic
> 4.0	Swelling	Very high	Montmorillonitic

Notes:

- Guidelines in Table 14.4 are used to decide the kind of foundations for civil engineering structures. One has to go for pile foundation, if free swell index (%) is more than 50.
- In Table 14.5 degree of expansion/swell potential is assessed independently with each parameter i.e. liquid limit, plasticity index, shrinkage index and FSI. The measurement procedures of some of these parameters are included in Chapter 12
- The guidelines may vary from one source to another in the literature. BIS guidelines are only included in this section.

SOIL CRACKS (TAPE AND ROD METHOD)

Cracking refers to the formation of cracks in drying soils due to shrinkage. It is a common feature of heavy clay soils including Vertisols (Ahmad, 1996). Cracks, in general refer to soil cracks that are open to surface. Sometimes, confusion prevails whether the cracks pertain to surface crusting or are the real cracks. To overcome this confusion, cracks 2 cm deep or 1 cm wide are referred as real cracks, otherwise not. The intensity of cracks is judged by the width, depth and spacing of cracks. Since the volume of cracks takes into account all these parameters, it is a good measure of crack intensity. The volume of the cracks is affected by the length of drying period, and the kind of soil and its clay content. Wetting in general reduces the volume of cracks. Clearly, crack dimensions are liable to change on short-term and seasonal time scales. The crack volume affects water infiltration rate, water required for irrigation especially the first irrigation and porosity distributions (Stewart et al., 2016). Commonly used methods to measure crack volume are:

- Tape and rod method
- The displacement method
- The Cast and excavate method
- Correlation methods

In the tape and rod or manual method, representative cracks are physically measured along a transect (Dasog et al., 1988, Ringrose-Voase and Sanidad, 1996). This method is simple, repeatable, and non-destructive, although labour intensive. In the displacement method, a water-filled bladder with instrumented standpipe is used to measure changes in crack width, which is reflected in water level variations (Stewart et al., 2012, Stewart and Abou Najm, 2017). The method is suitable only for measuring large cracks where it is possible to install the bladder. The cast and excavate method uses a hardening substance such as liquid latex to create three-dimensional casts of the soil cracks (Abou Najm et al., 2010). It is capable of revealing crack morphology and topology, which cannot be captured with other methods. Being destructive in nature, it is not possible to have repeat observations. Finally, several empirical relations that relate total crack area per unit soil surface with other soil parameters such as clay content, COLE and volumetric moisture content are used to calculate crack volumes (Yassoglou et al., 1994). Amongst the field methods, manual measurement of the geometry of cracks is commonly adopted to determine the intensity of the cracks. The limitations of this method is that it may be biased towards large cracks, requires assumptions of crack geometry to determine crack volumes and does not provide information on subsurface connections between creacks.

Apparatus: Frame: 1 m by 1 m with 0.1 m inside grid (may be of different sizes and shapes), Steel straps 6.35-mm wide and 0.79-mm thick of various lengths or steel wire of sufficient thickness, Flexible retracting tape

Procedure

1. Select an area for measurement of soil cracks
2. Depending upon size of the area and cracks geometry, select sample area, which may vary from 4 m^2 to 400 m^2. The sample area may be taken along a transect or in a random manner (Dasog et al., 1988)
3. Measure the crack length and width approximately every 10 cm using a flexible retracting tape unless a significant change is noticed
4. Measure depth of the crack relative to the soil surface to the nearest mm with steel straps or steel wire. Do not put much pressure on the strap/wire so that it does not penetrate below the actual depth of the crack
5. Calculate the crack volume assuming triangular geometry for each crack and sum them all to calculate the total volume.

Calculations

The formula used to calculate crack volume, V of each node is (Rivera, 2008):

$$V = \frac{1}{3} \times L \times \left[\left(\frac{1}{2} W_1 D_1 \right) + \left(\frac{1}{2} W_2 D_2 \right) + \sqrt{ \left\{ \left(\frac{1}{2} W_1 D_1 \right) + \left(\frac{1}{2} W_2 D_2 \right) \right\} } \right] \quad ...(14.15)$$

Various parameters for the node used in the equation are given in Fig. 14.4.

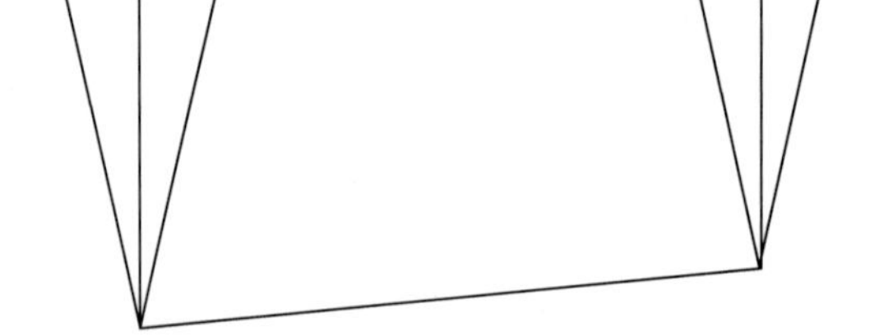

Fig. 14.4. *Crack geometry and dimensions of the crack for volume calculations*

PREFERENTIAL FLOW PATHS (MACRO-PORES)

Preferential flow refers to the phenomena where fluid, gas, and/or solutes move through preferred pathways, thereby bypassing large portions of a porous system. Macro-pores such as cracks, worm holes, fractures and root holes function as preferential flow paths for water and pollutants. Consequently, it is important to know the degree and extent of these macro-pores to judge percolation, ease of drainage, and the quantity of water required for irrigation. Dyes and paints have been used to investigate preferential flow paths (Alaoui and Goetz, 2008). Some dyes (e.g., methylene blue) exhibit high adsorption in organic-rich soils, therefore selection of suitable dyes is an important element in delineating flow pathways using this method. A dilute white paint solution (acrylic fiber resin emulsion) has been found to have minimal adsorption in forest soils (Noguchi, et al., 1999). The method described herein is based on the use of white vinyl paint.

Apparatus: White vinyl water paint, Bucket, Water, Plaster of Paris, scoop, Trowel, Spatula, Folding tape, Camera, and Sketch book

Procedure

1. Make a pit 40 cm wide, 50 cm long and 5 cm deep, at a suitable place in the field. Dilute 8 L of white vinyl paint with 80 L water. Pour it into the pit
2. Next day, make another pit 70 cm wide, 100 cm long, and 100 cm deep (Fig. 14.5). One of the long sides of this pit should coincide with the lengthwise center line of the first pit (Kohno, 1987). With a garden trowel and spatula, clean the face of the wall of the pit that lies along this line, so that the areas turned white by permeation of the paint are clearly visible
3. Sketch these white macro-pores correctly in the sketch book. Also record their depth, their shapes and sizes, concentration and degree of dryness of the paint, and any other marked characteristics of the soil profile. Photographs of the macro-pores may be taken
4. The concentration of the paint is an index of judging the horizontal continuity of the macro-pores. A high concentration reveals a low horizontal continuity of the macro-pore, and a low concentrate indicates the opposite. If the paint remains liquid in a certain zone, the vertical continuity of the macro-pore is high above the zone and low below the zone
5. To determine the horizontal continuity of macro-pores, pour a dilute solution of white paint into a hollow dug about 50 cm away from the pit and observe how it exudes from the side wall into the pit.

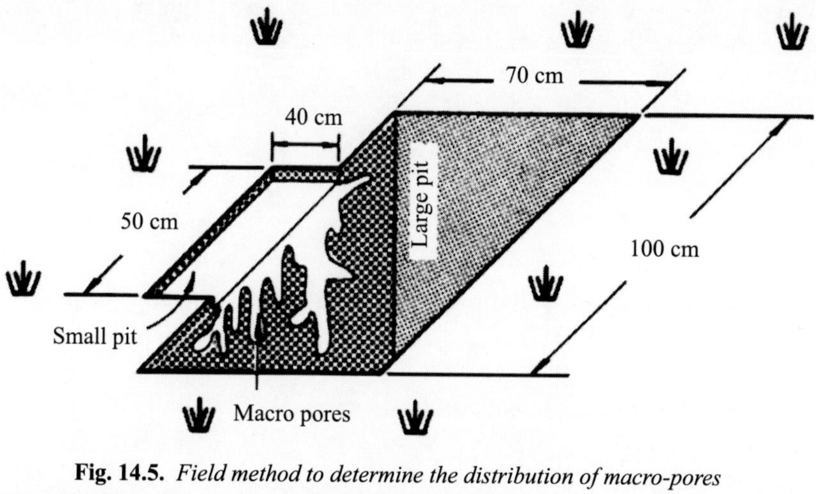

Fig. 14.5. *Field method to determine the distribution of macro-pores*

Precautions

- Determine the optimum dilution of paint required before proceeding to conduct the field test. The paint should be able to permeate rapidly but still have good contrast between white macro-pores and darker soil matrix
- If macro-pores are present only in the sub-soil, remove the surface soil and pour the diluted paint directly into the sub-soil
- Pour plaster of Paris into macro-pores to obtain a rough measurement of pore volumes and shapes. This action is required only if such details are required for the investigation.

COEFFICIENT OF LINEAR EXTENSIBILITY (COLE)

The soil shrink-swell potential is primarily a function of clay mineralogy, particle size distribution or percent clay and fine clay particles (Wilding, 1998). It is critical to characterize the swelling potential of the soils to mitigate the adverse effects of expansive soils. Coefficient of Linear Extensibility (COLE) is used as the controlling parameter to characterize swelling potential of expansive soils (Grossman et al., 1968). The COLE is a laboratory-based value that quantifies the swelling and shrinkage potential of a soil layer. COLE is calculated by using the difference in volume of the soil ped when moist and dry. Another method of measuring COLE is $COLE_{rod}$, which is performed with the use of a saturated soil paste moulded into rods trimmed to various lengths. These are then dried and their length measured to calculate $COLE_{rod}$.

Grossman et al. (1968) Method

COLE is calculated by extracting cores and measuring change in circumference before and after drying.

Equipment and Reagents: Insect-mounting or collection pins, Caliper or 0.1-mm ruler, Distilled water

Procedure

1. Wet the soil core or clod to field capacity. Place 2 pins at a minimum of 5 cm apart. Place pins on vertical face (relative to soil surface, place one pin below the other). It helps to have the measurements integrated over the depth
2. Measure distance between pins when soil core or clod is wet
3. Measure distance between pins when soil core or clod is air-dry.

Calculations

$$COLE_{ped} = (L_w - L_d)/L_d \quad ...(14.16)$$

Here $COLE_{ped}$ is the coefficient of linear extensibility by clod or core method, L_w is the distance between pins when wet (cm) and L_d is the distance between pins when dry (cm). Report COLE on whole-soil basis, cm/cm.

Note: Do not place pins on horizontal surface.

Schafer and Singer (1976) Method

Equipment and Reagents: Spatula, Paper cups-250 g, Sieve 10-mesh (2 mm), Caliper or 0.1-mm ruler, Plastic syringe, 25-cm^3 with 1-cm diameter orifice, Distilled water

Procedure

1. $COLE_{rod}$ is determined on the soil passing through 2-mm sieve
2. Fill 250 g cup about half full of soil (100 g). Add water and mix until a paste slightly drier than saturation is obtained
3. Allow paste to equilibrate for 24 hr and readjust to the appropriate water content, if necessary. Paste should glisten slightly but should not flow when tilted (Bower and Wilcox, 1965). Surface of paste should become smooth once the cup is repeatedly tapped on a table
4. Remove the plunger of the plastic syringe and load the paste into the syringe
5. Slowly extrude minimum 3 rods (6 to 10 cm length) onto a smooth surface. Trim the rod ends perpendicular to the drying surface
6. Measure and record the length of each rod without disturbing the trimmed ends
7. Air-dry the rods for 24 to 48 hr and re-measure the length of the rods
8. One can draw more rods and use some of them to determine $COLE_{rod}$ at other moisture contents such as 33 kPa (10 kPa for sandy soils) by equilibrating the rods on a pressure plate extractor. Few rods can be used to determine $COLE_{rod}$ at oven-dry moisture content. Calculations can accordingly be made and reported for the given moisture content

Calculations

$$COLE_{rod} = (L_w - L_d)/L_d \qquad ...(14.17)$$

Here $COLE_{rod}$ is the coefficient of linear extensibility by the rod method, L_w is the moist rod length, cm, L_d is the dry rod length at a given moisture content, cm. Linear Extensibility Percent, LEP_{rod} is then given as:

$$LEP_{rod} = COLE_{rod} \times 100$$

Report COLE as cm/cm on < 2 mm basis for a given moisture content. The results obtained by this method significantly correlate with COLE determined on natural soil clods ($p < 0.001$, $R^2 = 0.83$).

USDA–SCS (1971) Method (Whole Soil)

COLE values of <2 mm soil fractions are adjusted to whole-soil basis as follows.

$$COLE_{whole} = COLE(< 2mm) \times [1 - V(> 2\text{ mm})] \qquad ...(14.18)$$

$$COLE_{whole} = COLE\ (< 2mm) \times V(< 2\text{ mm}) \qquad ...(14.19)$$

Here COLE (< 2mm) is COLE of < 2-mm fraction, V > 2 mm is the volume fraction of > 2 mm fraction, V (< 2 mm) is the volume fraction of < 2 mm fraction.

Applications

While $COLE_{ped}$ and $COLE_{rod}$ are used to assess the relative swell- shrink potential of the soils, $COLE_{std}$ is used to assess the swell-shrink potential of a soil. $COLE_{std}$ characterizes the variation of soil volume from 1/3 atm water retention (i.e. field capacity) to oven-dry conditions. From the laboratory measured values $COLE_{std}$ is calculated as:

$$COLE_{std} = (V_{1/3\ atm} - V_{dry})^{1/3} - 1 \qquad ...(14.20)$$

Here $V_{1/3\ atm}$ is the soil volume at 1/3 atmosphere water retention and V_{dry} the soil volume at oven-dry conditions. The range of soil swell-shrink potential can be distinguished from the COLE values (Taboada, 2003):

Soil swell-shrink potential	$COLE_{std}$
Low	< 0.03
Moderate	0.03 – 0.06
High	0.06 – 0.09
Very high	> 0.09

STRUCTURAL STABILITY OF SOILS

Childs (1940, 1942) introduced a method for assessing the stability of soil aggregates. This method studies the disruptive forces associated with quick wetting on the soil structure. Childs described the comparative changes in aggregate structure, following quick and slow wetting, by means of the changes in pore size distribution. This technique is based on the interpretation of the high-energy part of the drainage moisture characteristic curve, high energy referring to matric potential from slightly greater than zero to approximately -60 cm of water (Laryea et al., 1997). The extent of aggregate disintegration due to quick wetting is evaluated by means of a series of moisture characteristics curves. Alternatively, the values of the equivalent pore neck radius for the most common pore can also be obtained by plotting the slope of the moisture characteristic curve versus matric potential (differential of moisture characteristic). The concentration of pores is shown by the presence of a peak in the differential of moisture characteristic at a matric potential ψ, which is related to the equivalent pore neck radius, r. The pore size distribution in turn is also measured by the moisture release characteristics of the two samples at low tensions (Joshua, 1990). If the soil aggregates are fully stable and do not collapse on fast wetting, the pore size distribution in the fast and the slow wetted samples should be nearly equal and their ratio will be close to unity. If the soil aggregates are completely unstable, then they will fully collapse on fast wetting and the ratio will almost be zero. Any intermediate value in the ratio indicates intermediate aggregate stability between these two extremes. This approach to evaluate the structural stability of soil is described in this section.

Procedure

The procedure reported by Joshua (1990) uses Haines apparatus to conduct this test. Four test apparatus are required for the two treatments with two replications each.

1. Prepare soil sample for slow and quick wetting by sieving air-dry soil through a nest of 2 mm and 1 mm round-holed sieves. Use the material passing the 2 but retained on 1-mm sieve
2. For slow wetting, pour a known mass of air-dry soil, and by tapping gently form a layer 1 cm thick on the surface of sintered glass Buchner funnel maintained at -10 cm suction. Let the system remains in this position for about 24 hr to wet the soil slowly. Next day, flood the soil slowly by increasing the matric potential from -10 to zero. The moisture characteristic curve of the soil is determined as explained in Chapter 10. Determine the outflow in steps of 5 cm suction up to 25 cm
3. For quick wetting, pour a known mass of soil to form a layer 1 cm thick on the surface of sintered glass Buchner funnel maintained at -10 cm suction (Joshua, 1990). Raise the burette quickly to fully flood the soil for about 20 s. Lower to zero tension to drain the excess water. The drainage moisture characteristic is determined as per the procedure outlined in Chapter 10. Determine the outflow in steps of 5 cm suctions up to 25 cm for this case as well.
4. Plot the average cumulative outflow of the two replications versus tension (only up to 25 cm) imposed on both the slow and fast wetted soils on the same graph (Fig. 14.6). The area under the curves for slow and fast wetted soils is determined.

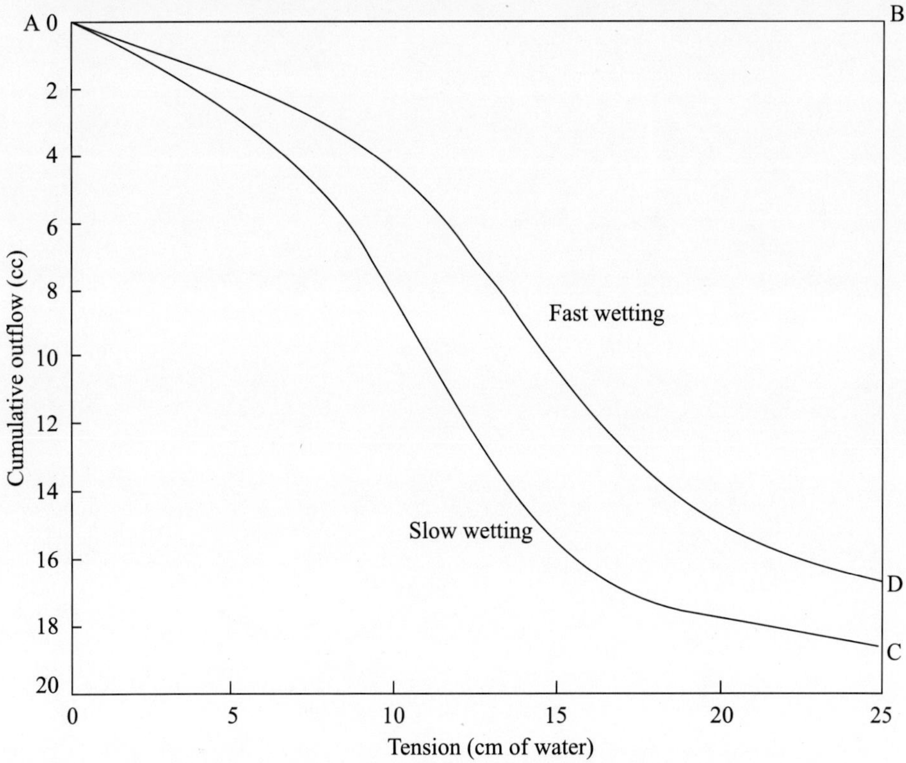

Fig. 14.6. *Plots of cumulative outflow of slow and fast wetted soils*

Structural stability index = Area of the fast wetted curve / Area of the slow wetted curve

Structural stability index = Area ABDA/Area ABCA ...(14.21)

A structurally stable soil will show values nearer to unity.

Note: High structural stability index may be obtained for sandy soils, which have a large proportion of single grain sizes in this range. Thus, the volume fraction of single grains in this size range should be reported so that it is possible to interpret the results in a realistic manner.

Chapter **15**

Soil Carbon

Carbon in soils is present in two forms namely soil inorganic carbon (SIC) and soil organic carbon (SOC), a measurable component of soil organic matter (SOM). The sum of the two forms constitutes the total carbon of the soil. The former consists of mineral forms mainly as carbonate minerals either from weathering of parent material or from the reaction of the soil material with atmospheric CO_2. SOM, on the other hand, is all organic materials found in soils irrespective of their origin or state of decomposition. It is the very foundation for healthy and productive soils, an essential constituent of soils for developing ecologically sound agriculture. Although, organic matter makes up only a small percentage of most soils, may be in the range of 1–6%, it is known to be a great modifier of almost all soil properties, physical, chemical and biological (Murphy, 2015). Few of the physical properties influenced by organic matter are: bulk density, infiltration, soil structure, aggregate stability, sealing and crusting, moisture holding capacity etc. Besides, it influences the effects of chemical amendments, fertilizers (nutrient cycling and nutrient availability) and is a great modifier of the cation exchange capacity (CEC) and buffer capacity (BC) of soils. It also influences the biological health of the soils by influencing the diversity and activity of soil organisms. Studies have shown that if soil organic carbon (SOC) levels fall below 1%, then it has an adverse effect on wide range of soil properties. Total soil carbon is probably the largest store of terrestrial carbon, soils' capacity to store carbon being much larger than the pools of carbon in the atmosphere and vegetation. Soil acts as a carbon sink being a part of the global carbon cycle. Considering this role and the importance of organic matter in the soils, testing SIC and SOC has become a routine test for soil physicists, chemists and biologists

INORGANIC CARBON

Carbonates are typically present in neutral to alkaline soils, but solid phase carbonates are also found in some acid environments in the form of nodules. Calcium carbonate equivalent (CCE) is determined, if the soil pH is above 7 or if effervescence is observed after treatment with 1M HCl. Soil carbonates are typically measured by dissolving carbonates in acid solution and then determining either H^+ consumption, or Ca and Mg production, or CO_2 production as per the following reactions.

$$CaCO_3 + 2H \rightarrow Ca^{2+} + CO_2 + H_2O \quad ...(15.1)$$

$$CaMg(CO_3)_2 + 4H^+ \rightarrow Ca^{2+} + Mg^{2+} + 2CO_2 + 2H_2O \quad ...(15.2)$$

Calcium Carbonate Equivalent (Titrimetric Method)

Principle

This method, as already stated, is based on acid neutralization of carbonates using hydrochloric acid. After neutralization of the soil sample, CCE is measured using the alkaline to back titrate the consumed H^+ (FAO, 2020a).

Apparatus and Reagents: Precision balance, Burette 50 mL, Volumetric pipettes 10 mL, 20 mL, Erlenmeyer flasks, 250 mL, Volumetric flasks, 1000 mL, Watch glass, Hot plate

Deionized or distilled water

- **Hydrochloric acid (HCl), 1 N:** Dilute 98.33 mL concentrated HCl (32%, sp. gr. 1.16) in deionized water, mix well, let it cool and bring to 1 L
- **Sodium hydroxide (NaOH), 1 N:** Dissolve 40 g NaOH in deionized water, transfer to a 1 L flask, let cool, and bring to volume
- **Sodium carbonate solution (Na_2CO_3), 1 N:** Dissolve 53 g anhydrous Na_2CO_3 in deionized water and bring to 1 L volume. Do not keep longer than one week
- **Phenolphthalein indicator [$C_6H_4COOC(C_6H_4$-4-OH$)_2$]:** Dissolve 0.5 g phenolphthalein indicator in 100 mL ethanol (ethyl alcohol)
- **Methyl-orange indicator [4-$NaOSO_2C_6H_4N$:NC_6H_4/-4-$N(CH_3)_2$]:** Dissolve 0.1 g methyl-orange indicator in 100 mL deionized water.

Procedure

1. Air-dry sample passed through a 10 mesh sieve is grounded to a very fine powder using a wooden mortar
2. Weigh 1 to 10 g (±0.001 g) of soil sample and pour into a 250-mL Erlenmeyer flask
3. Using a volumetric pipette, add 20 mL of standardized 1 N HCl to the flask
4. Cover the Erlenmeyer flask with a watch glass and boil the soil-acid mixture for 5 min
5. Add 50–100 mL deionized water using a graduated cylinder. Allow it to cool
6. Add 2 or 3 drops of phenolphthalein indicator
7. Titrate with 1 N NaOH solution while swirling the flask. Take the reading when a faint pink colour develops.

Calculations

$$CaCO_3 \text{ equivalent (\%)} = [(V_h N_h - V_n N_n)/\text{g of soil}] \times 0.05 \times 100 \quad \text{...(15.3)}$$

Here V_h, N_h, V_n, and N_n are the volume and normality of HCl and NaOH respectively.

Notes:

- The estimate of carbonate may be somewhat high because other constituents of the soil may also react with the acid to various degrees
- Add alcohol to samples that have a high content of calcium carbonate to avoid/minimize foaming
- Typically, 20 mL 1 N HCl dissolves up to 1 g $CaCO_3$. Therefore, 20 mL of acid solution can be expected to dissolve all of the carbonate in 1 g of soil

Calcium Carbonate (Gravimetric loss of CO_2)

Apparatus and Reagents: Pipette, Erlenmeyer flask, Hydrochloric acid, 3 N

Procedure

1. Pipette 10 mL of hydrochloric acid in an Erlenmeyer flask, stopper with a cork and weigh, W_1
2. Weigh 1-10 g of soil (W_2) so that the soil contains about 0.1 to 0.3 g of calcium carbonate
3. Add soil to flask little at a time to prevent excessive frothing
4. After the effervescence has subsided, replace the stopper loosely and swirl the flask. Continue swirling at frequent intervals and weigh after each swirling
5. Once the weight does not change much in the 30 min period, the reaction may be taken as complete
6. Remove the stopper and swirl the flask for 10-20 s to remove any carbon dioxide that may still remain in the flask. Weigh the content with stopper on, W_3

Carbon equivalent (%) = (weight of CO_2 lost × 227.4)/weight of the soil sample ...(15.4)

Weight of CO_2 lost = $W_1 + W_2 - W_3$

CALCIUM CARBONATE EQUIVALENT (VOLUMETRIC CALCIMETER)

In this method, the volume of released carbon dioxide is measured at constant temperature and pressure after treating the sample with hydrochloric acid. The change in water level in the system is a direct measure of the mass of CO_2 evolved and hence the soil carbonate decomposed (Allison and Moodie, 1965, FAO 2020b).

Apparatus and Reagents: Volumetric calcimeter comprising 250 mL reaction flask, 10 mL cup, three-way stopcock, burette, and tubing connections (Fig. 15.1), Magnetic stirrer, Electronic balance, Analytical balance, Volumetric dispenser 10.00 mL, volumetric flask 1000 mL

- **Hydrochloric acid (HCl), 4 N:** Add 393.3 mL of concentrated HCl (32%, sp. gr. 1.16) to 400 mL of deionized water. Dilute to a total volume of 1 L.
- **Reagent grade $CaCO_3$**

Procedure

1. Air-dry sample passed through a 10-mesh sieve is further grounded to very fine powder using a suitable mortar and pestle
2. Put 0.5 to 5.0 g ± 1 mg of soil sample to the reaction flask and add 20 mL deionized water
3. Add 7 mL of acid solution to the cup
4. Install the sample flask in the system. Open the three-way stopcock to the atmosphere. Adjust the liquid level of the measuring burette to exactly 0 mL by adjusting the height of the levelling bulb (Fig. 12.1)
5. Close the system to the atmosphere with the three-way stopcock
6. Lower the levelling bulb about 2 cm

7. Simultaneously begin adding HCl solution from the cup to the sample by tilting the flask and begin lowering the levelling bulb. The level of the liquid in the levelling bulb should be kept 1 to 2 cm below the level of the liquid in the measuring burette
8. Turn on the magnetic stirrer (slow stirring rate). If a magnetic stirrer is not available, shake continuously by hand
9. When the level in the gas burette stabilizes, then equalize the liquid levels in the levelling bulb and the measuring burette
10. Read and record the volume of CO_2 evolved
11. Weigh 0.1 g dry, fine-grained, reagent-grade $CaCO_3$ into the reaction flask
12. Repeat the steps without adding soil and measure the CO_2 evolved
13. Repeat the procedure with only 20 mL of water as blank (can be avoided)

Calculations

$$CaCO_3\ (\%) = 100\ [m_2\ (V_1 - V_3)/m_1(V_2 - V_3)]/[1 + m] \quad ...(15.5)$$

Here m_1 is the mass of the test soil, g, m_2 is the mean mass of the calcium carbonate standard, g, V_1 is the volume, mL of carbon dioxide produced by reaction of the test soil, V_2 is the mean volume, mL of carbon dioxide produced by the calcium carbonate standard, V_3 is the volume change, mL in the blank determination (this value can be even negative), m is the water content of the air-dry sample, expressed as a fraction by mass of the oven-dry sample. The result is rounded to a whole number.

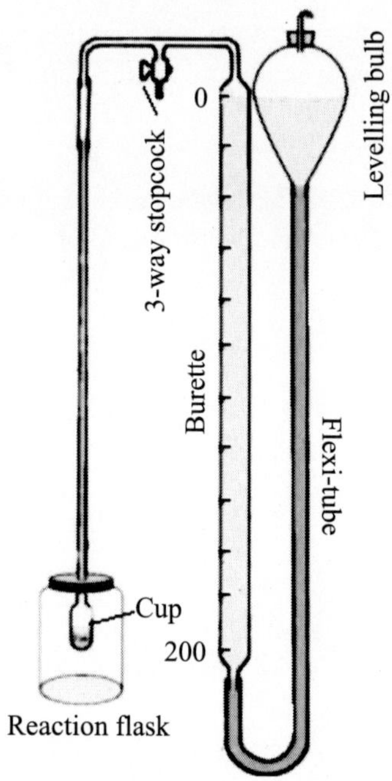

Fig. 15.1. *Volumetric calcimeter to determine soil inorganic carbon*

Notes:

- More water needs to be added to soil samples having high organic matter content. The dilution helps to decrease the oxidation of organic matter
- Foaming can be minimized by adding alcohol to samples having a high content of calcium carbonate
- Add $FeCl_2$ or $SnCl_2$ (5% by weight) to minimize the oxidation of organic matter and the subsequent production of CO_2
- The samples used for carbonate analysis should be well grounded and homogenized to minimize errors because carbonate minerals are not uniformly distributed in the soil
- Most carbonates in the soil decompose rapidly in dilute acid. However, dolomite decomposes slowly in cold acid. Therefore, soils containing dolomite should be allowed longer reaction time to ensure complete decomposition of carbonates.

Puri (1930) Method

Apparatus and Reagents: Conical flask, Pipette, Burette, Volumetric flask, Hot plate

- **0.5 N H_2SO_4**
- **Reagent grade $CaSO_4$**
- **0.1 N $AlCl_3$ $6H_2O$:** Dissolve 8.05 g of aluminum chloride in a small quantity of water and dilute to 1 L with distilled water
- **Bromothymol blue (1% alcoholic solution):** Dissolve 1 g of bromothymol blue in 100 mL of alcohol
- **Bromocresol green (1% alcoholic solution):** Dissolve 1 g of bromocresol green in 100 mL of alcohol

Procedure

1. Weigh 10 g of soil and stir it with 100 mL of distilled water in a conical flask
2. Add 0.2-0.5 g of $CaSO_4$ and bring the suspension to boil and add 10 mL of 0.1 *N* $AlCl_3.6H_2O$ solution
3. Shake the suspension and add 10 drops of bromothymol blue and bromocresol green indicators and note down the colour. A green colour develops, if $CaCO_3$ is present (more than 1%). A golden yellow colour indicates the absence of $CaCO_3$
4. If presence of carbonate is indicated, bring the contents of the flask to boiling point and titrate against 0.5*N* H_2SO_4. Boil for a minute or two after each addition and allow settling for a couple of minutes
5. Continue the titration till golden yellow colour persists on boiling for a couple of minutes. Note down the volume of 0.5 *N* H_2SO_4 used for titration.

Since 1 mL of 0.5 *N* H_2SO_4 used is equal to 1 mL of 0.5 *N* $CaCO_3$ or 0.5 meq of $CaCO_3$ or 0.025 g $CaCO_3$

$$CaCO_3 \text{ in soil (\%)} = (0.025 \times Y \times 100)/\text{Weight of the soil}$$
$$= 2.5\ Y/\text{weight of soil} \quad ...(15.6)$$

Here Y is the amount of 0.5 *N* H_2SO_4 used.

Interpretation of Results

$CaCO_3$ (%)	Soil rating
< 5% $CaCO_3$	Fit for all crop
5-10% $CaCO_3$	Fit for all crops except citrus
>10% $CaCO_3$	Unfit for all crops

Inorganic Carbon

Inorganic carbon (%) = $CaCO_3$ (%) × 0.012

ORGANIC CARBON

Oxidizable Organic Carbon (Walkley-Black, 1934)

Apparatus and Reagents: Conical flask, Thermometer

- **Potassium dichromate (1 *N* $K_2Cr_2O_7$) solution**: Dissolve 49.04 g of analytical grade $K_2Cr_2O_7$ (dried at 105°C for 2 hr) in distilled water and make up the volume to 1 L
- **Concentrated H_2SO_4**, not less than 96% (specific gravity 1.84)
- **Ortho-phosphoric acid** (85%) or **sodium fluoride** (NaF)
- **Diphenylamine indicator:** Dissolve 0.5 g diphenylamine indicator in a mixture of 100 mL concentrated sulphuric acid and 20 mL distilled water. Store in an amber colour bottle
- **Ferrous ammonium sulphate [0.5 N $FeSO_4(NH_4)_2SO_4.6H_2O$] solution** viz., **Mohr's salt**: Dissolve 392 g analytical grade ferrous ammonium sulphate (FAS) in distilled water, add 15 mL concentrated sulphuric acid and make volume to 2 L with distilled water.

Procedure

1. Take 2.0 g of soil (ground and sieved through 0.5-mm sieve) in a 250 mL conical flask
2. Add 10 mL of 1 *N* $K_2Cr_2O_7$ solution through a pipette and shake gently to mix the contents
3. Add 20 mL of concentrated sulphuric acid while swirling the flask slowly. Lot of heat of dilution is produced, reaction being exothermic in nature
4. Cool the contents to room temperature in dark. Note that the flask should be kept on a dry tile or asbestos sheet, being hot
5. Add 0.5 g NaF or 5.0 mL of ortho-phosphoric acid, 100 mL of distilled water and 10 drops of diphenylamine indicator solution. Shake vigorously to mix
6. Then titrate against 0.5 *N* FAS solution till the colour changes from violet to bright bottle green through blue
7. Note the volume of FAS solution used for titration
8. Carry out a blank titration.

Note: If this method is used to estimate organic carbon in saline soils, some errors might occur as Cl^- ions present in saline soils will also be oxidized by dichromate ions. Thus, it will underestimate the oxidizable organic carbon. Therefore, these soils should first be treated with silver sulphate/mercuric sulphate/mercuric oxide associated with potassium dichromate and sulphuric acid to precipitate chlorides. For this purpose 1.25 g of silver sulphate is dissolved in 100 mL of sulphuric acid. This reagent is used instead of only sulphuric acid for tests in saline soils.

Calculations

$$\text{Total organic carbon (\%)} = [N \times (V_1 - V_2)/ \text{ Weight of soil}] \times 0.39 \quad ...(15.7)$$

Here, N is the normality of FAS solution, V_1 is the volume of FAS used for the sample (mL) and V_2 is the volume of FAS used for the blank (mL).

Since 1 mL of 1 N $K_2Cr_2O_7 = 3 \times 10^{-3}$ g carbon, the factor 0.39 is given as $3 \times 10^{-3} \times 100 \times 1.3$. The factor 1.3 is used assuming 77% recovery because organic matter is not fully oxidized with this procedure.

$$\text{Organic matter (\%)} = \text{Total organic carbon (\%)} \times 1.724 \quad ...(15.8)$$

Here, 1.724 is the Van Bemmelen factor as explained in the following paragraph.

The organic carbon content in mineral surface soils ranges from 1.2 to 3.5%. Since soil organic carbon averages about 58% of organic matter, the organic matter in soils generally ranges from about 1 to 6% (% OM = % C × 1.724. The factor 1.724 is given by 100/58). A close relationship also exists between organic carbon and nitrogen in soils. Most organic materials have on an average of about 5% nitrogen. An approximate value of the soil nitrogen percentage can, therefore be obtained by multiplying the soil organic matter percentage by 0.05.

Soil Organic Matter by Loss on Ignition

The principle of this method states that all OC can be converted to CO_2 at a chosen temperature. Thus, the weight loss is proportional to the amount of SOM in the sample (Konen et al., 2002). The loss on ignition (LOI) method uses this approach to determine OM content by weighing the sample before and after ashing (Storer, 1984; Bisutti et al., 2004).

Apparatus: Oven, Muffle furnace or oven - capable of maintaining a temperature of 500°C, Porcelain crucibles, Balance with accuracy of 0.001 g, Scoop - 1 cm^3 capacity

Procedure

1. Take the weight of empty crucible (W_c). Add soil to the crucible and weigh accurately so that it contains 2 g of the soil sample (W_{cs})
2. Dry the soil in the crucible in an oven set at 105°C for 1 hr
3. Remove the crucible and allow it to cool for 15 min
4. Weigh the crucibles with the dried sample (W_{ds}).
5. Place the crucible in the muffle furnace equilibrated at 430°C for 2 hr

6. Remove the crucibles and allow it to cool for 1 hr
7. Weigh the crucible and the ashed sample and record its weight (W_{ca}).

Calculations

Moisture (%)

$$\text{Moisture (\%)} = 100\,(W_{cs} - W_{ds})/(W_{cs} - W_c) \quad ...(15.9)$$

Loss on ignition (%)

$$\text{LOI (\%)} = \text{Total organic carbon (\%)}$$
$$= 100(W_{ds} - W_{ca})/(W_{ds} - W_c) \quad ...(15.10)$$

Here LOI is weight percent loss on ignition, W_c is the weight of empty crucible, W_{cs} is the weight of crucible plus soil before drying at 105°C, W_{ds} is the weight of crucible plus soil after drying at 105°C and W_{ca} is the weight of crucible plus ashed soil.

$$\text{Organic matter, OM (\%)} = 1.724 \times \text{LOI} \quad ...(15.11)$$

This conversion factor can vary for one soil to another soils, but 1.724 provides a reasonable estimate of SOM for most purposes.

$$\text{SOC (kg/ha)} = \text{SOC (\%)} \times \text{BD (g/cm}^3) \times \text{SD (cm)} \quad ...(15.12)$$

Here, SOC is the soil organic carbon (%), BD is the bulk density (g/cm^3) and SD is the soil sampling depth (cm). Using the conversion factor of 1.724, the amount of SOM would be

$$\text{SOM (kg/ha)} = \text{SOC (kg/ha)} \times 1.724.$$

Colorimetric Method (Datta et al., 1962)

Apparatus and Reagents: Spectrophotometer, Conical flasks, Pipettes - 2, 5 and 10 mL

- Standard potassium dichromate 1 N
- Concentrated sulphuric acid containing 1.25% Ag_2SO_4
- Sucrose (AR quality)

Preparation of standard curve

Sucrose is used as a carbon source in this method. Take 1 mg to 20 mg sucrose in 100 mL flasks. Add 10 mL standard $K_2Cr_2O_7$ and 20 mL of concentrated H_2SO_4 in each flask. Swirl the flask and leave for 30 min. A blank is prepared in the same manner without adding sucrose. Adjust the spectrophotometer to zero with the blank. Green colour developed in the solutions with sucrose is read on spectrophotometer at 660 nm. The reading so obtained is plotted against mg of sucrose as carbon source (carbon = weight of sucrose × 0.42 because carbon content of sucrose is 42%) or against mg C directly. A standard curve prepared by Motsara and Roy (2008) is shown in Fig. 15.2 (Ministry of Agriculture, 2011).

Procedure

1. Take 1 g of soil in 100 mL conical flask
2. Add 10 mL of 0.1667M $K_2Cr_2O_7$ and 20 mL of concentrated H_2SO_4 containing 1.25% Ag_2SO_4

3. Stir the reaction mixture and allow to stand for 30 min
4. The green colour of chromium sulphate so developed is read on a spectrophotometer at 660 nm after setting the blank at zero

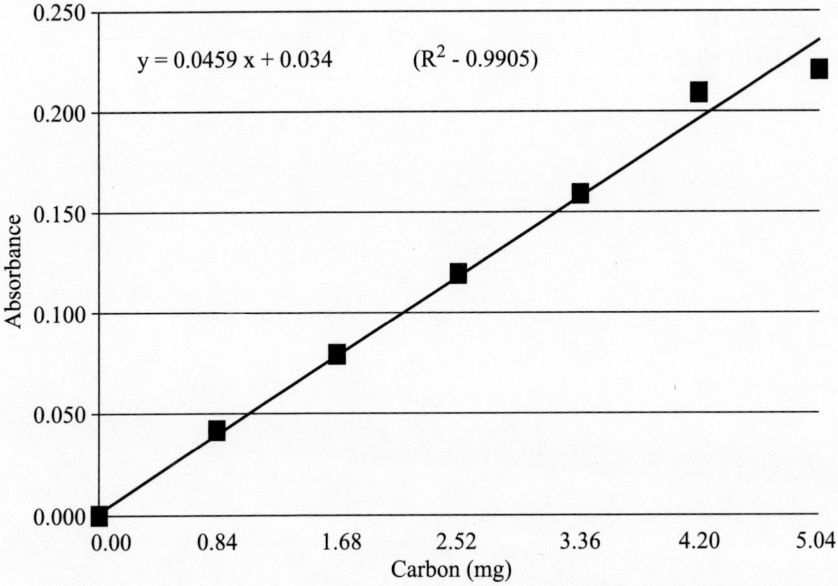

Fig. 15.2. *A calibration curve for spectrophotometer method of carbon determination*

Calculation

The carbon content of the sample is found directly from the standard curve.

C(%) = mg C observed × 100/1000 (observed reading is for 1 g soil, expressed as mg)

Percent OM = %C × 1.724

Total Carbon or Total Organic Carbon

Carbon/Hydrogen/Nitrogen/Sulphur (CHNS) analyzer is used to measure the total carbon while total organic carbon (TOC) analyzer is used to measure the total organic carbon of soil.

1. Finely ground air-dry soil sample (say 50-70 mg) is wrapped in tin boats (Supplied by the manufacturer), weighed and placed in the instrument
2. The direct values of total carbon and total nitrogen are obtained. Knowing the exact weight of the soil, express the results in %
3. A similar procedure is used to estimate total organic carbon of the soil by using the instrument TOC analyzer.

Interpretation of Results

Organic carbon	Soil rating
< 0.5 % (5 g/kg of soil)	Low
0.5 to 0.75% (5 to < 7.5 g/kg of soil)	Medium
> 0.75 % (>7.5 g/kg of soil)	High

ANNEXURE I

PHYSICO-CHEMICAL CHARACTERISTICS OF WATER

Water, essential for sustenance of life on the earth, is a clear, tasteless, odourless liquid and a binary compound. It can occur in three different forms namely gaseous, liquid and solid. Its dissolving capability to dissolve almost everything is both a boon and a curse. While its physical, chemical and biological properties change with dissolution of substances, continuous application of waters can greatly influence the soil's physical, chemical and biological properties. Some of the properties of water, which influence various physical properties of soils are given as under.

Property	**Value**
Molecular weight	18.015 g/mole
Molar volume	55.5 moles/L
Boiling point (BP)	100°C at 1 atm
Freezing point (FP)	0°C at 1 atm
Triple point temperature	273.16 K at 4.6 torr
Surface tension	73 dynes/cm at 20°C
Vapour pressure	0.0212 atm at 20°C
Heat of vaporization	40.63 kJ/mol
Heat of fusion	6.013 kJ/mol
Heat capacity	4.22 kJ/kg-K
Dielectric constant	78.54 at 25°C
Viscosity	1.002 centiPoise at 20°C
Density	1 g/cc at 4°C
Specific heat	4180 J/kg-K
Heat conductivity	0.60 W/m-K (T=293 K)
Melting heat	3.34×10^5 J/kg
Evaporation heat	22.6×10^5 J/kg
Critical temperature	647 K
Critical pressure	22.1×10^6 Pa
Speed of sound	1480 m/s (T=293 K)
Relative permittivity	88 at 0°C (32 °F) and 80 at 20°C
Liquid compressibility	480×10^{-12} m^2/N at 10°C
Coefficient of thermal expansion	256.32×10^{-6} K^{-1}at 25°C
Index of refraction (relative to air)	1.31 for ice 1.33 for water

Relationship of ρ, μ and υ of water and air (T = 10°C, p = 1 atm)

Parameter	Water	Air	Unit
Density, ρ	1000	1.37	Kg/m^3
Dynamic viscosity, μ	1.3×10^{-2}	1.8×10^{-4}	Poise
Kinematic viscosity, υ=μ/ρ	1.3×10^{-5}	1.3×10^{-5}	m^2/s

The variation of some of these properties with temperature is given in Table A-1. The general effect of some chemical properties of water is reported in Table A-2.

Table A-1. *Properties of water as a function of temperature*

Temp	Density	Specific weight	Isothermal compressibility	Dynamic viscosity	Vapour pressure	Surface tension
(°C)	(kg/m^3)	(kN/m^3)	$10^9 \times (1/pa)$	(mPa-s/ centiPoise)	(kPa)	(N/m)
0	999.84	9.806	0.501	1.7921	0.6113	0.07564
10	999.70	9.804	0.478	1.3077	1.2281	0.07422
20	998.21	9.789	0.458	1.0050	2.3388	0.07275
30	995.65	9.765	0.446	0.8007	4.2455	0.07118
40	992.22	9.731	0.441	0.6560	7.3814	0.06956
50	988.03	9.690	0.440	0.5494	12.344	0.06791
60	983.20	9.642	0.443	0.4688	19.932	0.06619
70	977.78	9.589	0.449	0.4061	31.176	0.06442
80	971.82	9.530	0.457	0.3635	47.373	0.06261
90	965.35	9.467	0.468	0.3165	70.117	0.06075
100	958.40	9.399	0.480	0.2838	101.325	0.05885

Table A-2. *Chemical properties that have significant influence on soil physical properties*

Property	Effect
Electrical conductivity (EC)	Hydraulic properties increase with increasing EC, flocculation, high osmotic pressure limits water availability to plants
Sodium adsorption ratio (SAR)	Hydraulic properties decrease with increasing SAR, causes dispersion and deflocculation, reduces aggregate stability, increases the bulk density of both the surface crust and underlying soil, collapse of soil structure
Residual sodium carbonate (RSC)	Hydraulic properties decrease with increasing RSC, increased dispersion and deflocculation, reduces aggregate stability, increases the bulk density of both the surface crust and underlying soil, collapse of soil structure, swelling and dispersion of organic matter, reduced water availability to plants, temporary water logging conditions, high sediment production in runoff
Mg/Ca ratio	For other parameters being same, high ratio decreases hydraulic properties, causes deflocculation

ANNEXURE II

SUB-ORDINATE DISTINCTIONS WITHIN MASTER HORIZONS AND LAYERS

Letter symbol	Meaning
a	Highly decomposed organic material, fiber content after rubbing is less than 17% (by volume)
b	Identifiable buried genetic horizons in the mineral soil
c	Significant accumulation of concretions or nodules with cementation. The common cementing agent may be iron, aluminium, manganese, or titanium
d	Physical root restriction, either natural or manmade such as dense basal till, plough pans, and mechanically compacted zones
e	Organic material of intermediate decomposition in which rubbed fiber content is 1/6 to 2/5 of the volume
f	Frozen soil in which the horizon or layer contains permanent (not seasonal) ice
g	Strong gleying in which iron has been reduced and removed during soil formation or in which iron has been preserved in a reduced state because of saturation with stagnant water
h	Illuvial accumulation of organic matter in the form of amorphous, dispersible organic matter-sesquioxide complexes
i	Slightly decomposed organic material in which rubbed fiber content is more than 2/5 of the volume
j	Accumulation of jarosite
jj	Evidence of cryoturbation includes irregular and broken horizon boundaries, sorted rock fragments, and organic soil materials occurring as bodies and broken layers within and/or between mineral soil layers
k	Accumulation of pedogenic carbonates, commonly calcium carbonate
m	Continuous or nearly continuous cementation or induration of a soil matrix by carbonates (km), silica (qm), iron (sm), gypsum (ym), carbonates and silica (kqm), or salts more soluble than gypsum (zm)
n	Accumulation of sodium on the exchange complex sufficient to yield a morphological appearance of a natric horizon
o	Residual accumulation of sesquioxides
p	Ploughing or other disturbance of the surface layer by cultivation, pasturing or similar uses
q	Accumulation of secondary silica

r	Weathered or soft bedrock including saprolite, partly consolidated soft sandstone, siltstone or shale; or dense such that roots penetrate only along joint planes and are sufficiently incoherent to permit hand digging with a spade
s	Illuvial accumulation of sesquioxides and organic matter in the form of illuvial, amorphous, dispersible organic matter-sesquioxide complexes if both organic matter and sesquioxide components are significant and the value and chroma of the horizon are >3.
ss	Presence of slickensides
t	Accumulation of silicate clay that either has formed in the horizon and is subsequently translocated or has been moved into it by illuviation
v	Plinthite, which is composed of iron-rich, humus-poor, reddish material that is firm or very firm when moist and that hardens irreversibly when exposed to the atmosphere under repeated wetting and drying
w	Development of colour or structure in a horizon but with little or no apparent illuvial accumulation of materials
x	Fragic or fragipan character that results in genetically developed firmness, brittleness, or high bulk density. Some part of the layer is physically root-restrictive
y	This symbol indicates an accumulation of gypsum
z	Accumulation of salts that are more soluble than gypsum

ANNEXURE III

ALTERNATIVE PROCTOR TEST METHODS (ASTM, 2012)

ASTM D698 uses a 4-inch-diameter mould. Soil is compacted in three layers with 25 blows of a 5.5 lb hammer falling from a height of 12 inches. It gives a compaction force of 12,375 ft-lbf/ft^3. The Modified Proctor test, ASTM D1557 uses the same mould, but soil is compacted in 5 layers with a 10 lb hammer falling through 18 inches using 25 blows on each layer. It gives a compaction force of about 56,250 ft-lbf/ft^3. In both the tests, 6 inches mould is allowed. To have about the same compaction force of standard Proctor and of modified Proctor, the number of blows per layer is increased to 56. The details are listed as under.

	Standard Proctor ASTM 698			**Modified Proctor ASTM 1557**		
	Method A	**Method B**	**Method C**	**Method A**	**Method B**	**Method C**
For test sample, use soil passing sieve	Sieve No.4	3/8" Sieve	¾" Sieve	Sieve No.4	3/8" Sieve	¾" Sieve
Mould	4" diameter	4" diameter	6" diameter	4" diameter	4" diameter	6" diameter
No. of layers	3	3	3	5	5	5
No. of blows/layer	25	25	56	25	25	56

GLOSSARY OF TERMS

A

Activity: It is the ratio of the plasticity index to the clay fraction.

Adhesion: The molecular attraction between surfaces (unlike molecules) that holds substances together. For example water adheres to soil particles.

Aeration (soil): The process of exchange of air in soil with air from the atmosphere. Air in a well-aerated soil is quite similar in composition to the air in the atmosphere. Contrarily, air in a poorly aerated soil contains considerably higher levels of carbon dioxide and low oxygen content. Also the amount of air-filled pores in the soil, expressed as the volume difference between total porosity and actual soil moisture should not be less than 10% for good plant growth.

Aggregate: A unit of soil structure generally < 20 mm in diameter, these are formed by natural processes.

Aggregate stability: The ability of soil aggregates to withstand disruptive forces, whether such forces are imposed mechanically (as during tillage or trampling) or by the action of water (causing swelling, slaking, and dispersion of clay).

Aggregation: A process in which primary soil particles, sand, silt, clay are bound together, usually by natural forces or by microbial activity or by substances derived from root exudates. Soil aggregates are naturally arranged to form soil peds, units of soil structure. From an agronomic point of view, most important soil aggregates are in the range 3-1 mm. Also see clods.

Air-filled porosity: The proportion of a soil's volume that is filled with air at a given time; a low air-filled porosity indicates poor aeration and restricted drainage.

Air void ratio: It is the ratio of the volume of the air space to the volume of solids in a soil mass.

Amendment, soil: A substance (e.g., lime, gypsum, or peat) added to amend soil to its pH or imporve physical properties.

Angle of repose: The angle of repose also known as the critical angle of repose of a granular material is the steepest angle of descent relative to the horizontal plane on which a material can be piled without slumping. The material on the sloping face is at the verge of sliding at this angle. It ranges from 0° to 90°.

Anoxia: It is an extreme form of hypoxia in which the level of oxygen in the root zone environment is totally depleted.

Arable land: Agricultural land that is cultivable for crop production.

Aspect: The direction in which a slope faces.

Atmosphere, technical: A unit of pressure abbreviated as at, 1 atmosphere = 98066 Pa and 0.98066 bar.

Atmosphere, standard: A unit of pressure abbreviated as atm, 1 atmosphere = 101325 Pa.

Atterberg limits: An engineering classification of soil consistence at different moisture contents.

Available soil water: It refers to the total amount of water in the root zone available for evapotranspiration, usually expressed in mm.

Available water capacity (available moisture capacity): The capacity of soils to hold water that can be used by most plants. It is the difference between the amount of soil water at field moisture capacity and the amount of soil water at the permanent wilting point. It is commonly expressed as mm of water per m depth of soil.

B

Bar: A unit of pressure. One bar equals 14.5 pounds per square inch (psi) or about 0.987 standard atmospheres or 10^5 Pa or 100 kPa.

Basic infiltration rate: The nearly constant infiltration rate of the soil achieved after a reasonable time depending upon the soil texture.

Bedrock: Mass of solid rock beneath the soil. If it is close enough to the surface to weather into soil, it is the parent material.

Blocky: A type of soil structure. Soil particles in this kind of structure are so arranged that their shapes look like small cubes with sharp or rounded edges.

Boulders: A more or less rounded block or fragment of rock of average dimension 300 mm or greater. Boulders are normally rounded as these are carried or rolled along by water or ice.

Bound water: The thin layer of water tightly adsorbed on to the surface of clay particles, not available to plants.

Breakthrough curve: A plot of the relative solute concentration at the outflow boundary of a soil column as a function of the volume of effluent.

Bulk density: A measure of the weight of the soil per unit volume (g/cc), usually expressed on an oven-dry (110° C) basis.

C

Calcareous soil: A soil with enough calcium carbonate, which effervesces when treated with aqueous hydrochloric acid.

Caliche: A layer or horizon cemented as a result of calcium carbonate depositions. It usually occurs somewhere in the soil profile but may also occur at the surface in eroded soils.

California bearing ratio: It is the ratio of the force per unit area required to penetrate a soil mass with a circular piston of 50 mm diameter at the rate of 1.25 mm/min to that required for the corresponding penetration in a standard material.

Capillarity: Forces between water and soil surfaces in the small (capillary) pores. Also, the tendency of a liquid to enter into narrow pores within a porous body, due to the combination of the cohesive and adhesive forces former being within the liquid (surface tension) and the latter between the liquid and the solid.

Capillary fringe: The zone above the free water table where some or all of the capillary interstices are filled with water; this water is continuous with the water in the saturated zone but is held above that zone by capillarity against gravity. Water is held at less than 1 atmospheric pressure.

Capillary rise: The upward movement of water from a free water table due to adhesion of water in soil capillaries and the cohesion of water molecules.

Capillary water: Water held as a film around soil particles and spaces between particles. Surface tension holds capillary water in the soil against a tension usually greater than 60 cm. It can still move upwards against gravity.

Carbon cycle: It is a sequence or cycle of transformations of carbon dioxide. It is first converted to organic forms by photosynthesis or chemosynthesis. It is then recycled through the biosphere and ultimately returned to its original state i.e. carbon dioxide through the processes of respiration or combustion.

Characteristic curve of soil moisture: A graph of the water content in a unit mass (mass wetness) or unit volume of soil (volume wetness) plotted as a function of matric suction. If the same information is measured during desorption, it is also called the moisture retention or release curve.

Chiseling: The process of tillage using an implement having one or more soil-penetrating points, to shatter or loosen hard compacted layers to a depth below normal plow depth. A process employed to improve water movement in soils.

Chroma: One of the three variables of colour. The relative purity of a colour directly related to the dominance of the determining wavelength.

Clay: Soil particle smaller than 0.002 mm or 2μm, having high specific area to influence soil colloidal properties as well as stability of soil structure, also a soil texture class.

Clay coating/film: Named variously as clay skins, clay flows, illuviation cutans, or argillans, it is coatings of clay on peds surfaces, mineral grains and lining pores.

Clay loam: A soil texture class. See soil texture.

Clay minerals: Clay-sized hydrous aluminium silicates with large interlayer space that are capable of holding significant amounts of water and other substances; these are capable of swelling and shrinking because of having a large surface area; examples are montmorillonite, smectite and kaolinite.

Clods: A lump or mass of soil, aggregates produced by tillage or logging.

Clogging: A process that blocks pores and reduces water flow in soils. Mechanisms can be physical, chemical, and biological.

Coarse fragments: If round, mineral or rock particles 2 mm to 25 cm in diameter; if flat, mineral or rock particles 15 to 38 cm long.

Coarse textured soil. A soil in USDA soil textures of loamy fine sand or coarser (loamy sand or sand).

Coating: A layer of a substance that completely or partially covers the surface of the soil material. It comprises clay, calcite, gypsum, iron, organic material, salt, etc.

Coefficient of curvature: The parameter estimated using the gradation curve through sieve analysis. It is the ratio of the square of d_{30}, grain diameter 30% finer to the d_{60} times d_{10}, the diameter 60% and 10% finer.

Coefficient of linear extensibility (COLE): A measure of the potential one-directional change in the volume of a soil with change in soil moisture content.

COLE: An abbreviation of the coefficient of linear extensibility.

Cohesion: The internal mutual bonding of like molecules or particles of a particular substance, imparting strength to a body composed of that substance. Water molecules cohere to one another.

Cohesion less soils: Any soil that contains elements that do not stick together. For example in engineering, sands having little clay or silt.

Cohesive soils: Fine grained, low strength and easily deformable soils that have a tendency for particles to adhere. Air-dry soils have considerable strength and significant cohesion when submerged.

Colloid: Most active constituents of mineral and soil organic matter having a diameter of 0.1 to 0.001μm are often called soil colloids. The particle sizes are within, or approach, colloidal dimensions.

Columnar: A type of soil structure in which soil particles are arranged into tall vertical columns often with rounded tops.

Compaction: A process of rearranging soil particles to decrease pore space and increase bulk density. It produces a compacted, dense soil layer.

Complex slope: Irregular or variable slope. Planning or constructing terraces, diversions, and other water-control structures on a complex slope are difficult.

Concretions: Grains, pellets, or nodules of various sizes, shapes, and colours consisting of concentrated compounds or cemented soil grains. Their composition is unlike that of the surrounding soil. Commonly occurring compounds in concretions are calcium carbonate and iron oxide.

Conditioners (soil): Polymeric substances added to the soil to promote aggregate stability by gluing particles together within the aggregates as well as by coating the surfaces of aggregates.

Consistence (soil): The feel of the soil and the ease with which a soil lump can be crushed with the fingers. Some of the commonly used terms to describe consistence are: loose, friable, firm, plastic, sticky, hard, soft, and cemented.

Consistency index: It is the ratio of the liquid limit minus the natural moisture content of the soil to its plasticity index.

Consolidation (soil): The process by which a saturated body of soil is compressed, resulting in reduction of pore volume by expulsion of water. Tests of consolidation are intended to predict the gradual settlement/subsidence, of soil under prolonged loading such as that due to a heavy structure.

Contact angle of a liquid on a solid: A measure of the attraction or affinity (adhesion) between a liquid (e.g., water) and a solid surface (e.g., soil particles). A contact angle of zero implies that a drop of the liquid placed on the solid would spread over it (hydrophilic). At the opposite extreme, a contact angle of 180^0 implies a non-wetting of the surface (hydrophobic).

Crust: A surface layer of soil, from a few millimeters to 2.5 cm thick, caused by rain drops impact, clay dispersion and clay translocation. When dry, it is much more compact, hard, and brittle than the material just under it.

Crust formation: The development of a dense, hard, and relatively impermeable layer at the soil surface, due to the breakdown of soil aggregates. A crust inhibits aeration, infiltration, and the emergence of germinating seedlings.

Cutans: These are the modification of the soil texture, or soil structure, or fabric of the natural surfaces (particle, pore, or ped) in soil materials as a result of illuviation.

D

Darcy's law: The rate of flow of a fluid (e.g., water) in a porous body (volume of fluid per unit time per unit area perpendicular to the flow direction) is proportional to the hydraulic gradient. The constant of proportionality is called the hydraulic conductivity, which depends on the properties of the fluid as well as of the porous medium.

Decalcification: A process to remove calcium carbonate or calcium ions from the soil through leaching.

Deflocculation: See dispersion

Degradation (soil): The deterioration of soil productivity by processes such as erosion, organic matter depletion, leaching of nutrients, compaction, breakdown of aggregates, water logging, and/or salinization.

Densification: Methods that rearrange the soil particles resulting in increased density of the soil. It may restrict root growth and inhibit soil aeration.

Density of soil solids: Mass of soil solids per unit volume of the solids (excluding the volume of the voids). See particle density.

Diagnostic horizon: See horizon.

Disperse: The process in which aggregates are separated into individual soil particles. Also, a soil in which clay particles readily form a colloidal solution. Dispersed soils exhibit low permeability, tend to shrink, crack, become hard on drying, slake and become plastic on wetting.

Disperse system: A system in which at least one of the phases is subdivided into numerous minute particles, which together present a very large interfacial area per unit volume. Examples are colloidal sols, gels, emulsions, aerosols, and soils.

Dispersion: Dispersion is the breakdown of soil clay minerals into single particles in solution. Dispersion occurs when the bonds in a soil are not strong enough to keep the particles together. Soils having a high proportion of exchangeable sodium, or being too low in organic matter show high dispersion. Working the soil in too wet or too dry condition can also enhance its tendency to disperse on wetting. The opposite is flocculation.

Dispersion coefficient: It is defined as the ratio of water dispersible clay to the total clay expressed as percent.

Dispersion ratio: Ratio of the soil materials < 0.005 mm after limited mechanical dispersion without dispersant to the total material < 0.005 mm.

Drainage: Drainage of water from the soil, either through natural or artificial channels. Surface drainage refers to the removal of water from the soil surface. Subsurface drainage on the other hand is the removal of water from below the soil surface, a process used to lower the water table.

E

Edaphic: Of or pertaining to the soils; is the condition resulting from or influenced by factors inherent in the soil or other substrata, rather than by climatic factors.

Effective diameter: The diameter taken from soil gradation curve for which the 10% particles are finer is known as the effective diameter.

Efflorescence: The accumulation of dissolved substance (usually simple salts) at the soil surface due to evaporation.

Eh: Measurement and reporting of the redox potential of a solution is given as Eh.

Electrical resistance blocks: Commonly known as gypsum blocks, are porous bodies containing electrodes, placed in the soil to equilibrate with soil moisture. The electrical resistance measured with such blocks can be calibrated against independent measurements of soil wetness or soil water potential.

Eluviated horizon (E horizon): A horizon from which minerals, clay, and/or organic matter have been leached.

Eluviation: The movement of dissolved or suspended material from the upper horizons.

Erosion: The removal of soil from the land surface by erosion agents such as water, wind, ice, and gravity. It may also include other natural or manmade agents that are capable of abrading, detaching and removing soil particles or rock material from one point and deposit the same elsewhere.

F

Feldspar: Group of hard crystalline minerals comprising aluminum silicates of sodium or calcium or barium.

Field capacity: It is the upper limit of plant available water and is defined as the state of the soil when the downward movement of water due to gravitational forces has ceased. It is the bulk water content retained at 5 to 33 kPa of hydraulic head or suction pressure depending on soil type. This state is normally encountered about 24-72 hr after saturation or heavy rain.

Fifteen-atmosphere percentage: The moisture percentage, dry-weight basis, of a soil sample that has been wetted and brought to equilibrium in a pressure-membrane apparatus at 15 atm. The moisture at this level approximates the lower limit of water available for plant growth.

Fine texture: A broad group of textures consisting of, or containing, large quantities of fine fractions, particularly silt and clay. These soil includes sandy clay, silty clay, and clay texture classes. When used in reference to particle-size classes in USDA and FAO soil taxonomy, these soils have 35 to 60% clay.

Fines: The portion of a soil, which is finer than 0.075 mm.

Flocculation: The process of smaller individual particles of soil, especially clay, joining together into larger flocs. Whereas the presence of calcium tends to flocculate clays, and maintain structure, sodium tends to disperse clays in solution (See dispersion). The term flocculation and dispersion is commonly used to describe the two processes.

Friable: A term applied to soils having crumbling texture, somewhere between sand and clay. A handful of soil when pressed hard, forms a lump but the lump easily disintegrates.

G

Genesis, soil: The mode of origin and formation of soil. Genesis normally refers to processes or soil-forming factors in operation during Solum formation. Also, mode of formation of true soil from the unconsolidated parent material.

Geomorphology: Science of landforms that studies the evolution of the Earth's surface and interprets landforms as records of geological history.

Gley soil: Soil formed under naturally wet or waterlogged conditions as evidenced by gray colours stemming from the reduction of ferric iron to the ferrous state under anaerobic conditions.

Gradation: A process to describe the distribution of different particle sizes contained in a soil sample. Grading is usually expressed through the particle size distribution curve.

Gradient: The rate at which a variable changes with distance, as the concentration gradient, potential gradient, and temperature gradient. Also used in the context of land slope.

Granular: A type of soil structure in which soil particles are arranged into shapes resembling granola.

Gravel: Rounded or angular fragments of rock from 2 mm to 7.6 cm in diameter. An individual piece is known as pebble.

Gravimetric: Pertaining to weighing. Gravimetric methods are used to determine amounts of a substance by direct or indirect weighing. The most common use is in gravimetric soil moisture determination.

Gravitational water: The water that flows freely through soils in response to gravity.

Groundwater: That portion of the water below the surface of the ground at a pressure equal to, or greater than that of the atmosphere.

Gypsum blocks: See electrical resistance blocks.

H

Hardpan: A cemented soil horizon, or layer either naturally or through anthropogenic activities. The soil material is cemented by iron oxide, silica, calcium carbonate, or other substances. Also, a hardened layer of soil below the zone of cultivation caused by cementation (chemical process) and/or compaction of soil particles by tillage equipment. Hardpans inhibit root growth and trap water, resulting in a waterlogged soil layer.

Heat conduction: The propagation of heat within a body by internal molecular motion, driven by a temperature gradient. The quantitative evaluation of heat conduction is made with the help of an equation known as Fourier's Law.

Heat conductivity: A measure of the ease with which a material or object conducts or transmits heat.

Horizon- diagnostic: Dominant soil property or material defines the name of horizon, e.g. gypsic horizon having distinct calcium sulphate (gypsum: $CaSO_4$) enrichment; genetic horizon depending on the type of pedogenesis.

Horizon, soil: A layer of soil, approximately parallel to the surface, having similar characteristics but differs at least in one property say texture, colour or any other property of adjacent horizons above or below it.

Hue: The dominant spectral colour, one of the three colour variables in Munsell colour system.

Humification: Conversion of carbon of organic residues into humic substances through biochemical and abiotic processes.

Humus: Organic compounds in soil, not included are the non-decayed plant and animal tissues, or their partially decomposed products, or the soil biomass; the term is often used synonymously with soil organic matter. Its structure is amorphous having no clear shape or structure, specific weight is low and surface area high.

Hydraulic conductivity: A measure of the ease (or difficulty) with which a liquid (usually water) flows through the soil in response to a given potential gradient. The volume of water flowing per unit area per unit of time (flux) per unit gradient of hydraulic potential. See Darcy's law.

Hydraulic diffusivity: The ratio between the flux of water and the gradient of soil wetness. This term is somewhat misleading, since it does not refer to diffusion as such but to convection. The term is taken from the analogy to the diffusion equation (Fick's law), stating that the rate of diffusion is proportional to the concentration gradient.

Hydraulic gradient: Drop in hydraulic head per unit distance of flow.

Hydraulic head: The elevation with respect to a standard datum at which water stands in a riser or manometer connected to the point in question in the soil. This includes elevation, pressure, and the velocity heads, if the terminal opening of the sensing element is pointed upstream. Hydraulic head has the dimensions of length (L).

Hydromorphic soils: Soils formed under conditions of poor drainage and water logging in marshes, swamps, seepage areas or flats.

Hydrophobicity: The tendency of a surface or of an assemblage of particles to resist wetting by water, caused by the obtuse contact angle between water and the surface. Most mineral soils having a natural affinity for water are hydrophilic. The surface zone of some soils, however, may be hydrophobic (water-repellent), especially when coated with oily resins.

Hysteresis: The dependence of an equilibrium state on the direction of the process leading up to it. A prime example is the difference between the soil moisture characteristic curves measured in sorption (initially dry soil is gradually wetted) and in desorption (initially saturated soil is gradually dried).

Hygroscopic Water: Moisture content of the soil due to water adsorbed from the atmosphere, it is held tightly and is not available to plants but can be used by the micro-organisms.

I

Illuviation: The process of materials moving from one horizon and their deposition in another horizon of the same soil; the movement is usually from an upper to a middle or lower horizon. Movement can also occur laterally.

Impermeable: A term often used with dense soil horizons through which water moves extremely slowly. Unable to transmit water.

Infiltration: The downward movement of water into the soil surface or any other material. The infiltration rate is the velocity of water at which it enters into the soil expressed in units of mm/hr. Contrast it with percolation, which is the movement of water through soil layers or material.

Infiltration capacity: It is the maximum rate at which water infiltrates into a soil under a given set of conditions.

Infiltration rate: The rate at which water when impounded on the surface as a shallow depth penetrates the surface of the soil (with no lateral movement) at any given instant, usually expressed in cm/hr. The rate is limited either by the infiltration capacity of the soil or the rate at which water is applied at the surface.

Isobath: An imaginary line or a line on a map connecting all points having the same depth to the water table below the soil surface.

L

Land: Terrestrial bio-productive system that comprises soil, vegetation, other biota, and the ecological and hydrological processes that operate within the system.

Leaching: A process to remove soluble materials from one layer of the soil to another through application of excess water.

Liquid limit: The water content at which the soil changes from a plastic to a liquid state.

Liquidity index: The ratio expressed as a percentage of the natural water content of soil minus its plastic limit to its plasticity index.

Loess: An aeolian material, predominantly silt sized particles transported and deposited by wind forming a mineral-rich layer.

M

Macro-pores: Large pores significantly larger than the inter-granular or inter-aggregate soil pores, often formed by roots and small soil animals and worms, pores > 0.08 mm in diameter.

Map scale: It depicts a relationship between distance on the map and the corresponding distance on the ground (e.g. 1:10,000 means 1 cm on the map equals to 10,000 cm or 100 m on the ground).

Marsh: A wetland, a poorly drained transition zone between water and land dominated by grasses.

Massive: A soil that has no structure. Soil particles are completely stuck together.

Matric potential: A negative potential resulting from adhesive and capillary forces due to the soil matrix. These forces attract water in soil and lower its potential energy compared with free water not in contact with soil. It is always negative until saturation where it is zero.

Mature soil: A well developed soil usually having clearly defined horizons.

Mechanical analysis: A procedure in which the array of particle sizes of a soil is determined quantitatively, by means of sieving, sedimentation, and/or microscopy.

Medium textured soil: A name given to a group of soils in USDA classification i.e. very fine sandy loam, loam, silt loam, or silt.

Mesh: Sieve size given by the number of holes per linear inch in the sieve.

Micron: 1/1000 of a mm.

Micro-pores: Pores that are small compared with macro-pores. Usually found within structural aggregates, also referred as capillary pores.

Milligram per liter (mg/L): Amount of the substance in mg in one liter of water. It is equal to parts per million (ppm).

Minerals: The inorganic particles in soils that have weathered from rocks.

Mineral soil: Soil having mainly mineral material and low in organic material. The bulk density of these soils is more than that of organic soils.

Moderately fine textured: A group comprising clay loam, sandy clay loam and silty clay loam soil textures.

Monolith: It represents a vertical slice of soil taken from a vertical face of the soil profile pit or section. Although smaller in size, it represents the arrangement of soil horizons in the same proportions as the original profile.

Modulus of rupture test: A procedure to characterize the cohesive strength of dry soil briquettes. This test is mainly applied to characterize soil crusts.

Moisture constants: The amount of water that a soil contains at each of pre-defined equilibrium points are known as soil moisture constants.

Mottling, soil: Irregular spots of different colours that vary in number and size. Mottling generally indicates poor aeration and impeded drainage. Mottles are the spots or blotches of colour(s) in a soil that differ from that soil's dominant colour.

Munsell colour system: Soil colour is measured by comparing with a colour chart system, known as Munsell colour system. The colour is specified by the relative degrees of three simple variables i.e. hue (wavelength), value (degree of lightness or darkness), and chroma (purity or strength). It is written in the form: 10YR 6/4, which means the hue of the soil colour is 10YR, its value is 6, and chroma is 4.

N

Neutron moisture meter: A device to measure soil water content, consisting of a probe that includes a radioactive source of fast neutrons and a detector of slow neutrons. The density of the cloud of slowed neutrons around the probe is proportional to the volumetric wetness of the soil.

O

ODR: A short name for oxygen diffusion rate. See oxygen diffusion rate.

Organic material: Materials related to organic compounds, consisting of decaying plant or animal material that accumulates or is applied at the surface. Some common materials are: FYM, compost, green manure, and animal manure.

Organic soil: Organic soil is a soil that contains a significant amount of organic material recently derived from plant remains. As an example, Peats are organic soils with more than 75% organic matter.

Organisms: Living things such as bacteria, fungi, plants, or animals.

Osmosis: A process by which molecules of a solvent move through a semi-permeable membrane from a solution of low concentration into a solution of higher concentration to equalize the concentration of solute on both the sides of the membrane.

Osmotic effect: Reduction in the ability of plants to extract water from the soil as the osmotic pressure of the soil solution increases with increasing salinity.

Osmotic potential: The osmotic potential of the constituent water in soil at a specified temperature is the amount of useful work per unit mass of pure water, in J/kg, that must be done to transfer reversibly and isothermally an infinitesimal quantity of water from a pool of pure free water to a pool of free soil solution (identical in composition with the soil liquid phase at the point under consideration, and at the same height). Osmotic stress/tension/suction is identical to osmotic pressure. It can be interpreted as the negative of the negative osmotic pressure, in Pascal.

Osmotic pressure: The pressure that must be applied to a solution that is in contact with pure water via a semi-permeable membrane (permeable to the molecules of water but not to those of the solute) to prevent the net diffusion of pure water into the solution through that membrane.

Oxygen diffusion rate: The oxygen diffusion rate (ODR) expresses the mobility of oxygen in the soil, being of significance to the availability of oxygen to the plant roots. A parameter widely used to express the aeration status of the soil.

P

Pan: A compact, dense layer in the soil that impedes the movement of water and the proliferation of roots. For example, hard pan, fragi-pan, clay pan, plough pan, and traffic pan.

Parent material: Mineral or rock material on and/or from which soils are formed during pedogenesis, one of the five major soil forming factors.

Particle density: It is the density of the solid particles that collectively make up a soil sample. Expressed in g/cm^3, the values range from 2.55 to 2.70 g/cm^3 with a commonly adopted value of 2.65 g/cm^3.

Ped: The natural structural unit formed when soil particles (sand, silt, and clay) bind together.

Pedogenesis: The process of soil formation and development by soil forming factors, namely climate (mainly temperature and precipitation), parent material, living organisms (plants and biota), topography, time, water and man.

Pedon: A three-dimensional body of soil. Lateral dimensions are large enough to permit the study of horizon shapes and relations.

Penetrometer: An instrument for assessing soil strength in the field. It consists of a narrow probe, a cone driven into the soil. The soil's resistance to penetration is measured in terms of the energy expended per unit depth.

Percolation: The downward flow of water through saturated or nearly saturated layers of the soil profile, generally due to gravity (suction gradients being negligible).

Permanent wilting percentage: The moisture content of soil in percentage at which plants wilt and fail to recover turgidity.

Permeability: Permeability is defined as the capacity of soil to allow water to pass through it i.e. quantity of water flowing through a unit of soil surface under a pressure of 1 unit hydraulic gradient. Permeability is measured in cm/hr.

Permeability (intrinsic): Pertaining to the relative ease with which a porous medium can transmit a liquid under a hydraulic or potential gradient. It is a property of the porous medium only being independent of the nature of the liquid. It is given by the relation: k= Kυ/g, such that k is the intrinsic permeability, K is the hydraulic conductivity, υ is the kinematic viscosity of the fluid and g acceleration due to gravity. It has the units of L^2.

pF: It is defined as the logarithm of the absolute value of soil matric potential i.e. pF = log (ψ_m) where ψ_m is the matric potential in cm.

Physiological drought: A temporary daytime state of drought in plants due to the losses of water by transpiration being more rapid than uptake by roots although the soil may have an adequate water supply. Such plants usually recover during the night.

Plasticity: The property of soil which allows it to deform beyond the point of recovery without cracking or appreciable volume change.

Plasticity index: Given by the numerical difference between the liquid limit and the plastic limit. It is the range of moisture content within which soil remains plastic.

Plastic limit: The moisture content at which soil changes from semi-solid to plastic.

Platy: A type of soil structure. Flat plate like arrangement of soil particles.

Ploughing (tillage): Mechanical cultivation of agricultural soils to different depths by the plough making them suitable for cultivation.

Plough pan: A compact layer formed in cultivated soils, just below the tilled topsoil, as a result of the pressures exerted by tillage implements and tractor wheels.

Ponding: Standing water on soils in closed depressions. Unless artificially drained, the water can only be removed by percolation or evapotranspiration.

Pores: The space between soil particles, which is filled with water or air. See porosity.

Pore size distribution: The volume fractions of the various sizes of pores in a soil.

Porosity: The volume percentage of the total bulk of soil not occupied by solid particles. It is the volume of water and air that can be held in the soil. It is also the volume of pores in a sample divided by the sample volume.

Preferential flow: The rapid movement of water through preferential pathways such as cracks, fissures and root hoes etc.

Pressure: A measure of force per unit area, SI unit of Pa is equal to 1 N/m^2.

Pressure-membrane apparatus: An instrument to measure the soil moisture characteristic. Pressurized air is used to extract water incrementally from initially saturated soil samples resting on a pressure membrane instead of porous plate as in pressure-plate apparatus.

Pressure-plate apparatus: An instrument to measure the soil moisture characteristic in the range of 1 to 15 or more bars of tension. Pressurized air is used to extract water incrementally from initially saturated soil samples resting on a porous plate.

Prismatic: A kind of soil structure. Soil particles are arranged into shapes that resemble columns.

Profile, soil: See soil profile.

Puddle: Processes, mostly physical, that destroy the structure of the surface soil. It can result from the impact of raindrops, poor cultivation with implements and trampling by animals or tractor wheels.

Q

Quartz: SiO_2. An important rock-forming mineral, abundantly available in many soils, is resistant to weathering.

R

Redox potential: A quantitative measure of the reducing or oxidizing power of a medium or solution. The redox potential of anaerobic soils is usually determined by measuring the difference in potential between an inert platinum electrode and a reference electrode (a saturated calomel or Ag/AgCl solution).

Refraction: The bent in the light ray, when it enters into the light denser medium to high denser medium at an angle, is known as the refraction.

Refractive index: It is defined as the ratio of the speed of light in a vacuum to the speed of light in the substance.

Refractometry: Method of measuring the refractive index of substances.

Regolith: The layer of unconsolidated solid material, such as weathered rock, soil and superficial deposits overlying solid rock.

Relief: Relief is defined as the relative difference in elevation levels on the land.

Remote sensing: It is the acquisition of information about an object or phenomenon without making physical contact with the object.

Remoulded: Soil whose natural structure has been modified or disturbed by manipulation.

Root zone: The part of the soil that can be penetrated by plant roots. Also, the area of soil and oxygen surrounding the roots of a plant.

S

Sand: Soil particles between 0.05 mm and 2 mm (in some countries lower size limit is 0.06 mm). It is also a texture class name for coarse soil materials.

Saturated: The condition of a soil when all of its pores are filled with water. It is also used in the context of a condition in which cations of the specified kind occupy all the exchange capacity.

Saturated flow: The movement of water in a soil with all its pores filled with water, or when the matric tension is zero.

Saturation extract: The soil solution extracted from a soil, which is at saturation percentage.

Secondary mineral: A mineral resulting from the decomposition of a primary mineral or from the re-precipitation of the products of decomposition of a primary mineral.

Sedimentation: A technique for determining particle size distribution, wherein particles settle through the water at a rate proportional to their size.

Seepage: The loss of water from a pond or reservoir by percolation into and through the soil. An alternative use of the term is to describe the emergence of water from an exposed stratum of a saturated soil.

Semi-permeable membrane: A membrane that permits the passage of solvent molecules but not the molecules or ions of certain solutes. When such a membrane is placed between solutions of different concentration, solvent molecules move from the low concentration solution to high concentration solution through osmosis. See osmosis.

Shrinkage limit: The maximum water content at which any reduction in water content will not cause a decrease in the soil's volume.

Shrink-swell: A soil character, the soil shrinks when dry and swells when wet. Swelling increases soil volume while shrinking decreases soil volume.

Silt: Soil particles in the size range of 0.002 mm and 0.05 mm (in some countries, upper size limit is 0.06 mm), also used as a texture class name for medium and medium-fine soil materials.

Single-grained: A soil that has no structure. The soil particles in these soils are not bound to each other in any way. As an example beach sand.

Size separate: Individual mineral particles, typically referring to the less than 2.0 mm in diameter–fine earth fraction–and ranging between specified size limits. Sand, silt, and clay are common groupings of soil size separates.

Slaking: It is the breakdown of a lump of soil into smaller fragments on wetting, resulting from swelling of clay and consequently bursting out of the entrapped air.

Slope: Characterizes the tilt or inclination of a surface or a landscape, the inclination of the land surface from the horizontal. Percentage slope is the vertical distance divided by horizontal distance multiplied by 100.

Smectite: A family of expanding 2:1 clay minerals having high surface area and exchange capacity.

Soil aeration: The process by which air in the soil is replaced by air from the atmosphere. The soil air in a well-aerated soil is very similar in composition to the atmospheric air. Poorly aerated soils, on the other hand contain a much higher percentage of carbon dioxide and a corresponding lower percentage of oxygen than the atmospheric air.

Soil air: The air within the soil; the soil atmosphere.

Soil biology: A scientific discipline dealing with living components of soils, such as bacteria, fungi, protozoa, nematodes, arthropods and earthworms as well as mammals.

Soil chemistry: A scientific discipline dealing with chemical properties of soils. It also deals with the influence of fertilizers, pesticides and the other chemical substances applied on or into the soil on soil's behaviour and fertility.

Soil classification: Also termed soil taxonomy, is the scientific discipline dealing with grouping of soils into soil morphological units or soil types, according to similar or comparable soil forming properties.

Soil colour: Soil colour is one of the indicators of soil status, depends on factors such as the amount and state of organic matter, presence of iron oxide, and the amount of air and water in soil pores. Munsell soil colour charts are used to describe soil colour.

Soil compaction: Process that results in a decrease in the volume of voids between soil particles or aggregates. It is reflected in an increase in bulk density.

Soil cracks: Openings in horizontal (mm or several cm) and vertical (cm or several m) orientation, arising from swelling and shrinking processes. While heavy clay soils are more susceptible to crack formation, sandy soils either do not crack or cracks are very small and unstable.

Soil degradation: Negative processes often accelerated by human activities (improper soil use and cultivation practices, building areas) that leads to deterioration of soil properties and functions or destruction of soil as a whole, e.g. compaction, erosion, salinization.

Soil depth: Depth of the soil profile from the top to bedrock or to any other layer that may obstruct roots proliferation especially in vertical direction.

Soil-forming factors: The biological, chemical, and physical processes that act under the influence of the soil-forming factors to create soils.

Soil gradation curve: A graph of grain size plotted on a horizontal logarithmic axis against percentage finer or coarser on an arithmetical vertical axis. A point on the curve gives the percentage by weight of material finer/coarser in size than that at the given point on the graph.

Soil hardening: Soil hardening means destruction of soil structure resulting in hardening of the tillage layer adversely impacting the air permeability and seriously affecting the normal growth of crop roots.

Soil inorganic carbon: It is mineralized forms of carbon present in the soil, such as calcium carbonate. It is more stable than most organic carbon because it is not food or fuel for micro-organisms.

Soil micro-organisms: Represented by protozoa, viruses, bacteria, fungi and algae.

Soil moisture stress: The sum of the soil moisture tension and the osmotic pressure of the soil solution. It is the suction or negative pressure to which water must be subjected to be at equilibrium through a semi-permeable membrane with the solution in soil.

Soil moisture tension: The equivalent negative pressure or suction of water in soil.

Soil monitoring: Observation and measurement of selected soil properties and functions at pre-decided intervals may be regular or irregular, mainly to study changes in soil conditions.

Soil morphology: The branch of soil science dealing with the description, using standard terminology, of *in-situ* spatial organization and physical properties of soil regardless of potential use. Also, form and arrangement of pedological features, that describe the physical makeup of the soil, including the texture, structure, porosity, consistence, colour, and other physical, mineral, and biological properties of the various horizons, including the thickness and arrangement of those horizons in the soil profile.

Soil organic carbon (SOC): A parameter that includes all of the carbon forms for dissolved (DOC: dissolved organic carbon) and total organic compounds (TOC: total organic carbon) in soils.

Soil organic matter (SOM): It is the organic matter component of soil, consists of plant and/or animal derived materials at various stages of decomposition.

Soil physics: A discipline in agricultural sciences that deals with physical properties of soil such as density, porosity, water retention and permeability, hydraulic conductivity etc. In a general form, it is the branch of soil science that deals with the state and transport of matter and with the state and transformations of energy within the soil, as well as between the soil and the adjacent domains

Soil-Plant-Atmosphere continuum: A concept that recognizes that the soil, plant, and ambient atmosphere constitute a physically integrated, dynamic system in which various flow processes involving energy and matter occur simultaneously and interdependently like links in a chain.

Soil porosity: See porosity.

Soil profile: A vertical section of soil horizons from upper layer to the parent material or to the water table, showing the configuration of soil horizons used as a basis for soil classification.

Soil quality: Soil quality is the capacity of a specific kind of soil to function, within natural or managed ecosystem boundaries, to sustain plant and animal productivity, maintain or enhance water and air quality, and support human health and habitation.

Soil separates: Mineral particles less than 2 mm in equivalent diameter ranging between specified size limits. According to USDA, the names and sizes of separates are: very coarse sand - 2.0 to 1.0 mm, coarse sand -1.0 to 0.5 mm, medium sand - 0.5 to 0.25 mm, fine sand - 0.25 to 0.10 mm, very fine sand - 0.10 to 0.005 mm, silt - 0.05 to 0.002 mm and clay - <0.002 mm.

Soil shrinkage: Soils that shrink upon drying, a characteristic of the clay minerals of these soils. See shrink-swell.

Soil structure: The arrangement and organization of the particles in the soil, whether single-grained (unattached particles variously stacked) or associated in more or less stable aggregates. It is also described as the arrangement of soil particles into clusters (peds) of various shapes that resemble balls, blocks, columns, or plates.

Soil survey: The systematic study of the soil of an area including classification and mapping of the properties and the distribution of various soil units.

Soil taxonomy: The U.S. Department of Agriculture system of soil classification.

Soil texture: Numerical proportion (% by weight) of sand, silt and clay in a soil. Usually characterized as coarse (predominance of sand particles), medium (predominance of silt particles), or fine (predominance of clay particles).

Solum: The part of a soil profile comprising A, E, and B horizons (above the C-horizon) in which processes of soil formation are active.

Spatial variability: The non-uniform distribution of specifiable attributes, properties and processes of an area, whether random or systematic (structured), and the range of their variation expressed by means of statistical (probabilistic) criteria.

Specific gravity: The ratio of the weight or mass of a given volume of a substance to that of an equal volume of a reference substance. Water is the reference for liquids and solids.

Specific surface: The total surface area of particles per unit mass of the particles or per unit volume of the soil, usually expressed as m^2/g or m^2/m^3.

Stokes law: It relates the velocity of a spherical particle settling in an aqueous suspension under the influence of gravity as a function of the particle radius and density, and viscosity and density of the fluid.

Structure, honeycomb: An arrangement composed of soil particles having a comparatively loose, stable structure resembling a honeycomb.

Structural stability: The ability of the soil's aggregated structure to resist the forces tending to cause its disruption, such as slaking by water or grinding and compaction by machinery.

Structure, soil: See soil structure.

Sub-soil (B horizon): The soil horizon rich in minerals that eluviated or leached down, from the horizons above it. Not present in all soils.

Sub-soiling: Breaking up a compact sub-soil by pulling a special chisel plough through the soil.

Subsurface diagnostic horizons: One or more soil horizons below the surface horizon with specific properties as described in the soil taxonomy (e.g., the argillic horizon).

Surface soil: Synonym topsoil. The layer of soil occurring on the surface.

Swamp: Wetland, permanent inundation of large areas of land by shallow water bodies. It is similar to marsh but with more woody plants. Also similar to bog but with better drainage.

T

TDR: Abbreviation for time-domain reflectometry. A technique for measuring soil wetness, consisting of parallel metal rods inserted into the soil. When a step pulse of electricity is sent through the rods, the received signal is related to the dielectric properties of the medium. The reading depends mainly on the fractional volume of water present.

Tensiometer: A device for measuring soil-water matric potential in place. Applicable to the tension range of 0 to 1 bar.

Texture: See soil texture.

Textural class: Classifications of texture into groupings, typically named for the primary constituent particle size or a combination of the most abundant particles sizes.

Thermal conductivity: The flux of heat (the amount of heat conducted across a unit cross-sectional area in unit time) per unit temperature gradient.

Thermal regime of a soil profile: The variation of soil temperature with time for various depths in the profile.

Thermocouple psychrometer: A device to measure the relative humidity of the air phase in the soil.

Tidal flats: Nearly flat areas, periodically covered by tidal waters, not suitable for agricultural use.

Tillage: The mechanical stirring or turning of the soil profile. In agriculture, horticulture, and landscaping, tillage is used to prepare soil for seeding, to incorporate organic matter or chemicals, or to control weeds.

Tilth, soil: A qualitative parameter based on experience, the physical state of the soil that determines its suitability for plant growth taking into account texture, structure, consistence and pore space.

Time-domain reflectometry: See TDR.

Topsoil: The surface soil horizon (A) which is modified when cultivated, and designated Ap. See also surface soil.

U

Undisturbed sample: A soil sample that has been obtained with ample precautions to minimize any disturbance to the sample.

Uniformity coefficient: It is the ratio of d_{60} to d_{10}, where d_{60} and d_{10} are 60% and 10% finer on the soil gradation curve.

Unsaturated flow: The movement of water in the soil that is not fully saturated, pressure in the pore-water is negative.

V

Vadose water: Water in the vadose zone.

Vadose zone: Called the unsaturated zone or the aerated zone, the unsaturated soil that overlies the permanent water table.

Value: One of the character of the Munsell colour system indicating the lightness of a colour. On a scale of 10, 0 indicates pure black and 10 pure white.

Vapour pressure: Pressure exerted by a vapour being in equilibrium with the liquid or solid form, or both, of the same substance. As an example the partial pressure of water vapour in the atmosphere. It depends on the pressure and temperature of the atmosphere, as well as on the state of water in a water-containing body (such as soil) at equilibrium with the atmosphere.

Vertisols: Soils rich in expansive clay (montmorillonite). When subject to alternating cycles of wetting and drying, as in a semi-arid region, such soils tend to heave and then to settle and form wide, deep cracks, as well as slanted sheer planes extending deep into the soil profile.

Viscosity: The resistance of a fluid to shear (to a force causing the adjacent layers of the fluid to slide over each other). It is visualized as the fluid's internal friction.

Void ratio: Fractional volume of voids (pores) per volume of soil solids.

Volumetric heat capacity: The volumetric heat capacity of a material is the heat capacity of a sample of the substance divided by the volume of the sample. Informally, it is the amount of energy that must be added, in the form of heat, to one unit of volume of the material in order to cause an increase of one unit in its temperature

W

Water potential: A measure of the energy state of water, the sum of gravity, pressure, matric, and solute components in the soil, relative to that of pure water at atmospheric pressure and a standard elevation.

Water retention: The ability of soil to hold water for a period that is longer than infiltration, normally 48 hr in a freely draining soil. Soil texture, organic matter content and bulk density, all influence water retention.

Water retention curve: The graphical relationship between soil water content (by mass or volume) and the soil-water matric potential (the energy required to remove the water).

Water-stable aggregates: A soil aggregate which is stable to the action of water such as falling rain drops or agitation as in wet sieving analysis.

Water table: The top surface of a body of groundwater, the surface being at atmospheric pressure. All groundwater beneath the water table is at a pressure greater than atmospheric, whereas soil moisture above it is at a pressure smaller than atmospheric.

Weathering: The breakdown and changes in rocks and sediments at or near the Earth's surface caused by biological, chemical, and physical agents or their combinations.

Well graded: Refers to soil material consisting of coarse grained particles that are well distributed over a wide range in size or diameter.

Wetland: A transitional area between aquatic and terrestrial ecosystems remaining inundated or saturated with water for long enough periods to produce hydric soils and support hydrophytic vegetation.

Wetting front: The advancing boundary between dry soil and wetted soil during infiltration.

Wettability: The tendency of the soil to accept water (in contrast with the opposite tendency of some soils to repel water, called hydrophobicity). Water applied to a wettable soil typically forms an acute contact angle, and is drawn into the soil pores.

Wilting point, permanent: Soil moisture content when the rate of absorption of water by plant roots is too slow to maintain plant turgidity resulting in permanent wilting of the plants. The average moisture tension at the outside surface of the moisture film around soil particles when permanent wilting occurs is 1500 kPa.

REFERENCES

Abou Najm, M. R., Jabro, J. D., Iversen, W. M., Mohtar, R. H. and Evans, R. G. 2010. New method for the characterization of three-dimensional preferential flow paths in the field. Water Resour. Res. 46: doi:10.1029/2009WR008594.

Agrawal, R. P. and Sharma, D. P. 1979. On the modification of Richards modulus of rupture equipment. Annals of Arid Zone. 18: 68-74.

Agriculture Victoria, 2019. Practical Note: Soil Colour. http://vro.agriculture.vic.gov.au/dpi/vro/vrosite.nsf/pages/soilhealth_prac_soil_colour. 2 p. opened on 16.08.2019

Ahmad, N. 1996. Occurrence and distribution of Vertisols. Dev. Soil Sci. 24: 1–41.

Akalan, İ. 1967. Toprak fiziksel özellikleri ve erozyon. Ankara Üniversitesi Ziraat Fakültesi Yıllığı 3-4: 490-503. [in Turkish].

Alaoui, A. and Goetz, B. 2008. Dye tracer and infiltration experiments to investigate macropore flow. Geoderma. 144: 279–286.

Al-Hashemi, H. M. B. and Al-Amoudi, O. S. B. 2018. A review on the angle of repose of granular materials. Powder Technology. 330: 397-417.

Allison, L. E. 1956. Soil and plant responses to VAMA and HPAN soil conditioners in the presence of high exchangeable sodium. Proc. Soil Sci. Soc. Am. 20: 147-151.

Allison, L. E. and Moodie, C. D. 1965. Carbonate. In: Methods of Soil Analysis Part 2. Black, C. A. (Ed.). American Society of Agronomy, Madison. 1379-1396.

ASAE. 1999. Procedures for Using and Reporting Data Obtained with the Soil Cone Penetrometer. EP542 (R2013). American Society of Agricultural and Biological Engineers, Ml, USA. 5 p.

Ashworth, J., Keyes, D., Kirk, R. and Lessard, R. 2001. Standard procedure in the hydrometer method for particle size analysis. Communications in Soil Science and Plant Analysis. 32: 633-642.

ASTM. 2000. Standard Test Method for Determination of Thermal Conductivity of Soil and Soft Rock by Thermal Needle Probe, D 5334 – 00. Annual Book of ASTM Standards, Vol. 04.09. ASTM, West Conshohocken, PA, USA.

ASTM. 2001. Standard Test Method for Measuring the Angle of Repose of Free Flowing Mold Powders. Method No. C1444-00; ASTM, 2001. American Society of Testing and Materials (ASTM), West Conshohocken- PA, USA. 2 p.

ASTM. 2009. Standard Test Methods for Laboratory Compaction Characteristics of Soil Using Standard Effort (12,400 ft-lbf/ft^3 (600 kN-m/m^3)). Designation: D 698 – 07. American Society of Testing and Materials (ASTM) International, West Conshohocken, PA, USA. 13 p.

ASTM. 2012. Standard Test Methods for Laboratory Compaction Characteristics of Soil Using Modified Effort (56,000 ft-lbf/ft^3 (2,700 kN-m/m^3)). Designation: D1557 – 12. American Society of Testing and Materials (ASTM) International, West Conshohocken, PA, USA. 14 p.

Atterberg, A. 1911. Die plastizität der tone. Intematl. Mitt. Bodenk. 1: 10-43.

Bahmani, O. and Palangi, S. 2016. Evaluation of pedotransfer functions for estimating the soil water retention points. Eurasian Soil Science. 49: 652–660.

Bandyopadhyay, K. K., Aggarwal, P., Chakraborty, D., Pradhan, S., Garg, R. N. and Singh, R. 2012. Practical Manual on Measurement of Soil Physical Properties. Division of Agricultural Physics, Indian Agricultural Research Institute, New Delhi. 62 p.

Barasa, E. 2014. Standard Operating Procedures Method for Analysing Soil Samples for Soil Moisture Characteristics. World Agroforestry Centre, Nairobi, Kenya. Code: METH05V01. 10 p.

Baver, L. D., Gardner, W. H. and Gardner, W. R. 1972. Soil Physics. 4th ed. John Wiley & Sons, New York.

Beatty, S. M. and Smith, J. E. 2014. Infiltration of water and ethanol solutions in water repellent post wildfire soils. Journal of Hydrology. 514: 233-248.

BIS. 1970. Indian Standard Classification and Identification of Soils for General Engineering Purposes (First Revision). IS: 1498 – 1970 (Reaffirmed 2002). Bureau of Indian Standards, New Delhi. 26 p.

BIS. 1972. Indian Standard: Specification for Soil Thermometers. IS 6592-1972 (Reaffirmed 2002). Bureau of Indian Standards, New Delhi. 20 p.

BIS. 1972. Standard Test for Soils: Determination of Shrinkage Factors. IS 2720 (Part VI) - 1972. (Fifth Reprint, 1993 Reaffirmed 2001). Bureau of Indian Standards, New Delhi. 12 p.

BIS. 1977. Methods of Test for Soils: Determination of Free Swell Index of Soils (Third Reprint,1997. IS: 2720 (Part XL). Bureau of Indian Standards, New Delhi. 5 p.

BIS. 1980a. Indian Standard Code of Practice for Design and Construction of Pile Foundations Under-reamed Piles. IS: 2911 (Part III) - 1980 (First Revision) Fifth Reprint, 1998. Bureau of Indian Standards, New Delhi. 33 p.

BIS. 1980b. Methods of Test for Soils: Determination of Water Content - Dry Density Relation using Light Compaction. IS 2720-7: Part 7: 1980. Bureau of Indian Standards, New Delhi. 10 p.

BIS. 1980c. Indian Standard: Methods of Test for Soils: Part III- Determination of Specific Gravity. IS : 2720 - 1980 (First Revision). Reaffirmed, 2002. Bureau of Indian Standards, New Delhi. 8 p.

BIS. 1983. Methods of Test for Soils: Determination of Water Content - Dry Density Relation using Heavy Compaction. IS-2720 (Part-8):1983 (Reaffirmed- May 2015). Bureau of Indian Standards, New Delhi.

BIS. 1984. Indian Standard: Specification for Test Sieves: Wire Cloth Test Sieves. Part I.1984 (Reaffirmed- 1998). Bureau of Indian Standards, New Delhi. 13 p.

BIS. 1985. Method of Test for Soils: Determination of Liquid and Plastic Limits. IS: 2720 (Part5)-1985. Bureau of Indian Standards, New Delhi. 16 p.

BIS (Bureau of Indian Standards). 1986. Indian Standard Methods of Tests for Soils: Laboratory Determination of Permeability. (IS 2720 part 17, Reaffirmed 2002). Bureau of Indian Standards, New Delhi. 13 p.

BIS (Bureau of Indian Standards). 1987. Indian Standard Methods of Tests for Soils: Laboratory Determination of Permeability of Granular Soil (Constant Head). (IS 2720 part 36, Reaffirmed 2002). Bureau of Indian Standards, New Delhi. 6 p.

Bisdom E. B. A., Dekker L. W. and Schoute J. F. T. 1993. Water repellency of sieve fractions from sandy soils and relationships with organic material and soil structure. Geoderma. 56: 105–118.

Bisutti. I., Hike, I. and Raessler, M. 2004. Determination of total organic carbon- an overview of current methods. Trends in Analytical Chemistry. 23: 716-726.

Black, C. A. 1965. Methods of Soil Analysis. Part 1 Physical and Mineralogical Properties, American Society of Agronomy, No:9. Madison, WI, USA.

Blake, G. R. 1965. Particle density. In: Black, C. A. (Ed.) Methods of Soil Analysis: Part I. Madison: ASA. 371-373.

Blake G. R. 2008. Particle density. In: Chesworth W. (Ed). Encyclopedia of Soil Science. Encyclopedia of Earth Sciences Series. Springer, Dordrecht.

Blake, G. R., and Hartge, K. H. 1986. Bulk density. In: Klute, A. (Ed.) Methods of Soil Analysis. Part 1. 2nd ed. Agron. Monogr. 9. ASA and SSSA, Madison, WI. 363–375.

Blázquez, C. S., Martín, A. F., Nieto, A. M. and González-Aguilera, D. 2017. Measuring of thermal conductivities of soils and rocks to be used in the calculation of a geothermal installation. Energies. 10: 795-813.

Bouwer, H. 1966. Rapid field measurement of air-entry value and hydraulic conductivity of soil as significant parameters in flow system analysis. Water Resources Research. 2: 729-738.

Bouwer, H., Back, J. T. and Oliver, J. M. 1999. Predicting infiltration and ground water mounds for artificial recharge. J. Hydro Eng. ASCE. 4: 350-357.

Bouyoucos, G. J. 1965. Plaster of Paris block electrical measuring unit for making a continuous measurement of soil moisture under field conditions. In: Wexler, A. (Ed.), Humidity and Moisture: Measurement and Control in Science and Industry. Reinhold, New York. 4: 105-111.

Bowers, S. A. and Hanks, R. J. 1962. Specific heat capacity of soils and minerals as determined with a radiation calorimeter. Soil Science. 94: 392-396.

Bower, C. A. and Wilcox, L. V. 1965. Soluble salts. In: Black, C.A. et al., (Eds.). American Society of Agronomy, Madison. 933-940.

Bradford, J. M. 1986. Penetrability. In: Methods of Soil Analysis: Part 1 - Physical and Mineralogical Methods, 2nd ed., A. Klute (Ed.), American Society of Agronomy, Madison, WI.

Brady, N. C. and Weil, R. R. *1984.* The Nature and Properties of Soils. MacMillian Publishing Company, New York. 881 p.

Brady, N. C. and Weil, R. R. 2002. Nature and Properties of Soils.13th ed. Prentice Hall, Englewood, Nl.

Brezeale, E. and McGeorge, W. T. 1946. A new technique for determining wilting percentage of soil. Soil Science. 68: 371-374.

Briggs, L. J. and McLane, J. W. 1907. The Moisture Equivalent of Soils. Bureau Soils Bull. 45. US Department of Agriculture. Washington, DC.

Briggs, L. J. and Shantz, H. L. 1912. The Wilting Coefficients of Different Plants and its Determination. USDA Agri. Bureau Plant Ind. Bull. 230. US Department of Agriculture. Washington, DC

Bristow K. L. 2002. Thermal conductivity. In: Methods of Soil Analysis. Dane, J. H. and Topp, G. C. (Eds). Soil Science Society of America, Inc., Madison. 1209–1226.

Bruce, R. R. and Klute, A. 1956. The measurement of soil water diffusivity. Soil Sci. Soc. Am. Proc. 20: 458 - 462.

Bryan, R. B. 1968. The development, use and efficiency of indices of soil erodibility. Geoderma. 2: 5-25.

Buckland, G. D., Harker, D. B. and Sommerfeldt, T. G. 1986. Comparison of methods for determining saturated hydraulic conductivity and drainable porosity of two southern Alberta soils. Can. J. Soil Sci. 66: 249-259.

Bullock, P., Fedoroff, N., Jongerius, A., Stoops, G. and Tursina, T., 1985. Handbook for Soil Thin Section Description. Waine Res. Pub., Wolverhampton.

Campbell, G. S. and Shiozawa, S. 1992. Prediction of hydraulic properties of soils using particle size distribution and bulk density data. Proceedings of the International Workshop on Indirect Methods for Estimating Hydraulic Properties of Unsaturated Soils. M. T. van Genuchten et al. (Eds.) Univ. of Calif., Riverside. 317-328.

Carnes, A. 1934. Soil crust - method of study their strength, and a method of overcoming their injury to cotton stands. Agric. Engg. 15: 167-169.

Carrillo, M. L. K., Letey, J. and Yates, S. R. 1999. Measurement of initial soil-water contact angle of water repellent soils. Soil Science Society of America Journal. 63: 433–436.

Carter, D. L., Mortland, M. M. and Kemper, W. D. 1986. Specific surface. In: Klute, A. (Ed), Methods of Soil Analysis, Part 1. Physical and Mineralogical Methods. Agronomy No. 9, ASA, SSSA Publ., Madison, WI. 413-423.

Carter, M. R. 1990. Relative measures of soil bulk density to characterize compaction in tillage studies on fine sandy loams. Can. J. Soil Sci. 70: 425–433.

Cassel, D. K. and Klute, A. 1986. Water potential: tensiometry. In: Klute, A. (Ed): Methods of Soil Analysis: Part 1. Physical and Mineralogical Methods, 9. 2nd ed. Madison, Wisconsin USA, American Society of Agronomy - Soil Science Society of America. 563-596.

Cerato, A. B. and Lutenegger, A. J. 2002. Determination of surface area of fine-grained soils by the ethylene glycol monoethyl ether (EGME) method. Geotechnical Testing Journal. 25: Paper ID# GTJ200210035_253. 1-9.

Channarasappa, K. R. 1965. Tests of a Suggested Piezometric Method for Determining Hydraulic Conductivity of Saturated Soils. All Graduate Plan B and other Reports. 1136. https://digitalcommons.usu.edu/gradreports/1136

Chen, G., Pan, J., Han, B., and Yan, H. 1999. Adsorption of methylene blue on montmorillonite. Journal of Dispersion Science and Technology. 20: 1179–1187.

Chepil, W. S. 1952. Improved rotary sieve for measuring state and stability of dry soil structure. Soil Science Society of Am. J. 16: 113-117.

Childs, E. C. 1940. The use of soil moisture characteristics in soil studies. Soil Science. 50: 239-252.

Childs, E. C. 1942. Stability of clay soils. Soil Science. 53: 69-72.

Chopra, S. L. and Kanwar, J. S. 1991. Analytical Agricultural Chemistry. Kalyani Publishers, New Delhi.

Chossat, J. C. 1987. Relation between hydraulic conductivity and drainable porosity. In: Fifth National Drainage Symposium, Proceedings, ASAE, Chicago. International ST Book, 403: 81–92.

Christiansen, J. E. 1962. The pumped-well method of determining transmissibility of soils for drainage design. Paper presented at the meeting of the Irrigation and Drainage Division, American Society of Civil Engineers Convention, Houston, Texas.

Clarke, B. G., Agab, A. and Nicholson, D. 2008. Model specification to determine thermal conductivity of soils. Proceedings of the Institution of Civil Engineers - Geotechnical Engineering. 161: 161-168.

Crouch, R., Reynolds, K. C., Hicks, R. W. and Greentree, D. A. 2007. Soils and their use for earthworks. In: Soils – Their Properties and Management. Charman, P. E. V. and Murphy, B. W. (Eds.). Oxford University Press, Melbourne. 367–393.

Dane, J. H. and Hopmans, J. W. 2002. Water retention and storage. In: Dane, J. H. and Topp, G. C. (Eds.). Methods of Soil Analysis- Part 4, Physical Methods. Soil Science Society of America, Madison. 1692 p.

Dasog, G. S., Acton, D. F., Mermut, A. R. and De Jong, E. 1988. Shrink-swell potential and cracking in clay soils of Saskatchewan. Canadian Journal of Soil Science. 68: 251-260

Datta, N. P., Khera, M. S. and Saini. T. R. 1962. A rapid colorimetric procedure for the determination of the organic carbon in soils. J. Indian Soc. Soil Sci. 10: 67-74.

Davis, J. L. and Chudobiak, W. J. 1975. *In-situ* meter for measuring relative permittivity of soils. Geol. Surv. Can. Pap. 75–1A: 75–79.

Day, P. R. 1965. Particle fractionation and particle size-analysis. In: Black, C.A. et al., (Eds.) Methods of Soil Analysis. Agronomy No. 9, Part 1. American Society of Agronomy, Madison, WI. 545–567.

Dekker L. W. and Ritsema C. J. 1995. Fingerlike wetting patterns in two water-repellent loam soils. Journal of Environmental Quality. 24: 324–333.

Dekker L.W., Ritsema C. J., Oostindie, K., Moore, D. and Wesseling, J. G. 2009. Methods for determining soil water repellency on field-moist samples. Water Resources Research. 45: 1-6.

Department of Sustainable Natural Resources. 2015. Soil Survey Standard Test Method Water Repellency. https://www.environment.nsw.gov.au/resources/soils/testmethods/rep.pdf. opened on 16.08.2019.

Dieleman, P. J. 1974. Deriving soil hydrologic constants from field drainage tests. In : Drainage Principles and Practices. Vol. 3. International Institute for Land Reclamation and Improvement, Wageningen, The Netherlands. 329- 350.

Dirksen, C. 1999. Soil Physics Measurements. Catena Verlag, Reiskirchen, Germany, 154 p.

Diserens, E. 1934: Beitrag zur Bestimmung der Durchlässigkeit des Bodens in natürlicher Bodenlagerung. Schweiz. Landw. Monatstr. 12.

Doerr, S. H. 1998. On standardizing the 'water drop penetration time and the molarity of an ethanol droplet' techniques to classify soil hydrophobicity: a case study using medium textured soils. Earth Surface Processes and Landforms. 23: 663–668.

Donghao, M., JiaBao, Z., Jianbin, L., and Quanjiu, W. 2016. An improved method for determining Brooks–Corey model parameters from horizontal absorption. Geoderma. 263: 122-131.

Dumm, L. M. 1954. Drain spacing formula. Agricultural Engineering. 35: 726-730.

Dunin, F. X. 1976. Infiltration: its simulation for field conditions. In: Rodda, J.C. (Ed.), Facets of Hydrology, Wiley-Interscience, New York. 199-227.

Ebong, S. T., Attai, E. S. and Joshua, E. O. 2016. Measurement of thermal conductivity and specific heat capacity of three major geomorphological units in Akwa Ibom state, Nigeria. British Journal of Applied Science & Technology. 12: 1-7.

Emerson, W. W. 1967. A classification of soil aggregates based on their coherence in water. Australian Journal of Soil Research. 5: 47–57.

FAO. 2006. Guidelines for Soil Description. Food and Agriculture Organization of the United Nations. Rome.

FAO. 2020a. Standard Operating Procedure for Soil Calcium Carbonate Equivalent: Titrimetric Method. Food and Agriculture Organization of the United Nations, Rome.

FAO. 2020b. Standard Operating Procedure for Soil Calcium Carbonate Equivalent: Volumetric Calcimeter Method. Food and Agriculture Organization of the United Nations, Rome.

Feng, G., Wu, L. and Letey, J. 2002. Evaluating aeration criteria by simultaneous measurement of oxygen diffusion rate and soil-water regime. Soil Science. 167: 495 -503.

Fiedler, S., Vepraskas, M. J. and Richardson, J. L. 2007. Soil redox potential: its importance, field measurements, and observations. Advances in Agronomy. 94: 1-54.

Field, D. J., McKenzie, D. C. and Koppi, A. J. 1997. Development of an improved Vertisol stability test for SOILpak. Australian Journal of Soil Research. 35: 843-852.

Freeze, R. A. and Cherry, J. A. 1979. Groundwater. Prentice-Hall, Inc., Englewood Cliffs, New Jersey. 604 p.

Gardner, W. R. 1959. Solutions of the flow equation for the drying of soils and other porous media. Soil Sci. Soc. Am. Proc. 23: 183– 187.

Gardner, W. R. 1970. Field measurement of soil water diffusivity. Soil Sci. Soc. Am. Proc. 34: 832-833.

Goodwin, I. 2009. Gypsum blocks for measuring the dryness of soil. Future Farming Systems Research. AG0294. http://agriculture.vic.gov.au/agriculture/farm-management/soil-and-water/soils/gypsum-blocks-for-measuring-the-dryness-of-soil. Opened on 14.08.2019.

Grevers, M. C. J. 1991. Modulus of Rupture Test for determining Soil Strength of Surface Soils. Report M-111, Saskatchewan Institute of Pedology, University of Saskatchewan, Saskatoon, SK. 49-54.

Groenevelt, P. H., Grant, C. D. and Murray, R. S. 2004. On water availability in saline soils. Australian Journal of Soil Res. 42: 833-840.

Grossman, R. B., Brasher, B. R., Franzmeier. D. P. and Watker, J. L. 1968. Linear extensibility calculated from natural-clod bulk density measurements. Soil Sci. Soc. Am. Proc. 32: 570-573.

Gupta, I. C., Yaduvanshi, N. P. S. and Gupta, S. K. 2012. Standard Methods for Analysis of Soil Plant and Water. Scientific Publisher, India, Jodhpur. 224 p.

Gupta, R. P. and Dakshinamoorthi, C. 1980. Procedures for Physical Analysis of Soil and Collection of Agro-meteorological data. Report ICAR, New Delhi.

Gupta, S. K. 2019. Drainage Engineering: Principles and Practices. Scientific Publishers, Jodhpur. 531 p.

Haines, W. B. 1930. The hysteresis effect in capillary properties and the modes of moisture distribution associated therewith. J. Agric. Sci. 20: 96–105.

Halbertsma, J. M. and Veerman, G. J. 1994. A New Calculation Procedure and Simple Set-Up for the Evaporation Method to Determine Soil Hydraulic Functions. Report 88. DLO Winand Staring Centre, Wageningen, The Netherlands. 25 p.

Hanks, R. J. and Thorp, F. C. 1956. Seedling emergence of wheat as related to soil moisture content, bulk density, oxygen diffusion rate and crust strength. Proc. Soil Sci. Soc. Am. 20: 307-310.

Hanks, R. J. and Thorp, F. C. 1957. Seedling emergence of wheat, grain sorghum and soybeans as influenced by soil crust strength and moisture content. Proc. Soil Sci. Soc. Am. 21: 357-359.

Hardy, F. 1923. The physical significance of the shrinkage coefficient of clays and soils. J. Agr. Sci. (England). 13: 243-264.

Hazelton, P. and Murphy, B. 2016. Interpreting Soil Test Results: What Do All the Numbers Mean? NSW Gov. Dept. of Natural Resources, CSIRO Publishing. 152 p.

Hesse, P. R. 1971. A Textbook of Soil Chemical Analysis. John Murray, London.

Hicks, R. W. 2007. Soil engineering properties. In: Soils – Their Properties and Management. Charman, P. E. V. and Murphy, B. W. (Eds.). Oxford University Press, Melbourne. 192-205.

Hooghoudt, S. B. 1936. Bepailing van den doorlaatfactor van den grond met behulp van pompproeven, (z.g. boorgatenmethode). Verslag Landbouwk. Onder zoek. 42: 449-541.

Horton, R. E. 1939. Analysis of runoff-plot experiments with varying infiltration capacity. Transactions American Geophysical Union. 20: 693-711.

Huang, M., Rodger, H. and Barbour, S. L. 2015. An evaluation of air permeability measurements to characterize the saturated hydraulic conductivity of soil reclamation covers. Can. J. Soil Sci. 95: 15-26.

Hutson, J. L. and Wagenet, R. J. 1992. LEACHM: Leaching Estimation and Chemistry Model—a Process-Based Model of Water and Solute Movement, Transformations, Plant Uptake and Chemical Reactions in the Unsaturated Zone, Version 3. Cornell University, Ithaca, NY, USA.

Indoria, A. K., Sharma, K. L., Sammi Reddy, K. and Srinivasa Rao, Ch. 2017. Role of soil physical properties in soil health management and crop productivity in rainfed systems-I: Soil physical constraints and scope. Current Science. 112: 2405-2414.

IRRI. 1987. Physical Measurements in Flooded Rice Soils: The Japanese Methodologies. International Rice Research Institute, Los Baños, Philippines. 65 p.

Iversen, B. V., Moldrup, P., Schjonning, P. and Jacobsen, O. H. 2003. Field application of a portable air permeameter to characterize spatial variability in air and water permeability. Vadose Zone J. 2: 618-626.

Jackson, M. L. 1958. Soil Chemical Analysis. Verlag: Prentice Hall, Inc., Englewood Cliffs, NJ. 498 p.

Jackson, M. L. 1973. Soil Chemical Analysis. Prentice Hall, New Delhi.

Jackson, T. J. 1990. Laboratory evaluation of a field-portable dielectric soil moisture probe. IEEE Transactions on Geoscience and Remote Sensing. 28: 241–245.

Johnson, A. I. 1991. A Field Method for Measurement of Infiltration. Paper 1544 F (2nd ed). US Geological Survey, Denever, Co, USA. 31 p.

Joshua, W. D. 1990. Methods of Measurement of Physical Properties of Soil. Soil Mapping and Advisory Services. AG/Bot/85/011. FAO, Gaborone, Republic of Botswana. 30 p.

Kandhal, P. S. and Parker, F., Jr. 1998. Aggregate Tests Related to Asphalt Concrete Performance in Pavements. National Cooperative Highway Research Program (NCHRP) Report 405, Transportation Research Board, National Research Council, National Academy Press, Washington, D.C., 109 p.

Keen, B. A. and Raczkowski, H. 1921. Relation between the clay content and certain physical properties of a soil. Journal of Agricultural Science. 11: 441-449.

Keller, T., Silva, A. P. D., Tormena, C. A., Giarola, N. F. B., Cavalieri, K. M. V., Stettler, M. and Arvidsson, J. 2015. SoilFlex-LLWR: linking a soil compaction model with the least limiting water range concept. Soil Use and Management. 31: 321–329

Kennedy, J. G. and Rush, E. S. 1968. Trafficability of Soils: Development of Revised Mobility Index Formula for Self - Propelled Wheeled Vehicles in Fine-Grained Soils. Technical Memorandum NO. 3-240. U. S. Army Engineer Waterways Experiment Station, Corps of Engineers, Vicksburg, Mississippi, USA.

Kersten, M. S. 1948. The thermal conductivity of soils. Proc. 28th Annual Meeting Highway Research Board. Washington, D.C. 28: 391-409.

Khakural, B. R. and Sharma, P. P. 1984. Laboratory Manual of Soil Science (Soil Physics, Morphology, Genesis and Classification). Institute of Agriculture and Animal Science, Rampur-Nepal. 70 p.

Kiehl, E. J. 1979. Manual de edafologia: Relações solo-planta. São Paulo, Ceres. 264 p.

King, P. M. 1981. Comparison of the methods for measuring severity of water repellence of sandy soils and assessment of some factors that affect its measurement. Australian Journal of Soil Research. 19: 275-285.

Kirkham, D. 1946. Proposed method for field measurement of permeability of soil below the water table. Soil Sci. Soc.Amer. Proc. 10: 58-68.

Kirkham, D. 1947. Field method of determination of air permeability of soil in its undisturbed state. Soil Sci. Soc. Amer. Proc. 11: 93-99.

Klute, A. and Dirksen, C. 1986. Hydraulic conductivity and diffusivity: Laboratory methods. Methods of Soil Analysis: Part 1. Agronomy Monograph No. 9, American Society of Agronomy, Madison, WI. 687-734.

Kohno, E. 1987. Physical Measurements in Flooded Rice Soils: The Japanese Methodologies. International Rice Research Institute, Los Baños, Philippines. 32-34.

Konen, M. E., Jacobs, P. M. C., Lee, B. B., Talaga, J. and Mason, J. A. 2002. Equations for predicting soil organic carbon using loss-on-ignition for North Central US soils. Soil Science Society of America Journal. 66: 1878-1881.

Kořenková L., Šimkovic I., Dlapa P., Juráni B. and Matúš P. 2015. Identifying the origin of soil water repellency at regional level using multiple soil characteristics: The White Carpathians and Myjavska Pahorkatina upland case study. Soil and Water Res. 10: 78–89

Kostiakov, A. N. 1932. On the dynamics of the coefficient of water percolation in soils and the necessity for studying it from a dynamic view for purpose of amelioration. Trans. 6th Cong. Int. Soc. Soil Sci. Russian Part A: 17-21.

Kroetsch, D. and Wang, C. 2008. Particle size distribution. In: Soil Sampling and Methods of Analysis Carter and Gregorich (Eds.). Canadian Society of Soil Science. Taylor & Francis, Boca Raton, FL.

Kukkonen, M. and Suppala, I. 1999. Measurement of Thermal Conductivity and Diffusivity *in- situ*: Literature Survey and Theoretical Modelling of Measurements. Geological Survey of Finland, POSIVA OY, Helsinki, Finland. 69 p.

Kumar, **S.,** Gupta, S. K. and Sewa Ram. 1994. Inverse techniques for estimating transmissivity and drainable pore space utilizing data from subsurface drainage experiments. Agric. Water Management. 26: 41-58

Laryea, K. B., Pathak, P., and Katyal, J. C. (Eds.). 1997. Measuring Soil Processes in Agricultural Research. Technical Manual no. 3. International Crops Research Institute for the Semi-Arid Tropics, Patancheru and Central Research Institute for Dryland Agriculture, Hyderabad, 100 p.

Leelamanie, D. A. L., Karube, J. and Yoshida, A. 2008. Characterizing water repellency indices: contact angle and water drop penetration time of hydrophobized sand. Soil Science and Plant Nutrition. 54: 179-187.

Lees, C. H. and Chorlton, J. D. 1896. On a simple apparatus for determining the thermal conductivity of cements and other substances used in the arts. Philosophical Magazine. 41: 1-5.

Lemon, E. R. and Erickson, A. E. 1952. The measurement of oxygen diffusion in the soil with a platinum microelectrode. Soil Soc. Am. Proc. 16: 160 -163.

Lemos, P. and Lutz, J. F. 1957. Soil crusting and some factors affecting it. Proc. Soil Sci. Soc. Am. 21: 485-491.

Letey, J. 1969. Measurement of contact angle, water drop penetration time, and critical surface tension. Proc. Symp. Water Repellent Soils. Univ. California, Riverside. 43-47.

Letey, J., Osborn, J. and Pelishek, R. E. 1962. The influence of the water-solid contact angle on water movement in soil. Hydrological Sciences Journal. 7: 75-81.

Letey, J., Osborn, J. and Pelishek, R. E. 1962. Measurement of liquid-solid contact angles in soil and sand. Soil Science. 93: 149-153.

Letey, J. and Stolzy, L. H. 1967. Limiting distances between root and gas phase for adequate oxygen supply. Soil Sci. 103: 404–409.

Light, T. S. 1972. Standard solution for redox potential measurements. Analytical Chemistry. 44: 1038-1039.

Liu, Z. 2011. Measuring the Angle of Repose of Granular Systems Using Hollow Cylinders. M. S. Thesis: University of Pittsburgh, Pennsylvania, USA.

Liu, Z., Yu, X. and Wan, L. 2014. Capillary rise method for the measurement of the contact angle of soils. Acta Geotechnica. DOI 10.1007/s11440-014-0352-x.

Livneh, M. 1987. The Correlation between Dynamic Cone Penetrometer Values (DCP) and CBR Values. Transportation Research Institute Publication 87-303. Technion-Israel Institute of Technology, Haifa, Israel.

Loveday, J. and Pyle, J. 1973. The Emerson Dispersion Test and Its Relationship to Hydraulic Conductivity. Technical Paper No. 15. CSIRO. Australia Division of Soils, Adelaide, Australia. 7 p.

Low, J. E. , Loveridge, F. A. and Powrie, W. 2017. Error analysis of the thermal cell for soil thermal conductivity measurement. Proceedings of the Institution of Civil Engineers: Geotechnical Engineering. 170: 191-200.

Lowe, M. 2019. Soil Water Repellency and Its Limitations on Water Infiltration. Ph. D. Thesis. University of Western Australia, Perth-WA. 286 p.

Lynn, W. C. and Pearson, M. J. 2000. The Color of Soil. The Science Teacher. In: file://D:\less_pln\color\color.htm opened on 19.08.2019.

McIntyre, D. S. 1970. The platinum microelectrode method for soil aeration. Adv. Agron. 22: 235–283.

McMullen, B. 2000. SOILpak for Vegetable Growers. NSW Agriculture, Warriewood NSW, Australia.

Means, R. E. and Parcher, J. V. 1965. Physical Properties of Soils. Prentice Hall of India (Pvt.) limited, New Delhi.

Mills, J. J., Murphy, B. W. and Wickham, H. G. 1980. A study of three simple laboratory tests for the prediction of shrink-swell behaviour. Journal of Soil Conservation NSW. 36: 77–82.

Ministry of Agriculture. 2011. Methods Manual: Soil Testing in India. Ministry of Agriculture, GoI, New Delhi. 208+ p.

Mitchell, J. K. and Kao, T. C. 1978. Measurement of soil thermal resistivity. Journal of the Geotechnical Engineering Division. 104: 1307-1320.

Miura, K., Maeda, K. and Toki, S. 1997. Method of measurement for the angle of repose of sands. Soils and Foundations. 37: 89-96.

Morrison, J. E., Jr., and Bartek, L. A. 1987. Design and field evaluation of a hand-pushed digital soil penetrometer with two cone materials. Transactions of the ASAE. 30: 646-651.

Motsara, M. R. and Roy, R. N. 2008. Guide to Laboratory Establishment for Plant Nutrient Analysis, FAO Fertilizer and Plant Nutrition, Bulletin 19. FAO, Rome.

Murdock, L., Gray, T., Higgins, F. and Wells, K. 1995. Soil Compaction in Kentucky. Cooperative Extension Service, AGR-161. University of Kentucky, Lexington, Kentucky, USA.

Murphy, B. 2015. Key soil functional properties affected by soil organic matter – evidence from published literature. IOP Conf. Series: Earth and Environmental Science: 25. doi:10.1088/1755-1315/25/1/012008.

Murphy, B. W. 1986. Water Repellence in Surface Soils of Central Western New South Wales. Soil Conservation Service of NSW. Technical Bulletin 22/86. Cowra Research Centre, Cowra, NSW.

Musavi-Jahromi, S. H. and Shiravand, R. 2012. Determination of hydraulic conductivity applying empirical formulae and physical modeling. Archives Des Sciences. 65: 15-21.

Naeimi, V., Scipal, K., Bartalis, Z., Hasenauer, S. and Wagner, W. 2009. An improved soil moisture retrieval algorithm for ERS and METOP scatterometer observations. IEEE Transaction on Geoscience and Remote Sensing. 47: 1999- 2013

Nakayama, H. 1983. Methods of measuring soil-bearing capacity. In: Advanced Rice cultivation, Irrigation, and Drainage Technology in Japan. Fuji Marketing Research Co., Ltd., Tokyo, Japan.

Nikam, P. J. 1989. Effect of evapotranspiration and variable drainable pore space on subsurface drainage design. M. Tech. Thesis. GB Pant University of Agriculture and Technology, Pantnagar

Noguchi, S., Tsuboyama, Y., Sidle, R. C. and Hosoda, I. 1999. Morphological characteristics of macro-pores and the distribution of preferential flow pathways in a forested slope segment. Soil Sci. Soc. Am. J. 63: 1413–1423.

Norman, L. E. J. 1959. The one-point method of determining the liquid limit of a soil. Geotechnique. 9: 1-8.

NRCS. 2012. Cone penetrometer. In: National Engineering Handbook, Part 631 Engineering Geology, Natural Resources Conservation Service (NRCS). US Department of Agriculture, Washington, DC.

Nusier, O. K. and Abu-Hamdeh, N. H. 2003. Laboratory techniques to evaluate thermal conductivity for some soils. Heat and Mass Transfer. 39: 119-123.

Odong, J. 2007. Evaluation of empirical formulae for determination of hydraulic conductivity based on grain-size analysis. Journal of American Science. 3: 54-60.

Olmstead, L. B. 1937. Some Moisture Relations of the Soils from the Erosion Experiment Stations. Technical Bulletin No. 562. United States Department Of Agriculture ,Washington, D. C. 45 p.

Olmstead, L. B., Alexander, L. T. and Middleton, H. E. 1930. A Pipette Method of Mechanical Analysis of Soils based on Improved Dispersion Procedure. Technical Bulletin No. 170. United States Department of Agriculture Washington, DC. 23 p.

Owe, M., de Jeu, R. and Walker, J. 2001. A methodology for surface soil moisture and vegetation optical depth retrieval using the microwave polarization difference index. Geoscience and Remote Sensing, IEEE Transactions, 39: 1643-1654.

Pandey, R. S., Bhattacharya, A. K., Singh, O. P. and Gupta, S. K. 1992. Drawdown solutions with variable drainable porosity. Journal of Irrigation and Drainage Engineering. 118: 382-396.

Petrenko, L. R. and Berezhnyak, Y. M. 2008. Soil Science: Practical Methods Manual. Publishing House NAU, Kyiv. 239 p.

Philip, J. R. 1957. The theory of infiltration: 4- sorptivity and algebraic infiltration. Soil Science. 84: 257-264.

Philip, J. R. 1969. The theory of infiltration. Advances in Hydroscience. 5: 215-296.

Phogat, V. K., Tomar,V. S. and Dahiya, R. 2015. Soil physical properties. In: Soil Science: An Introduction. Rattan et al. (Eds). Chapter 6: 135-171.

Piper, C. S. 1942. Soil and Plant Analysis. Hans Publishers, Bombay. 368 p.

Poelchau, H. S., Baker, D. R., Hantschel, Th., Horsfield, B. and Wygrala, B. 1997. Basin simulation and the design of the conceptual basin model. In: Petroleum and Basin Evaluation, Welte, D. H., Horsfield, B. and Baker, D. R. (Eds.). Springer, Berlin. 36–41.

Prakash, K. 2015. Geotechnical Engineering Laboratory Instruction Manual. Sri Jayachamarajendra College of Engineering, Mysuru, India. 56 p.

Prakash, K. and Sridharan, A. 2004. Free swell ratio and clay mineralogy of fine-grained soils. Geotechnical Testing Journal, ASTM. 27: 220–225.

Prakash, K., Sudheendra, S. and Sridharan, A. 2014. Hygroscopic moisture content: determination and correlations. J. Environmental Geo-techniques. 3: doi:10.1680/envgeo.14.00008.

Puri, A. N. 1930. A New Method for Estimating Total Carbonates in Soils. Bull. No. 206. Imp. Agri. Res. Institute, Pusa. 7 p.

Quirk, J. P. 1955. Significance of surface areas calculated from water vapor sorption isotherms by use of the B. E. T. equation. Soil Science Journal. 80: 423-430.

RAJAD, 1995. Analysis of Subsurface Drainage Design Criteria. Chambal Command Area Development & Canadian International Development Agency. Kota, Rajasthan.

Ranganatham, B. V. and Satyanarayana, B. 1965. A rational method of predicting swelling potential for compacted expansive clays. In: Proceeding of the 6th ICSMFE, Montreal. 1: 92–96

Rawat, K. S., Tripathy, V. K., Gupta, S. K. and Rao, K. V. G. K. 2001. Effect of drain placement in a low hydraulic conductivity zone in a two-layered soil profile on transmissivity of shallow aquifers. J. Ind. Soc. Soil Sci., 49: 399-406.

Reeve, R. C. 1965. Modulus of rupture. In: Black, C. A. (Ed.). Methods of Soils Analysis. Part 1. American Society of Agronomy, no. 9. Madison, WI. 466-471.

Rice, R., Milczarek, M. and Keller, J. 2014. A critical review of single ring cylinder infiltrometers with lateral flow compensation. 14th Biennial Symposium on Managed Aquifer Recharge. 31.07.2014 to 01.08. 2014, Orange, California.

Richards, L. A. 1953. Modulus of rupture as an index of crusting of soil. Proc. Soil Sci. Soc. Am. 17 : 321-323.

Richards, L. A. 1941. A pressure-membrane extraction apparatus for soil solution. Soil Science. 51: 377-386.

Richards, L. A. (Ed.). 1954. Diagnosis and Improvement of Saline and Alkali Soils. Handbook No. 60. US Dept. of Agriculture. US Salinity Laboratory, Riverside CA. 160 p.

Richard, L. A. and Fireman, M. 1943. Pressure plate apparatus for measuring moisture sorption and transmission by soils. Soil Science. 56: 395-404.

Ringrose-Voase, A. and Sanidad, W. 1996. A method for measuring the development of surface cracks in soils: Application to crack development after lowland rice. Geoderma. 71: 245–261.

Ritchie, J. A. 1963. Earthwork tunnelling and the application of soil testing procedures. Journal of Soil Conservation NSW. 19: 111–129.

Ritzema, H. P. 1994. Drainage Principles and Applications. ILRI Pub. 16. International Institute for Land Reclamation and Management (ILRI), Wageningen, The Netherlands. 1125 p.

Rivera, L. D. 2008. Comparing Methods of Estimating Crack Volume in Shrink-Swell Soils. Report Submitted to Texas A&M University as Undergraduate Research Scholar. Texas A&M University, Texas. 39 p.

Saha, A. K. 2008. Methods of Physical and Chemical Analysis of Soil. Kalyani Publishers, Ludhiana. 488 p.

Santamarina, J. C., Klein, K. A., Wang, Y. H. and Prencke, E. 2002. Specific surface: determination and relevance. Canadian Geotechnical Journal. 39: 233-241.

Schafer, W. M. and Singer, M. J. 1976. A new method of measuring shrink-swell potential using soil pastes. Soil Sci. Soc. Am. J. 40: 805-806.

Schindler, U. 1980. Ein Schnellverfahren zur Messung der Wasserleitfähigkeit im teilgesättigten Boden an Stechzylinderproben. Arch. Acker- u. Pflanzenbau u. Bodenkd. Berlin 24: 1–7.

Schindler, U., Durner, W., von Unold, G. and Müller, L. 2010. Evaporation method for measuring unsaturated hydraulic properties of soils: Extending the measurement range. Soil Sci. Soc. Am. J. 74:1071–1083.

Schindler, U. and Müller, L. 2006. Simplifying the evaporation method for quantifying soil hydraulic properties. J. Plant Nutrition Soil Sci. 169: 1–7.

Schoeneberger, P. J., Wysocki, D. A. and Benham, E. C. and Soil Survey Staff. 2012. Field Book for Describing and Sampling Soils. Version 3.0. USDA–NRCS, National Soil Survey Center, Lincoln, NE.

Schuler, R. T., Casady, W. W. and Raper, R. L. 2000. Soil compaction. In: Conservation Tillage Systems and Management. Iowa State University, Ames, IA. MWPS-45: 70-76.

Shah, P. H. and Singh, D. N. 2006. Methodology for determination of hygroscopic moisture content of soils. J. ASTM International. 3: 1 -14.

Silva, A. P. and Kay, B. D. 1997. Estimating the least limiting water range of soil from properties and management. Soil Science Society of America Journal. 61: 877–883.

Silva, A. P., Kay, B. D. and Perfect, E. 1994. Characterization of the least limiting water range. Soil Science Society of America Journal. 58: 1775–1781.

Singh, Alam. 1967. Soil Engineering in Theory and Practice. Asia Publishing Co. New Delhi. 803 p.

Smedema, L. K. and Rycroft, D. W. 1983. Land Drainage: Planning and Design of Agricultural Drainage Systems. Cornell University Press, Ithaca, New York. 375 p.

Soil Science Division Staff. 2017. Soil Survey Manual. Ditzler, C. , Scheffe, K. and Monger H.C. (Eds.). USDA Handbook 18. Government Printing Office, Washington, D.C.

Soil Science Society of America. 1956. Report of Definitions Approved by the Committee on Terminology. Soil Sci. Soc. America Proc. 20: 430-440.

Soil Survey Division Staff. 1993. Soil Survey Manual. Soil Conservation Service. U.S. Department of Agriculture Handbook 18. Washington DC.

Sridharan, A. and Prakash, K. 2000. Classification procedures for expansive soils. Proceedings of ICE: Geotechnical Engineering. 143: 235-240.

Sridharan, A., Rao, S. M. and Joshi, S. 1990. Classification of expansive soils by sediment volume method. Geotechnical Testing Journal, ASTM. 13: 375–380.

State of Queensland. 2015. Soil colour. https://www.qld.gov.au/environment/land/management/soil/soil-properties/colour. Opened on 16.08.2019.

Steenhuis, T. S., Rivera, J. C., Hernandez, C. J. M., Walter, M. T., Bryant, R. B. and Nektarios, P. 2001. Water repellency in New York state soils. International Turfgrass Society Journal. 9: 624-628.

Sterling, R. L. Assessment of apparent soil thermal conductivity. https://web.ornl.gov/sci/buildings/conf-archive/1992%20B5%20papers/020.pdf (Opened on 22.03.2020)

Stewart, R. D. and Abou Najm, M. R. 2017. Field measurements of soil cracks. Methods of Soil Analysis. 2: doi:10.2136/msa2015.0043

Stewart, R. D., Abou Najm, M. R., Rupp, D. E. and Selker, J. S. 2012. Measurement tool for dynamics of soil cracks. Vadose Zone J. 11: 2136/vzj2011.0048.

Stewart, R. D., Abou Najm, M. R., Rupp, D. E. and Selker, J. S. 2016. Modeling multi-domain hydraulic properties of shrink swell soils. Water Resources Res. 52: 7911–7930.

Stokes, G. G. 1851. On the effect of internal friction of fluids on the motion of pendulums. Transactions of the Cambridge Philosophical Society. 9:8–106.

Stibinger, J. 2014. Examples of Determining the Hydraulic Conductivity of Soils: Theory and Applications of Selected Basic Methods. University Handbook on Soil Hydraulics. J. E. Purkyně University in Ústí n. Labem. 72 p.

Storer, D. A. 1984. A simple high sample volume ashing procedure for determination of soil organic matter. Communications in Soil Science and Plant Analysis. 15: 759-772.

Structx.com. 2020. Critical Angle of Repose. https://structx.com/Soil_Properties_005.html. Opened on 19.04.2020.

Taboada, M. A. 2003. Soil shrinkage characteristics in swelling soils. In: Departamento de Ingeniería Agrícola y Uso de la Tierra, Facultad de Agronomía UBA, Buenos Aires, Argentina. LNS0418038. 471-486.

Tisdall, A. L. 1951. Comparison of methods of determining apparent density of soils. Australian Journal of Agricultural Research. 2: 349-354

Topp, G. C. and Binns, M. R. 1976. Field measurement of hydraulic conductivity with a modified air-entry permeameter. Can. J. Soil Sci. 56: 139-147.

Topp, G. C. and Davis, J. L. 1985. Measurement of soil water content using TDR: A field evaluation. Soil Sci. Soc. Am. J. 49: 19–24.

Topp, G. C., Davis, J. L. and Annan A. P. 1980. Electromagnetic determination of soil-water content—Measurements in coaxial transmission-lines. Water Resour. Res. 16: 574-582.

Topp, G. C. and Ferré, P. A. 1998. Measuring water content in soil using TDR: A state-of-the-art. In: Comparison of Soil Water Measurement using the Neutron Scattering, Time Domain Reflectometry and Capacitance Methods. International Atomic Energy Agency, Vienna, Austria. IAEA-TECDOC-1137:111-124.

Topp, G. C., Parkin, G. W. and Ferre′, P. A. 2008. Soil Water Content. In: Carter and Gregorich (Eds.). Sampling and Methods of Analysis (2nd ed). Canadian Soc. of Soil Science and Taylor & Francis Group, Boca Raton, FL. 939-961

Tuller, M. and Or, D. 2005. Water films and scaling of soil characteristic curves at low water contents. Water Resour. Res. 41: W09403

USBR, 1978. Drainage Manual: A Water Resources Technical Publication. US Department of Interior, Bureau of Reclamation, USA. 321 p.

USDA. 1962. Soil Survey Manual. *USDA* Handbook No. 18. Oxford and IBH Publishing Company, Calcutta.

USDA. 1987. Soil Mechanics Level I. Module 3 – USDA Textural Soil Classification, Study Guide. USDA, Soil Conservation Service. Stillwater, OK, USA.

USDA-NRCS. 2003. Soil compaction: Detection, Prevention and Alleviation. Soil Quality. Agronomic Technical Note No 17. USDA-NRCS Soil Quality Institute. Auburn, AL.

USDA–NRCS (U.S. Department of Agriculture, Natural Resources Conservation Service). 2000. Soil quality concerns: Hydrophobicity. Soil Quality Information Sheet. Available at http://www.nrcs.usda.gov/Internet/FSE_DOCUMENTS/nrcs142p2_051899.pd

USDA–SCS (U.S. Department of Agriculture, Soil Conservation Service). 1971. Handbook of Soil Survey Investigations: Field Procedures. USDA–SCS. GPO, Washington, DC

USSCS. 1988. Standards and Specifications for Conservation Practices – Standard on Subsurface Drain. Report No. 606. USDA-SCS, Washington, DC, USA. 606/1 -606/7.

Van Schilfgaarde, J. (Ed.). 1974. Drainage for Agriculture. Agronomy no. 17. American Society of Agronomy, Madison, USA. 700 p.

Veneman, P. L. M. and Pickering, W. 1983. Salt bridge for field redox potential measurements. Commun. Soil Sci. Plant Analysis. 14: 669-677.

Vepraskas, M. J. 2002. Redox Potential Measurements. Report, North Carolina State University, Raleigh, NC. 17 p.

Viji, R. and Prasanna, P. R. 2012. Assessment of water holding capacity of major soil series of Lalgudi, Trichy, India. Journal of Environmental Research and Development. 7: 393-398.

Walkley, A. J. and Black, I. A. 1934. Estimation of soil organic carbon by chromic acid titration method. Soil Science. 37: 29-38.

Wallach, R., Ben-Arie, O. and Graber, E. R. 2005. Soil water repellency induced by long–term irrigation with treated sewage effluent. J. Environ. Qual. 34: 1910–1920.

Washburn, E. W. 1921. The dynamics of capillary flow. Phys. Rev. 17: 273 -276.

Watkins, D. C. 1987. Evaluation of an Air-entry Permeameter for Use in Investigations of Groundwater Flow beneath Embankments. Report No. SR 102. Hydraulic Research, Wellingford, UK. 36 p.

Weber, W. L. 1950. Heat and temperature measurements. Prentice Hall Inc., New York. 323 – 326.

Wesseling, J. 1974. Crop growth and wet soils. In: Drainage for Agriculture, J. van Schilfgaarde (Ed). Agronomy Monograph 17. American Society of Agronomy Inc., Madison, Wisconsin, USA. 7–37.

Wilde, S. A. 1950. A simple soil permeameter. Agronomy journal. 42: 522.

Wilding, L. P., and Tessier, D. 1998. Genesis of Vertisols: shrink-swell phenomena. In: Vertisols: Their Distribution, Properties, Classification, and Management, Wilding et al. (Eds.). Texas A&M University, College Station, TX. 55-79.

Wind, G. P. 1968. Capillary conductivity data estimated by a simple method. Water in the Unsaturated Zone, Symp. Wageningen, The Netherlands. 1966. Proc. UNESCO/ IASH. 181-191.

Yadav, L. R., Yadav, S. S., Sharma, O. P. and Shivram, A. C. 2012. Practical Manual on Agrometeorology. S. K. Rajasthan Agricultural University, Campus Jobner. 60 p.

Van Wijk, W. R. (Ed.). 1963. Physics of Plant Environment. North Holland Publishing Company, Amsterdam.

Yassoglou, N., Kosmas, C. S., Moustakasa, N., Tzianis, E. and Danalatos, N. G. 1994. Cracking in recent alluvial soils as related to easily determined soil properties. Geoderma. 63: 289-298.

Yoder, R. E. 1936. Direct method of aggregate analysis and study of the physical nature of erosion losses. J. Am. Soc. Agron. 28: 337-351.

USEFUL INFORMATION AND CONVERSION FACTORS

Meter: The meter is the length of the path traveled by light in vacuum during a time interval of 1/299,792,458 second.

Second: The second is the duration of 9,192,631,770 periods of the radiation corresponding to the transition between the two hyperfine levers (F=4, m_F=0 to F=3, m_F=0) of the ground state of the cesium 133 atom.

Mass: The kilogram (kg) is equal to the mass of the international platinum prototype kept by Bureau International des Poids et Measures (BIPM) at Sevres. 1 kg = 1,000 g (grams)

There is no SI unit of 'weight'.

Weight means the force due to gravity acting on a mass, the mass (kg) must be multiplied by g (9.81 m/s^2) to give the force in Newton's (N).

Density: The mega gram per cubic meter (Mg/m^3) is the density unit adopted for soil mechanics. 1 Mg/m^3 = 1 g/cm^3 = 1,000 kg/m^3

Force: The Newton (N) is that force which, applied to a mass of 1 kilogram, gives it an acceleration of 1 m/s^2. 1 N = 1 kg m/s^2, 1 kN = 1,000 N = approximately 0.1 ton f

Pressure and stress: The Pascal (Pa) is the pressure produced by a force of 1 Newton applied, uniformly distributed, over an area of 1 m^2. The Pascal is the unit for pressure and stress, and is exactly equal to N/m^2.

Viscosity: 1 CentiPoises (cP) = 1 CentiStokes (cSt)/Specific gravity

Kelvin: The Kelvin, unit of thermodynamic temperature is the fraction 1/273.16 of the thermodynamic temperature of the triple point of water.

Ampere: The ampere is the constant current, which if maintained in two straight parallel conductors of infinite length of negligible circular cross-section, and placed 1 meter apart in vacuum, would produce between these conductors a force equal to 2×10^{-7} Newton per meter length

Mole: The mole is the amount of substance of a system which contains as many elementary entities as there are atoms in 0.012 kg of carbon 12 (about 6.022×10^{23} atoms). When the mole is used, specify the elementary entities, may be atoms, molecules, ions, electrons, other particles, or specified groups of such particles.

Candela: The candela is the luminous intensity in a given direction of a source that emits monochromatic radiation of frequency 540×10^{12} Hz and that has a radiant intensity in that direction of 1/683 watt per steradian.

Standard gravity, g: It is equal to 9.80665 ~ 9.81m/s^2

CONVERSION FACTORS (LENGTH, AREA, VOLUME, VELOCITY AND DISCHARGE)

Length				
inch	ft	cm	meter	km
1	0.0833	2.54	0.0254	2.54E-05
12	1	30.48	0.3048	3.05E-04
0.3937	0.0328	1	0.01	1.00E-05
39.37	3.28	100	1	0.001
39360	3280	1.000E+05	1000	1
Area				
in^2	ft^2	cm^2	m^2	ha
1	0.0069	6.452	6.452E-04	6.452E-08
144	1	929	0.0929	9.290E-06
.1550	1.07E-03	1	1.000E-04	1.000E-08
1550	10.764	1.0 E+4	1	1.000E-04
1.550E+07	1.07E+05	1.0E+8	10,000	1
Volume				
in^3	ft^3	gallon	liter	m^3
1	0.00058	0.0043	0.0164	1.639E-05
1728	1	7.4805	28.3168	0.0283
231	0.1337	1	3.7854	0.0038
61.0237	0.0353	0.2642	1	0.001
61024	35.314	264.17	1000	1
Velocity				
in/s	ft/s	cm/s	m/s	km/hr
1	0.0833	2.54	0.0254	9.144E-02
12	1	30.48	0.3048	1.0972
0.3937	0.0328	1	0.01	3.6E-02
39.37	3.28	100	1	3.6
39360	3280	1.000E+05	1000	3.6E+03
Discharge				
in^3/s	ft^3/s (cusecs)	gallon/s	m^3/s (cumecs)	L/s
1	0.00058	0.0043	0.0164	16.4
1728	1	7.4805	28.3168	2.831E+04
231	0.1337	1	3.7854	3.7854E+03
61.0237	0.0353	0.2642	1	1.0E+03
61024	35.314	264.17	1000	1.0E+06

Conversion Factors (Mass, Force, Power, Energy, and Surface tension)

Mass			
ounce	pound	kg	tonne (metric)
1	0.0626	0.0283	2.835E-05
16	1	0.4536	4.536E-04
35.2741	2.2046	1	0.0010
35,840	2,240	1,016	1.0160
35,274	2,205	1,000	1
Note: 1 ton = 1.016 tonne (Mg)			
Force			
Pound force	Newton	kgf	gf
1 lbf	4.448	0.4536	453.6
Power			
BTU/hr	ft-lb/min	Kcal/hr	kW
1	12.9695	0.2520	2.931E-04
0.0771	1	0.0194	2.260E-05
3.9683	51.4663	1	0.001163
2,544	33,000	641.196	0.7457 (1 HP)
33,470	434,089	8,434	9.8091
3,412	44,254	859.859	1
BTU = British thermal unit, gf = gram fore			
Energy			
BTU	kCal	hp-hr	kW-hr
1	0.25200	3.930E-04	2.931E-04
0.0013	0.0003	5.051E-07	3.766E-07
3.9683	1	0.0016	0.0012
0.9478	0.2388 (1 kJ)	3.725E-04	2.778E-04
2,544	641.19	1	0.7457
3,412	859.85	1.341	1
Surface tension			
Pound force/in	N/m	gf/cm	dyne/cm
1	175.1268	0.9807	0.001
0.0057	1.0	1.0197	1000

CONVERSION TABLE: PRESSURE

Units	Pascal (Pa)	Bar (bar)	Atmosphere metric (at)	Atmosphere standard (atm)	Torr (Torr)	Pound-force per square inch (psi)
1 Pa	1 N/m^2	10^{-5}	1.0197×10^{-5}	9.8692×10^{-6}	7.5006×10^{-3}	145.04×10^{-6}
1 bar	100,000	10^6 dyne/cm^2	1.0197	0.98692	750.06	14.5037744
1 at	98,066.5	0.980665	1 kgf/cm^2	0.96784	735.56	14.223
1 atm	101,325	1.01325	1.0332	1 atm	760	14.696
1 torr	133.322	1.3332×10^{-3}	1.3595×10^{-3}	1.3158×10^{-3}	≡ 1 Torr; ≈ 1 mmHg	19.337×10^{-3}
1 psi	6.894×10^3	68.948×10^{-3}	70.307×10^{-3}	68.046×10^{-3}	51.715	1
1 mm H_2O	9.8067	0.000098	1.0×10^{-4}	0.000097	0.0735	0.001421

Units to Measure Moisture Potential (Pressure) and Their Approximate Conversions

cm of water	Kilopascal (kPa)	Atmosphere or bars	Megapascal (MPa)	Jouls per kilogram (J/kg)	Pounds per square inch (lb/in^2)
100	10	0.1	0.01	10	1.45
15000	1500	15	1.5	1500	218

Conversion Table: Temperature

Degrees Celsius (C) to degrees Kelvin (K)	(C) + 273.15 = (K)
Degrees Celsius (C) to degrees Fahrenheit (F)	[(C) × 1.8] + 32 = (F)
Degrees Celsius (C) to degrees Rankine (R)	[(C) × 1.8] + 491.67 = (R)
Degrees Fahrenheit (F) to degrees Kelvin (K)	[(F) × 0.555556] + 255.37 = (K)
Degrees Fahrenheit (F) to degrees Rankine (R)	(F) + 459.67 = (R)
Degrees Kelvin (K) to degrees Rankine (R)	(K) × 1.8 = (R)
Degrees Rankine (R) to degrees Celsius (C)	[(R) - 491.67] × 0.555556 = (C)
Degrees Rankine (R) to degrees Kelvin (K)	(R) × 0.555556 = (K)

Conversion Table: Kinematic and Dynamic Viscosity

Kinematic viscosity	
1 cSt	1×10^{-6} m²/s
1 cm²/s	1×10^{-4} m²/s
1 ft²/h	2.58064×10^{-5} m²/s
1 ft²/s	9.2903×10^{-2} m²/s
1 in²/h	1.79211×10^{-7} m²/s
1 in²/s	6.4516×10^{-4} m²/s
1 m²/h	2.77778×10^{-4} m²/s
1 m²/s	1 m²/s
1 St	1×10^{-4} m²/s
Viscosity (dynamic)	
1 lbf s/ft²	47.88 N-s/m²
1 g/(cm s)	0.1 N-s/m²
1 Poise	0.1 N-s/m²
0.1 Poise	1 kg/m-s (N-s/m²)

SUBJECT INDEX